BLACK-BODY THEORY
AND THE
QUANTUM DISCONTINUITY
1894–1912

BLACK–BODY THEORY
AND THE
QUANTUM DISCONTINUITY
1894-1912

Thomas S. Kuhn

With a new Afterword

THE UNIVERSITY OF CHICAGO PRESS
Chicago & London

The University of Chicago Press, Chicago 60637
The University of Chicago Press, Ltd., London

© 1978 by Oxford University Press, Inc.
All rights reserved. Published 1978
University of Chicago Press Edition 1987
Printed in the United States of America

96 95 94 93   5 4 3

*Reprinted by arrangement with
Oxford University Press.*

Library of Congress Cataloging-in-Publication Data

Kuhn, Thomas S.
  Black-body theory and the quantum discontinuity, 1894–1912.

  Reprint. Originally published: Oxford, Oxfordshire : Clarendon Press ; New York : Oxford University Press, 1978. With a new afterword.
    Bibliography: p.
    Includes index.
    1. Black-Body radiation—History. 2. Quantum theory—History. I. Title.
[QC484.K83   1987]   530.1'2   86–30833
ISBN 0–226–45800–8 (pbk.)

For
Sarah, Liza, and Nat,
my teachers in discontinuity

PREFACE

This book is the outcome of a project I had not intended to undertake. An account of its genesis may therefore suggest the volume's purpose and simultaneously provide some clues to the nature of historical research. Early in 1972 a change in professional circumstances enabled me to begin detailed study of the history of quantum theory, a topic with which I had long been concerned but one of which my knowledge was for the most part superficial. At that time I intended to presuppose the first stage of the development of quantum concepts, for it had been much studied by extremely competent scholars.[1] Rather than begin at the beginning, as this volume does, I planned to prepare a monograph on the development of quantum conditions, a central theme in the evolution of the so-called old quantum theory and one which could provide a strategic overview of the development of that theory as a whole. Only against the background provided by that overview, I thought, could the emergence of matrix mechanics, wave mechanics, and electron spin during 1925 and 1926 be understood.

In a general way I was aware of the structure of the developments I wished to explore, and I also knew the climactic episodes with which my story would close: the inventions, during 1922 and 1923, of Landé's vector model of the atom and of Bohr's model of the periodic table. Nevertheless, I lacked one detail prerequisite to the start of focused research. I did not know when physicists first began to look for quantum conditions, when they first asked about the nature of the restrictions placed by the quantum on the motion of systems more general than Planck's one-dimensional harmonic oscillator. That question, I was aware, had been much discussed at the first Solvay Congress late in 1911, but I did not know when or how it had initially arisen, and I could not therefore tell where the story I wished to relate should begin. Neither the printed proceedings of the Congress nor the abundant secondary literature on the first decade of the development of quantum concepts provided clues.

After numerous weeks of fruitless search for an answer, I determined to attempt a less direct approach: I would work my way chronologically through Planck's relevant papers, readily accessible in his collected scientific works. Planck might not, of course, be the person who first conceived the need for generalized quantum conditions, but his first mention of that need would localize my problem in time and very probably, through context and accompanying citations, in space as well. As always at the beginning of a major research project, the time available seemed ample, and I did not therefore begin my search by reading Planck's famous quantum papers of 1900 and 1901, papers I had read many times before and thought I understood. Instead, I started with his earlier work on black-body theory, the first product of which had been published in 1895.

That reading program had, for me, an extraordinary result. Having assimilated Planck's classical black-body theory, I could no longer read his first quantum papers as I and others had regularly read them before.[2] They were not, I now saw, a fresh start, an attempt to supply an entire new theory. Rather they aimed to fill a previously recognized gap in the derivation of Planck's older theory, and they did not at all require that the latter be set aside. In particular, the arguments in Planck's first quantum papers did not, as I now read them, seem to place any restrictions on the energy of the hypothetical resonators that their author had introduced to equilibrate the distribution of energy in the black-body radiation field. Planck's resonators, I concluded, absorbed and emitted energy continuously at a rate governed precisely by Maxwell's equations. His theory was still classical.

Shortly afterwards I discovered that that same classical viewpoint was also developed, but far more clearly, in the first edition of Planck's well-known *Lectures on the Theory of Thermal Radiation*, delivered in the winter of 1905–06 and published late in the following spring. Even in the middle of 1906, neither restrictions on classically permissible energy nor discontinuities in the processes of emission or absorption were to be found in Planck's work. Those are, however, the central conceptual novelties we have come to associate with the quantum, and they have invariably been attributed to Planck and located in his work at the end of 1900. Only after studying the extended treatment of Planck's theory in the *Lectures* of 1906 was I quite able to believe that I was now reading his first quantum papers correctly and that they did not posit or imply the quantum discontinuity.

At that point, early in the summer of 1972, I temporarily suspended my attempt to locate the start of the search for quantum conditions. Instead, I began work on an article embodying my new reading of Planck. Gradually and against my will, that article became a book, partly because I found that understanding Planck's early black-body theories demanded an acquaintance with previously unexplored aspects of Boltzmann's statistical treatment of irreversibility and partly because I came to realize that I must explain how discontinuity had entered physics if it had not, as previously thought, come from Planck. Numerous revisions later, this book results.

In its final form the manuscript is divided into three parts, the last a brief epilogue. Part One is the story I had originally intended to tell in an article, but much extended, especially at the start, to provide the background material appropriate to a book. Chapter I opens with a sketch of the black-body problem, describes the development of Planck's research before he took that problem up, and explores the ways in which his earlier concern with the foundations of thermodynamics both motivated and shaped his approach to thermal radiation. It concludes with a sketch of the earliest stage of Planck's black-body research, culminating in 1896 with his presentation of differential equations for a radiation-damped resonator.

Chapter II is a long but essential digression on the development of Boltzmann's statistical treatment of irreversibility, which proved critically important to the route followed by Planck's research from the beginning of 1898. Chapter III, which describes the development of Planck's black-body theory from 1896 through 1899, presents the first of two distinct stages in his assimilation of Boltzmann's statistical approach. The second emerges in Chapter IV, which considers the direction taken by Planck's research during 1900 and 1901, years in which he invented his famous black-body distribution law and then provided the first two derivations for it. Chapter V, which concludes Part One, considers how Planck and his first readers understood his revised theory during the years from 1900 to 1906.

The four following chapters, which constitute Part Two, trace the emergence and assimilation of the concept of a discontinuous physics. Chapters VI and VII deal primarily with the work of Ehrenfest and Einstein, the two physicists who first recognized that Planck's black-body law could not be derived without restricting resonator energy to integral multiples of $h\nu$ or some equivalent non-classical step. Their demonstrations, both published in 1906, had little apparent impact, but the next, presented by Lorentz in 1908, is the presumptive cause of a rapid change in the attitude, at least of German physicists, towards the quantum. Chapter VIII considers the circumstances that led Lorentz to embrace the discontinuous version of Planck's black-body theory and describes the way in which other recognized experts on radiation—most notably Wien, Planck himself, and probably James Jeans—followed Lorentz's lead during 1909 and 1910. By the end of the latter year most of the theorists who had studied the black-body problem in depth were convinced that it demanded the introduction of discontinuity.

As that conviction was established, the black-body problem lost its central role in the development of quantum concepts, for it offered no clues to the source and nature of discontinuity. Further progress would depend on the investigation of other areas proposed for the quantum's application, and by the beginning of 1911 there had been many of these, though only one that was beginning to be taken seriously. That situation is the one described in Chapter IX, which sketches the development of other proposed quantum applications in the course of a survey of the state of the quantum in 1911 and 1912. Among other things, it suggests that during the first of these years, leadership in the investigation of the quantum very suddenly passed from the black-body problem to the previously neglected topic of specific heats at low temperatures. One by-product of that transfer was a far larger audience for the quantum, which was soon internationally known. Another provides the answer to the question from which my reluctant search of the black-body literature began. Because it transformed the locus of discontinuity from Planck's resonators to massive atoms and molecules, the specific heat problem is the primary source of the search for quantum conditions. The question of how to apply the quantum to multidimensional mechanical problems was not raised publicly until 1911, but it was then raised repeatedly and in a variety of forms.

That survey of the state of the quantum concludes Part Two, and a brief epilogue, constituting Part Three, closes this volume. Its subject is the so-called second theory of black-body radiation, developed by Planck during 1911–12 and definitively formulated in the second edition of his *Lectures*, which differs decisively from the first. Usually interpreted as a retreat towards classical theory and a sign of its author's conservatism, the second theory proves to be the first in which Planck found a place for discontinuity of any sort. Localizing discontinuity in what he later called "the physical structure of phase space," it was also a serious piece of physics, one that influenced a number of contemporaries, including Niels Bohr, and which was briefly a serious contender in the growing field of competing non-classical formulations of the interaction between radiation and matter. Because it simultaneously returns attention to the themes of Part One and illustrates the state of the quantum early in the second decade of this century, the second theory provides an appropriate ending for this volume. The black-body problem would not, for some years, carry physical theory further.

Though my own close involvement with the black-body problem began only in the spring of 1972, my concern with the development of the quantum theory is a decade older. It originated in my association during the years 1961–64 with Sources for History of Quantum Physics, an archival project that sought, both by interviewing participants and by making copies of original manuscripts, to preserve records on which future studies of the development of the subject might be based.[3] Because still living physicists were the primary object of that enterprise's attention, very few of the records it succeeded in preserving are directly relevant to the years dealt with in this volume. The project, however, also sought to locate relevant manuscripts already on deposit in European libraries. Virtually all the manuscript materials referred to below were located in the course of that library survey; in its absence, many of them would doubtless be unknown to me.

Equally important, though far less tangible, work on the project supplied much of the overview of the quantum theory's development which has set the concerns and guided the selection of materials for this volume. Though a historian may not work backwards from the end project of the development to be explored, he can scarcely work at all without a preliminary sketch of the terrain. I have been particularly fortunate in mine, for it was a cooperative project to which my principal assistants, John L. Heilbron and Paul L. Forman, made major contributions as did some of the physicists to whom the project introduced us. Footnotes will record the debts I can still detail, but will not thereby begin to suggest the extent of what I owe them.

More recent debts have been accumulated during the long course of this volume's preparation. Hans Kangro and Martin Klein have provided the basic previous accounts from which much of my work departs, in both senses of the word. Just because we have differed at key points of interpretation, I am especially grateful for their generosity in hearing and criticizing my views at an early stage of their development. Later, as my manuscript took shape, a number of colleagues in history of science offered significant suggestions about all or parts of it. John Heilbron, Russell McCormmach, Noel Swerdlow, John Stachel, and

Spencer Weart responded to one or another version of the whole. Jed Buchwald, Stephen Brush, Paul Forman, and Daniel Siegel criticized drafts of one or more chapters. For guidance through or around occasionally recondite problems of physical theory, I am indebted to discussions with John Bahcall, Freeman Dyson, Edward Frieman, and John Hopfield. Finally, three of my students or former students—Robert Bernstein, Bruce Wheaton, and Norton Wise—have studied the manuscript with care in the course of checking footnotes, quotations, translations, and bibliographical citations. Their critical contributions have gone well beyond the significant minutiae assigned to them, and Robert Bernstein has also taken responsibility for the index. All these people have helped me clarify my text and avoid errors of both commission and omission. Nevertheless, the canonical disclaimer is more than usually appropriate in this case: for residual problems in the present text, I alone am responsible.

Anyone engaged in work of this sort makes a nuisance of himself to librarians. I must specially acknowledge the patience and good cheer with which my depredations have been borne by the staffs of the Mathematics and Natural Science Library at the Institute for Advanced Study and the Mathematics-Physics Library at Princeton University. Much of the manuscript material upon which my narrative depends is deposited on microfilm in the Library of the American Philosophical Society, and I am grateful to Murphy Smith and his staff for making copies easily available to me. Other essential help with manuscripts has been provided by Dr. Tilo Brandis of the Staatsbibliothek Preussicher Kulturbesitz and his staff, by Dr. A. Opitz of the Deutsches Museum, and by E. van Laar of the Algemeen Rijksarchief at The Hague. For permission to reproduce materials which the custodians of these collections have made available to me, I am most grateful to: Frau Gerda Föppl, on behalf of the Wien heirs; Professor T. H. Von Laue; Dr. Otto Nathan, for the Einstein estate; Frau Dr. Nelly Planck; and Fru Pia de Hevesey. Dr. van Laar of the Algemeen Rijksarchief and Mrs. M. Fournier of the Museum Boorhaave have also transmitted authorization on behalf of their respective institutions.

To Helen Dukas, who has done so much to assemble and preserve the Einstein archive at the Institute for Advanced Study, I owe a special debt. Not only has she been a generous guide to the rich collection over which she presides, but, exposed by proximity to my repeated importunities, she has been a constant help on questions of German orthography and idiom. Other help of the same sort has from time to time been provided by my colleagues Albert Hirschman and Michael Mahoney as well as by occasional German visitors to the Institute for Advanced Study. And, at a time of great need, Victor Lange deciphered for me some key phrases in Gabelsberger shorthand scattered through the Ehrenfest research notebooks discussed in Chapter VI.

Work on this volume was begun during a one-semester leave from Princeton University, supported in part by the University and in part by the National Science Foundation under Grant S-1265.[4] The effectiveness of my work, then and since, has also been much enhanced by an association with the Institute for Advanced Study, first as a visitor and more recently as a part-time member. In the latter capacity my work was for two years supported in part by the National Endowment for the Humanities under Grant H-5426 and for another three by the National Science Foundation under Grants GS 42905x and SOC 74-13309. To all of these institutions, as well as to the patient secretarial staff of the School of Social Science at the Institute for Advanced Study, I am very deeply in debt. In arrangements for publication and for the book's final form, I have had much valued assistance from the staff of the Oxford University Press, especially Leona Capeless, who has provided the perceptive and firm, but nonetheless flexible and understanding, editorial criticism that I had previously concluded could not exist. Both in detail and tone the manuscript has greatly benefitted from her intervention.

My most extended and least tangible debt is to the members of my family. They bore patiently and usually cheerfully the dislocations of home and school life caused by my involvement with the archival project that first interested me in the history of quantum physics. Since its close, they have tolerated the preoccupations and inattentions which, in my case at least, seem the usual concomitant of scholarly effort. Sometimes they must have wondered whether the flame is worth the candle, but they have been supportive nonetheless. For that and a great deal else, I thank them.

Princeton, N.J.   T.S.K.
September 1977

NOTE TO THE PAPERBACK EDITION

The prospect of a new edition of *Black-Body Theory* gives me great pleasure. Paperback publication will make the book accessible to a wider and more casual audience than is likely to come to terms with the hardcover version. Republication also makes possible the inclusion within a single binding both of the original text and of a retrospective article about it, the latter reprinted here as an afterword. Prepared six years after the book's original publication, that afterword has multiple purposes. For those requiring guidance through the technical complexities of the text, it supplies a summary of the book's major points and of some reasons to take them seriously. In addition, it considerably clarifies the description, in chapter IV, of Planck's derivation of the black-body law and of the connection between my account of that derivation and the book's main thesis. Third, the afterword discusses a topic on which the book itself remains scrupulously silent: the relationship between the historical enterprise illustrated by this volume and the more abstract view of scientific development presented, especially, in *The Structure of Scientific Revolutions*. Understanding that relationship, it suggests, may also help in understanding aspects of the book's initial reception. I am most grateful to the editors of *Historical Studies in the Physical Sciences* for permission to reprint the article here, to the University of Chicago Press for their willingness to include it, and to the Oxford University Press for permitting this reissue of the book.

Boston, Mass.   T.S.K.
November, 1986

CONTENTS

PART ONE. PLANCK'S BLACK-BODY THEORY, 1894–1906: THE CLASSICAL PHASE

I. PLANCK'S ROUTE TO THE BLACK-BODY PROBLEM . . . 3
   The Black-Body Problem 3
   Planck and Thermodynamics 11
   Planck and the Kinetic Theory of Gases 18
   Planck on the Continuum and Electromagnetism 29

II. PLANCK'S STATISTICAL HERITAGE: BOLTZMANN ON IRREVERSIBILITY 38
   Boltzmann's H-Theorem 39
   The First Interpretation of the H-Theorem 42
   Loschmidt's Paradox and the Combinatorial Definition of Entropy . 46
   The Conflation of "Molecular" and "Molar" 54
   Molecular Disorder 60
   Epilogue: Molecular Disorder and the Combinatorial Definition after 1896 67

III. PLANCK AND THE ELECTROMAGNETIC H-THEOREM, 1897–1899 72
   Cavity Radiation without Statistics 73
   The Entry of Statistics and Natural Radiation 76
   Planck's "Fundamental Equation" 82
   Entropy and Irreversibility in the Field 84

IV. PLANCK'S DISTRIBUTION LAW AND ITS DERIVATIONS, 1900–1901 92
   Planck's Uniqueness Theorem and the New Distribution Law . 92
   Recourse to Combinatorials 97
   Deriving the Distribution Law 102
   The New Status of the Radiation Constants 110

V. THE FOUNDATIONS OF PLANCK'S RADIATION THEORY, 1901–1906 114
   The Continuity of Planck's Theory, 1894–1906 115
   Natural Radiation and Equiprobable States 120

Energy Elements and Energy Discontinuity 125
The Quantum of Action and Its Presumptive Source 130
Planck's Early Readers, 1900–1906 134

PART TWO. THE EMERGENCE OF THE QUANTUM DISCONTINUITY, 1905–1912

VI. DISMANTLING PLANCK'S BLACK-BODY THEORY: EHRENFEST, RAYLEIGH, AND JEANS. 143
The Origin of the Rayleigh–Jeans Law, 1900–1905 144
Ehrenfest's Theory of Quasi-Entropies 152
The Impotence of Resonators 158
Complexion Theory and the Rayleigh–Jeans Law 167

VII. A NEW ROUTE TO BLACK-BODY THEORY: EINSTEIN, 1902–1909. 170
Einstein on Statistical Thermodynamics, 1902–1903. . . . 171
Fluctuation Phenomena and Black-Body Theory, 1904–1905 . 176
Einstein on Planck, 1906–1909 182

VIII. CONVERTS TO DISCONTINUITY, 1906–1910 188
Lorentz's Rome Lecture and Its Aftermath 189
Planck on Discontinuity, 1908–1910 196
The Consolidation of Expert Opinion: Wien and Jeans . . . 202

IX. BLACK-BODY THEORY AND THE STATE OF THE QUANTUM, 1911–1912. 206
The Decline of Black-Body Theory 207
The Emergence of Specific Heats 210
Quanta and the Structure of Radiation 220
The Quantum and Atomic Structure 226
The State of the Quantum 228

PART THREE. EPILOGUE

X. PLANCK'S NEW RADIATION THEORY. 235
Planck's "Second Theory" 235
Revising the *Lectures*. 240
Some Uses of the Second Theory 244
The Fate of the Second Theory. 252

NOTES 255
BIBLIOGRAPHY 323
AFTERWORD: Revisiting Planck 349
INDEX 371

BLACK-BODY THEORY
AND THE
QUANTUM DISCONTINUITY
1894–1912

Part One

PLANCK'S BLACK-BODY THEORY, 1894–1906: THE CLASSICAL PHASE

I
PLANCK'S ROUTE TO THE BLACK-BODY PROBLEM

Between late 1894 and the end of 1900 three lines of nineteenth-century scientific research were associated in novel ways by the work of the established German physicist Max Planck (1858–1947). An unexpected product of their interaction was the quantum theory, which, during the next three decades, transformed the classical physical theories from which it had developed. Part One of this volume describes the conception and gestation of that new theory during the years before 1906, a period in which Planck worked alone; Part Two considers its birth and early development from 1906 to 1912, when others reformulated the theory with a success sufficient to ensure its survival; Part Three, an epilogue, returns briefly to Planck in order to examine his initial constructive response to their apparently revolutionary reformulation. This opening chapter describes the problem to which Planck turned in the mid-1890s, discusses the concerns that led him to it, and examines the first stage of the research that followed.

The black-body problem

The research topic that led Planck to the quantum is the so-called black-body problem, usually known at the time as the problem of black radiation.[1] If a cavity with perfectly absorbing (i.e., black) walls is maintained at a fixed temperature T, its interior will be filled with radiant energy of all wavelengths. If that radiation is in equilibrium, both within the cavity and with its walls, then the rate at which energy is radiated across any surface or unit area is independent of the position and orientation of that surface. Under those circumstances, the energy flux reaching an infinitesimal surface $d\sigma$ from an infinitesimal cone of solid angle $d\Omega$ may be written $K \cos\theta \, d\Omega \, d\sigma$, where K is the intensity of the radiation and θ is the angle between the normal to $d\sigma$ and the axis of the cone $d\Omega$. Since radiation at a variety of wavelengths contributes to the total flux, the intensity may be more precisely specified by a distribution function K_λ such that K is given by $\int_0^\infty K_\lambda \, d\lambda$ and $K_\lambda \, d\lambda$ is the intensity due to radiation with wavelength between λ and $\lambda + d\lambda$. Determining and explaining the form of K_λ are the central components of the black-body problem, which originated in the work of Gustav Kirchhoff (1824–87).

During the winter of 1859–60 Kirchhoff announced the following theorem.[2] Let $d\sigma$ be an element of the interior surface of the wall of an arbitrary cavity, not necessarily black, and let $a_\lambda(T)$ be the fraction of the incident energy with wavelength between λ and $\lambda + d\lambda$ absorbed by that element when the cavity is maintained at temperature T. The rate at which energy in that range is absorbed by $d\sigma$ is then $\pi a_\lambda K_\lambda \, d\sigma$, the factor π being introduced by integration over $d\Omega$. Similarly, let $\pi e_\lambda(T) \, d\sigma$ be the rate at which energy in the same range is radiated into the cavity from $d\sigma$. Obviously, for equilibrium, total emission and absorption must be equal, or $\int_0^\infty a_\lambda K_\lambda \, d\lambda = \int_0^\infty e_\lambda \, d\lambda$. Kirchhoff was able to show, by considering a cavity with different materials in different walls, that the equality of emitted and absorbed energy must also apply separately to each infinitesimal wave-length range, i.e., that $a_\lambda K_\lambda = e_\lambda$. In addition, he demonstrated that, since K_λ is constant throughout the cavity, the ratio of e_λ to a_λ must be the same for all materials, however differently those materials may emit and absorb. Those results constitute Kirchhoff's radiation law:

$$\frac{e_\lambda}{a_\lambda} = K_\lambda(T),$$

where the intensity distribution K_λ is a universal function, dependent only on temperature and wavelength, not on the size or shape of the cavity or on the material of its walls. For a cavity with black walls, $a_\lambda = 1$ everywhere, and $e_\lambda = K_\lambda$. The radiation emitted by a black body is therefore identical, in its intensity distribution, to the equilibrium radiation contained in a cavity of any material for which $a_\lambda \neq 0$ at all wavelengths. The cavity may even have perfectly reflecting walls ($a_\lambda = 0$) provided that it somewhere contains a speck of dust which will, by absorption and re-emission, permit an initially arbitrary distribution of energy to approach equilibrium.

Beginning late in 1894, Planck undertook to explain that remarkable uniformity and, a few years later, to derive the form of the universal function $K_\lambda(T)$. By then, however, two other striking regularities of black radiation had been discovered, and these, especially the second, supplied essential background for his research. When Kirchhoff wrote on cavity radiation just after mid-century, he assumed only that radiant energy was propagated in waves, like light; little else could be taken for granted about it. Thirty years later, especially after Heinrich Hertz (1857–94) demonstrated the existence of electric waves in 1888, both visible and thermal radiation were increasingly assumed to be electromagnetic, with properties governed by Maxwell's equations. Consequences of those equations were first applied to black-body radiation in 1884 by the Austrian Ludwig Boltzmann (1844–1906). Then, in 1893, the year before Planck began his work on black radiation, Boltzmann's results were decisively extended by Wilhelm Wien (1864–1928), a recently licensed docent at the University of Berlin.

Boltzmann's initial objective was to show that recognition of the existence of radiation pressure could eliminate an apparent conflict between the second law of thermodynamics and the behavior of the recently invented radiometer.[3] Its pursuit led him to a powerful formulation of radiation thermodynamics. For equilibrium, the net flux of energy across the surface of any volume in a cavity's interior must be zero, a condition which Boltzmann showed could be satisfied only if the density u of radiant energy were related to its intensity K by the equation $u = 4\pi K/c$, with c the velocity of propagation. (The equation applies also to the distribution functions for energy density and intensity; u_λ must therefore, like K_λ, be a universal function of wavelength and temperature, and u must be a function of temperature alone.) It had previously been shown, in addition, that a plane wave perpendicularly incident on a reflecting or perfectly conducting surface exerts a pressure p equal to its energy density,[4] so that for isotropic radiation $p = u/3$. Taken together, these relations permit the direct application of thermodynamics to black radiation.

Let the radiation be confined in a cylinder of volume V closed by a reflecting piston. If radiation pressure does work, increasing the cylinder's volume by δV, then heat δQ must be added to maintain the temperature constant. By the first law of thermodynamics,

$$\delta Q = \delta U + p\,\delta V = \delta(uV) + \tfrac{1}{3} u\,\delta V = V\frac{\partial u}{\partial T}\delta T + \left(V\frac{\partial u}{\partial V} + \tfrac{4}{3} u\right)\delta V.$$

The expansion δV also changes the entropy of radiation S by an amount $\delta S = \delta Q/T$, where T is measured from absolute zero. By the second law of thermodynamics, δS must be an exact differential, so that $\partial^2 S/\partial V\,\partial T = \partial^2 S/\partial T\,\partial V$. Since u is a function only of T by Kirchhoff's law, straightforward manipulation yields the equations $du/dT = 4u/T$ and

$$u = \sigma T^4,$$

where σ is a universal constant. That relation between the energy density of black radiation and the cavity temperature had been proposed in 1879 by Josef Stefan (1835–93) as a likely extrapolation from preliminary experiments.[5] In the literature on black-body theory it is generally known as the Stefan-Boltzmann law.

That law is of no present importance, but the techniques developed in obtaining it are. Less than a decade after they were made public, Wien used them to derive a fundamental property of the distribution functions u_λ and K_λ.[6] Like Boltzmann, he dealt with radiation in a cylinder closed by a piston, but both his cylinder and piston were perfectly reflecting so that an arbitrary initial distribution of energy would be preserved unless the piston were moved. If the cavity volume were increased adiabatically, however, two effects would combine to alter the distribution. First, the energy in each wavelength range would be reduced as the corresponding radiation did work in moving the piston. Second, the wavelength of any radiation reflected from the moving piston would be increased by the Doppler effect, which would thus transfer the corresponding energy from one wavelength range to another.

Calling on the second law of thermodynamics, Wien showed that, if the radiation were initially at equilibrium with the cavity at a particular temperature, it would remain in equilibrium as the piston moved and the temperature rose or fell. (By introducing a suitably chosen radiation filter he was able to demonstrate that a departure from equilibrium would permit the direct conversion of heat to work.) Next, by a quantitative analysis of the redistribution of energy due to the Doppler effect and to work done by the piston, Wien showed how to compute the final distribution of energy from the initial one for a given intervening change in the cavity's volume. In the case of an equilibrium distribution, recourse to the Stefan-Boltzmann law enabled him also to specify the temperature corresponding to both the initial and final states. If the distribution function u_λ were known at one temperature, Wien could compute its form at any other.

Wien's result is called the displacement law because it shows how the curve for u_λ is displaced as the temperature of the cavity changes. In modern notation it takes the following simple form,

$$u_\lambda = \frac{4\pi}{c} K_\lambda = \lambda^{-5}\phi(\lambda T), \tag{1a}$$

where ϕ is an arbitrary function of a single variable. When, in consequence of Planck's work, frequency replaced wavelength as the standard independent variable, the displacement law assumed the more familiar form,

$$u_\nu = \frac{4\pi}{c} K_\nu = \nu^{-3}\phi(\nu/T), \qquad (1\mathrm{b})$$

where u_ν and K_ν are, respectively, the energy density and intensity in the frequency range ν to $\nu + \mathrm{d}\nu$. With ϕ unspecified the distribution law remained unknown, but Wien's result provided an important clue to its pursuit. What required specification had become a function of a single variable, no longer of two.

Physicists able to follow Wien's argument and to accept its premises presumably found his result persuasive. During the first decade of this century the displacement law rapidly become a standard tool. But it could scarcely have had that status at the time of its announcement in 1893. Arguments from the second law were not everywhere well understood; Maxwell's equations were only beginning to be widely known and used; the radiation from hot bodies was primarily the province of experimentalists, and their results were still preliminary in numerous respects. Postponing until later in this chapter consideration of the status of thermodynamics and electromagnetic theory in the 1890s, let us look briefly at the state of experimentation relevant to the distribution law.[7]

Suggestive observations date from William Herschel's discovery of the sun's infrared spectrum at the beginning of the nineteenth century, and they include measurements reported by J. H. J. Müller in 1858, John Tyndall in 1865, and A. P. P. Crova in 1880. But these experiments, like all those made before the mid-1880s, examined the spectra of only a few sources (the sun, gas flames, glowing filaments), all very hot and with temperatures only vaguely known. Measurements drawn from them provided little information in the infrared region and were, in any case, of questionable relevance to the properties of equilibrium radiation because the sources of radiation were not necessarily black. The first experiments which began to supply the sort of information needed to fix K_λ were those reported in 1886 by the American astronomer S. P. Langley (1834–1906). His objective was to determine the effect on solar radiation of its absorption and re-emission from the relatively cool surface of a planet.

Langley's radiator was copper, coated with lampblack, and he investigated the continuous spectrum to which it gave rise at a series of controlled temperatures below 1000 °C. At such temperatures, the detectable emission spectrum is confined to the infrared. To explore it, Langley improved the thermocouple, invented the bolometer, and skillfully calibrated a rock-salt prism for infrared wavelengths up to about 5 μ. Figure 1 reproduces one set of the curves which he obtained.[8] Qualitatively, they conform closely to all subsequent measurements, displaying temperature-dependent intensity maxima from which each curve declines asymptotically to zero with both increasing and decreasing wavelength. But their significance is primarily qualitative: only the three highest temperature curves have maxima in the region where Langley could establish reliable wavelengths. Published barely eight years before Planck took up the black-body problem, Langley's experiments are the mere beginning of the work on which the development and evaluation of quantitative black-body laws would depend.

It was, however, an important beginning, since it stimulated both experimentalists and theoreticians (doubtless including Wien) to pursue the determination of Kirchhoff's universal function. In 1887 the Russian W. A. Michelson (1860–1927) combined the Stefan–Boltzmann law with a speculative statistical hypothesis about the mechanism of emission to derive the radiation formula[9]

$$K_\lambda = b\lambda^{-6} T^{3/2}\, e^{-a/\lambda^2 T},$$

with a and b disposable constants. That equation, he showed, reproduced all the qualitative characteristics of Langley's experimental curves. But quantitatively it was not very satisfactory, a fact soon emphasized by H. F. Weber (1843–1912) of the Zurich Technische Hochschule, a physicist currently engaged in measuring the emission spectrum from carbon filament lamps.[10] After criticizing the theoretical basis of Michelson's derivation (including its reliance on the Stefan–Boltzmann law), Weber proposed an alternate formula based on his own and other experiments. His candidate for the Kirchhoff function required three disposable constants and took the form

$$K_\lambda = b\lambda^{-2}\, e^{hT-(a/\lambda^2 T^2)}.$$

When Wien, five years later, published the displacement law, his only reference to experiment was through Weber's law. Like his own law, Wien pointed out, Weber's required that the wavelength λ_m at which the intensity function reached its maximum be governed by the equation $\lambda_m T = $ Constant. Since the two laws were in other respects clearly

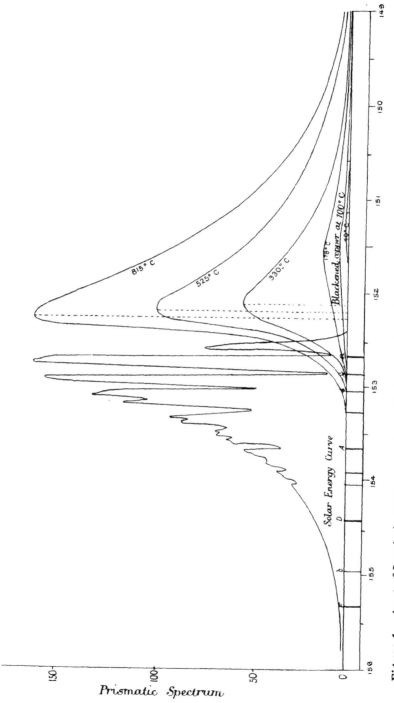

Figure 1. A set of Langley's curves comparing radiant energy from the sun with that from an experimental radiator of blackened copper. The horizontal axis is the length of the prismatic spectrum measured on an arbitrary linear scale.

incompatible, Wien's invocation of Weber's formula highlights the very limited authority of experiments on black radiation in 1893.

Three years later the situation had changed notably. Following the publication of Langley's work, a number of young experimentalists had set out to improve the sensitivity of bolometric measurements and to increase the range and precision of wavelength determinations in the infrared. One of them was Friedrich Paschen (1865–1947), then an Assistant at the Technische Hochschule in Hannover and at the beginning of what was to prove a distinguished career in spectroscopy. Paschen applied his improved instruments to the search for Kirchhoff's function with striking success. In 1895 he reported that the wavelength λ_m is, to a close approximation, inversely proportional to absolute temperature, thus providing direct evidence for the displacement law. Then, in the following year, an extension of his measurements led him to propose a new and especially simple form for the distribution function

$$K_\lambda = b\lambda^{-\gamma} e^{-a/\lambda T}.$$

Paschen's values for the constants were necessarily tentative, but γ appeared to lie in the range 5 to 6 with a mean value of 5.66.[11]

Paschen's radiation formula was first published, with his permission, in a paper by Wien, who had learned of it through correspondence and had at once seen its relation to a highly speculative derivation of his own, one he had previously refrained from publishing. A heated gas, Wien pointed out, can serve as the source of black radiation. In such a gas the number of molecules with velocities in the range between v and $v + dv$ is, by Maxwell's distribution law, proportional to $v^2 \exp(-v^2/\alpha^2)$ with α^2 proportional to the gas temperature T. If, in addition, one makes the far from natural assumption that both the wavelength and the intensity of the radiation from a given molecule are functions only of that molecule's velocity, then the distribution of radiation from the gas must take the form $K_\lambda = F(\lambda) \exp[-f(\lambda)/T]$. In that expression both F and f are unknown functions, derivable from the also unknown relations between wavelength and velocity, on the one hand, and between intensity and velocity, on the other. To specify them further, Wien noted that his formula would conform with the Stefan–Boltzmann and the displacement law only if $F = b\lambda^{-5}$ and $f = a/\lambda$. The result is the famous Wien distribution law,

$$K_\lambda = b\lambda^{-5} e^{-a/\lambda T}, \qquad (2)$$

a formula differing from Paschen's only in that it specifies the value of the constant γ. Unless $\gamma = 5$, Wien pointed out, Paschen's law is irreconcilable with the Stefan-Boltzmann law, itself an apparently unproblematic consequence of thermodynamics.[12]

As a product of theory, the Wien distribution law had, of course, little authority until Planck rederived it by a very different route in 1899. The hypothesis that both wavelength and intensity are functions only of the translational velocity of the emitting molecules was at best ad hoc. But the law was nevertheless unlikely to be merely wrong. It did conform to the requirement of the displacement law, and that law was derivable without resort to ad hoc hypotheses. Probably more important, it very closely resembled the law that Paschen had adduced from the best experiments made to date. A reduction in γ by less than 15 percent would make the two coincide. Further experiments might well bring about that reduction. Very soon they did.

By January 1899, Paschen's own investigations were directed to checking Wien's form of the law, and he soon reported that the value of γ declined from 6.4 to 5.2 as the emitter was changed from reflecting platinum to highly absorbent carbon. In February of the same year, Otto Lummer (1860–1925) and Ernst Pringsheim (1859–1917) provided fuller confirmation of Wien's form, using, for the first time, an experimental black cavity within which radiation could reach equilibrium before its intensity was measured. Other experts on infrared technique, including both Ferdinand Kurlbaum (1857–1927) and Heinrich Rubens (1865–1922), provided additional support, and Planck—still in 1899—supplied a magistral derivation from first principles.[13] Whatever the status of its derivation, Wien's law had triumphed. In the event, of course, that triumph was extremely brief. Early in 1900 the application of new long-wavelength infrared techniques to the newly deployed experimental cavities disclosed the law's limitation, with decisive effects on the subsequent development of physics. But that is another chapter, to be considered at an appropriate point below. No such outcome could have been foreseen when Planck's black-body research began or when, in 1899, it reached its first, apparently satisfactory, conclusion.

Planck and thermodynamics

Despite its brevity, the preceding sketch of the black-body problem discloses the three fields that were to interact consequentially within Planck's work. Two are obvious: thermodynamics and electromagnetic theory. The third, statistical mechanics, is the source of the Maxwell distribution to which both Michelson and Wien appealed when deriving their proposed distribution laws. Before 1900 Planck had made important contributions to all three, but they occupied very different places in his thought. Thermodynamics had been his first love, and his work in it was well known before he first turned, at age thirty-six, to electromagnetism. For him the latter's role was initially instrumental: Maxwell's equations provided conceptual tools with which to solve thermodynamic problems, particularly the problem of black radiation. Statistical techniques entered Planck's research later still and against much resistance. Though their appearance marks the first step on his path to immortality, Planck's resort to them was at the time an admission of failure. The remainder of this chapter will consider Planck's background in thermodynamics, his hopes for electromagnetic theory, and the beginning of his attempts to apply that theory to cavity radiation. In the process, his reasons for resisting statistics will be encountered as well.

Thermodynamics, as an abstract quantitative theory of the role of heat in macroscopic physical processes, was a relative novelty when Planck first encountered it in the 1870s. Its historical roots are traceable to the gas and steam-engine theories of the first third of the nineteenth century, but its formulation required a series of experimental and conceptual innovations that cluster at mid-century.[14] During the 1840s numerous thermal, chemical, electrical, and mechanical effects were recognized as qualitatively interconvertible without quantitative gain or loss. The generalization that captured those results was soon to be known as the law of conservation of energy, and it surpassed all earlier physical principles in the range of its concrete applications. Once conceptually assimilated, furthermore, it seemed so nearly inevitable that it was sometimes accorded a priori status. Indeed, the theorem that every cause must be quantitatively equivalent to its effect, else the universe will run down, had played a role in several of its initial enunciations.[15]

Conservation of energy is the first law of thermodynamics, and the second followed quickly, though by no means straightforwardly. In 1824 a young French engineer, Sadi Carnot (1796–1832), had derived a consequential set of theorems concerning heat engines from the assumption that heat is a special caloric fluid which does work as it passes from a higher to a lower temperature reservoir. Virtually unknown for more than a decade after its publication, Carnot's theory was revived, extended, and successfully applied to significant new problems in the 1840s, just as the law of energy conservation was gaining currency. Since Carnot's theory in its original formulation required that heat be conserved, the two were incompatible. But the disturbing conflict was resolved in 1850–51 when Rudolph Clausius (1822–88) and William Thomson (1824–1907) independently developed a new way of deriving Carnot's theorems. For that purpose they required a new physical principle, later usually known as the impossibility of perpetual motion of the second kind. Clausius gave it in the form: heat cannot of itself pass from a colder to a warmer body, the rest of the universe remaining unchanged. Thomson's formulation was equivalent: it is impossible to construct an engine that will raise a weight simply by cooling a heat reservoir. Those were the first statements of the second law of thermodynamics.[16]

Because these statements supply an observational basis for the second law, one of them, or an equivalent, still often appears early in books on thermodynamics. But as statements about what cannot occur in nature, their positive physical import is by no means transparent, and they are therefore rapidly displaced by some more perspicuous formulation, one that can be applied directly to a variety of physical situations. That sort of formulation was first systematically developed by Clausius in a series of papers which appeared between 1854 and 1865. All but the last were collected with supplementary commentary in the first edition of his *Mechanical Theory of Heat*, published in 1864 when Planck was a young schoolboy at Kiel. Republished in English in 1867 and in French in 1868, that book was the first monograph on thermodynamics, and many physicists learned the subject from it.[17] Its second edition, furthermore, was the field's first text. In its new preface, Clausius pointed out that, "The mechanical theory of heat in its present state of development already constitutes a wide-ranging and independent subject of study." He had therefore, he continued, reworked the contents of his papers so that the new edition of his book "would form a coherent developing whole and thus take the form of a textbook."[18] That version of Clausius's thermodynamics was published in 1876, and Planck was among the first to put it to its intended use.[19] Its formative influence on his career was great. Early in 1879 he submitted to the University of Munich a doctoral thesis based on Clausius's work but recommending a fundamental reformulation of his approach. As one might expect, Planck's subsequent research was the first to be affected by that recommendation.

Before examining Planck's reformulation of thermodynamics, let us look briefly at the route on which he had encountered Clausius's version.[20]

Born at Kiel in 1858, Planck was educated primarily in Munich, where his father assumed the University's chair of civil law in 1867. Very little science of any sort was included in his Gymnasium curriculum, but he did acquire some bits of physics from a much-admired mathematics teacher, and they impressed him deeply. In later life he particularly emphasized the impact—"like a sacred commandment"—of the conservation of energy. It was, he wrote, the first law he had learned "which possessed absolute validity, independent of man."[21†] Whether or not a product of hindsight, that memory of an early concern with laws of the greatest possible depth and generality indicates what particularly attracted him to thermodynamics.

For three years, beginning in the fall of 1874, Planck studied physics at the University of Munich, where he encountered the mechanical theory of heat though probably not the developed second law.[22] Next he spent a year at Berlin, where he attended the courses of Helmholtz and Kirchhoff. Both, he wrote in his *Scientific Autobiography*, attracted him greatly as men; in addition, his exposure to them and the Berlin circle "greatly expanded [his] scientific horizons." But, Planck continued,

> I must acknowledge that I gained little from the lectures.... Therefore, I could only still my need for continuing scientific education by reading works which interested me, and those naturally were ones relating to the energy principle. In this way I came upon the papers of Rudolph Clausius, whose clarity of expression and thought made a powerful impression. With growing enthusiasm I worked my way deeply into them. What I particularly admired was the exact formulation of the two laws of thermodynamics and the pioneering demonstration [*erstmalige Durchführung*] of the sharp separation between them. Previously, as a consequence of the material theory of heat, the opinion had been current that the transmission of heat from a higher to a lower temperature was like the sinking of a weight from a higher to a lower altitude, and this erroneous view was not easily suppressed.[23]

That encounter with Clausius determined Planck's choice of subject for a doctoral thesis, and the ideas he developed there helped to shape his approach to the black-body problem fifteen years later. An examination of what Clausius had done and of how Planck's approach to thermodynamics differed will show what was involved.

In 1850, as previously noted, Clausius had modified the foundations of Carnot's theory to reconcile it with energy conservation. The way in

---

† This sign (†) is attached to occasional notes to indicate that they contain substantive additions to or qualifications of the text, not simply citations and bibliographical discussion.

which he thereafter built on that modified foundation was, however, very much like Carnot's own. In particular, the primary thermodynamic systems considered by Clausius were always, like the idealized cylinder and piston imagined by Carnot, in thermal and mechanical interaction with their environment. In addition, all Clausius's formulations of the second law were statements about the behavior of such systems when carried through a closed cycle. In 1854 he gave the second law the form

$$\oint \frac{dQ}{T} \leq 0, \qquad (3)$$

where the equality sign applies only if the cycle is reversible, and where dQ is the quantity of heat absorbed by the system from its environment and T the absolute temperature at which that heat is absorbed.[24]

Clausius, of course, quickly went farther. His later papers considered the value of $\int dQ/T$ over open paths. In 1865 he introduced both the symbol S and the name entropy for the value of that integral:

$$S_1 = S_0 + \int_0^1 \frac{dQ}{T}, \qquad (4)$$

where the path from configuration 0 to configuration 1 must be reversible.[25] Applying the second law, equation (3), to that definition, he demonstrated that entropy must be a single-valued function of a system's configuration or state. Finally, from that property together with the first law in its standard thermodynamic form, he showed how to pass quickly to many of the now-familiar partial differential relations governing the macroscopic variables which characterize physical systems. But equation (3) or a close equivalent continued, for him, to represent the second law.[26]

Approaching his subject in the late 1870s, Planck found his starting point in Clausius rather than Carnot: it was equation (4), which defines entropy as a single-valued function of the state variables of a specified system. How, Planck asked, would entropy change with time as the corresponding system developed *by itself*, in thermal and mechanical isolation from its environment? The early pages of his thesis presented his answer in the form

$$S' - S \geq 0, \qquad (5)$$

where S' is the entropy at a later time, S at an earlier.

That equation was Planck's version of the second law. Though recognizing its mathematical equivalence to Clausius's form, equation (3),

Planck nevertheless insisted that it was conceptually clearer, more general, and more fundamental.[27†] Just as the first law governed the behavior of energy over time, so the second governed that of entropy. More important, by catching the parallel between the two absolute laws from which thermodynamics derived, the new formulation highlighted their decisive difference. The total energy of an isolated system must remain constant over time; its entropy can only increase or, in the ideal limiting case, remain constant. Equation (5) prohibited not simply the spontaneous passage of heat from a lower to a higher temperature but any process which would decrease the entropy of an isolated system. Viewed in this way, the second law rapidly became for Planck "The Principle of the Increase of Entropy."[28] Its function, emphasized by Planck in the opening paragraph of his doctoral thesis, was to determine the direction in which natural processes develop "so that a return of the world to a previously occupied state is impossible."[29]

Planck's reformulation of the second law has a presumptive bearing on the subsequent development of thermodynamics, but its significance here is more restricted. In its new form the law was especially well adapted to the study of equilibrium and the approach of a closed system to it. Planck made the point himself late in a paper published three years after the appearance of his thesis. Its title was "Evaporation, Melting and Sublimation," and its main text developed several significant thermodynamic theorems. Then, its concluding section began:

> All the previously enumerated theorems are strict consequences of a single proposition: that stable equilibrium corresponds to the maximum of entropy. That proposition, in turn, follows from the more general one that in every natural process the sum of the entropies of all participating bodies is increased. Applied to thermal phenomena this law is the most general expression of the second law of the mechanical theory of heat as I have [elsewhere] shown in detail.[30]

Black radiation is, however, a case of thermal equilibrium, and the preceding passage suggests how Planck would later approach it. If an arbitrary initial distribution of energy is injected into an insulated cavity, then the distribution will move towards equilibrium as energy is absorbed and re-emitted by any bits of black material the cavity contains. The approach to thermal equilibrium is irreversible, and entropy must therefore increase until equilibrium is achieved. If one had a formula for the entropy of radiation as a function of the field variables, then the black-body distribution function would be the one that maximized the total entropy of the radiation in the cavity. That is the approach Planck would begin to explore late in 1894, reaching a first conclusion in 1899. Clausius's formulation of the second law offered no equally apparent point of entry.

Planck is not likely, however, to have had the black-body problem in mind when he defended his thesis in 1879. During the fifteen years that followed, his published research continued much as it had begun. Many of his papers as well as a small book on conservation of energy[31] were intended to extend and clarify the foundations of thermodynamics, a field still widely misunderstood, especially by a prominent group of anti-mechanists known as the energeticists. Accepting the first law of thermodynamics as the fundamental principle of science, they hoped to reduce both matter and force to mere manifestations of energy. Discarding the distinction between reversible and irreversible processes, they believed they could derive a totally general version of the second law from the first.[32] Planck was referring to them when he later wrote of the difficulties in suppressing the view "that the transmission of heat from a higher to a lower temperature was like the sinking of a weight from a higher to a lower altitude."[33] Other early papers published by Planck dealt with applications of thermodynamics, initially to physical problems including saturation, change-of-phase, and equilibrium. Then, between 1887 and 1894, he turned increasingly to the exciting new field of physical chemistry just being opened by the pioneering research papers of Arrhenius and van't Hoff. Planck later emphasized, apparently with much justice, how little attention had been paid to at least the more basic aspects of his thermodynamic theory.[34] But that theory or its applications, more likely the latter, were sufficiently significant and well known to justify his appointment in 1885 to a special chair of mathematical physics at Kiel and then, in 1889, to the University of Berlin as Kirchhoff's successor.[35]

The move to Berlin brought Planck back, of course, not only to the center of German physics but more obviously to what would shortly become the world center for theoretical and experimental research on black radiation. Wien, Lummer, Pringsheim, Rubens, and Kurlbaum all worked there, either at the University, the Technische Hochschule, or the Physikalisch-Technische Reichsanstalt in nearby Charlottenburg. In such a setting Planck's turn from physical chemistry to radiation theory is not surprising. By the mid-1890s a physicist in Berlin could scarcely be unaware that Kirchhoff, Boltzmann, and Wien had firmly established the applicability of thermodynamic argument to radiation.

What Planck had begun to do by applying thermodynamics to chemistry, he could now reasonably expect to achieve for radiation theory as well. Actually, as the following pages will show, the program on which he embarked in 1894 had a far more ambitious goal, but he might never have pursued it if research on the black-body problem had not offered by-products of a more familiar sort.

Planck and the kinetic theory of gases

Planck's larger objective in taking up the black-body problem was to reconcile the second law with mechanics. By the mid-1890s, problems in the relation between the two were widely recognized and, in England, much discussed. But Planck had been aware of difficulties long before and, in 1881, had suggested the direction from which he expected their solution to emerge. In retrospect his suggestion seems extraordinarily implausible, but it is nevertheless the one that motivated his turn to black radiation thirteen years later. A first step in recovering its original cogency requires a digression on the early developmental stages of the kinetic theory of gases.

The belief that heat was the motion of material particles had dominated seventeenth-century science and had never thereafter been lost from view, not even during the thirty years, roughly 1790 to 1820, when most physicists subscribed to the caloric theory. But transforming the mode-of-motion viewpoint into a quantitative theory of significant scope depended on the development of a model of material aggregates to which mathematics could be easily applied. Ultimately, gas models proved to have the requisite simplicity, but not until the end of the eighteenth century. Previously, gases had been thought to be a distinct chemical species. (Before the eighteenth century air was its only member.) Their particles, which were usually conceived as space-filling, could only rotate or vibrate in place.[36†] Gas models thus presented would-be kinetic theorists with all the complexities still characteristic of liquids and solids. That situation changed only after Lavoisier's new chemistry had persuaded scientists that gases were simply a particular physical state of substances which could also exist in the solid and liquid form. Once it was recognized that steam, for example, was a gas like any other, easily imagined estimates showed that its molecules could occupy only a minuscule fraction of the volume filled by the gas. Initially, the newly empty space was thought to be filled with caloric fluid, a view which nicely accounted for the uniform properties of gases. After 1820, however, as the belief in caloric declined, gas models in which molecules traveled for some distance in straight lines between collisions were first conceived.[37]

Models of that sort proved quite persuasive after the recognition at mid-century of the interconvertibility of heat and work, and more of them then began to appear. The first to attract much attention, however, was contained in a pair of papers published by Clausius in 1857 and 1858.[38] He had been working on them for some time, refraining from publication so that their speculative micromechanics would not, by association, interfere with the reception and understanding of his purely thermodynamic work, and his papers are particularly rich in accomplishment and suggestiveness as a result of the delay. The first paper opened by showing that, if gas pressure is due to the mechanical impact of molecules on the container's walls, then $\frac{3}{2}pV = \frac{1}{2}nmv^2$, where p is pressure, V volume, and n the number of gas molecules, each with mass m and speed v. Comparing that result with the Boyle-Charles law, Clausius pointed out that absolute temperature must be proportional to the translational kinetic energy of a molecule.

That result had been derived several times before, but Clausius's papers quickly carried it farther. The first includes, for example, the earliest distinction in print[39] between the translatory motion of molecules, on the one hand, and their rotational and vibrational motion, on the other. It also suggests that the ratio of total kinetic energy to translatory energy must be fixed for a given gas, and it relates that ratio to the ratio of the specific heats at constant pressure and constant volume. The second paper introduces the concepts of mean free path and sphere of molecular action, and it applies them to explaining the observed slow rate of gaseous diffusion. Systematic development of the kinetic theory of gases dates from these papers, the second of which was published in the year Planck was born.

Clausius's papers quickly captured the interest of James Clerk Maxwell (1831–79), whose first contribution to kinetic theory was published in 1860.[40] That paper extended and improved Clausius's results in several significant respects, the most important being the application of statistical concepts to the distribution of molecular velocities in a gas. Clausius had known that those velocities varied greatly, both from molecule to molecule and, for an individual molecule, from time to time, but for purposes of computation he had used a fixed speed v. Adapting a standard argument from error theory,[41] Maxwell argued that the fraction of molecules with speeds between v and $v + dv$ must be given by

$$\frac{4}{\alpha^3 \sqrt{\pi}} v^2 \, e^{-(v^2/\alpha^2)} \, dv, \qquad (6)$$

where α is a constant easily shown to equal $\frac{2}{3}$ the mean value of v. The foundations of the argument by which Maxwell first arrived at that distribution law were subject to criticism, but he presented a much improved derivation in 1867,[42] and there have been numerous others since. Two of them, both due to Boltzmann, had an important influence on the development of Planck's black-body theory and will therefore be central topics in the next chapter.

From the start, the distribution law opened new avenues to Maxwell. He used it to show, for example, that a mixture of two gases could be in equilibrium only if the individual molecules of each had the same mean translational energy. Since two gases in equilibrium also have the same temperature, the results that Clausius had gained by comparing kinetic theory with the gas laws could be made more precise: absolute temperature is proportional to mean translational energy per molecule, the proportionality constant being independent of the gas; equal volumes of two gases at the same temperature and pressure must, as a consequence, contain the same number of molecules. Elsewhere in his 1860 paper Maxwell improved Clausius's computation of the mean free path and applied the result to the theory of viscosity, a quantity he was startled to find independent of pressure. Finally, after considering additional problems concerning diffusion and heat conduction, Maxwell treated the general question of equilibrium between complex molecules, concluding that the mean translational and rotational energies per molecule must be the same. Together with his result for mixed gases, this conclusion foreshadowed the equipartition theorem, a general result which both he and Boltzmann later derived. Very nearly the full range of problems that would occupy kinetic theorists for a generation had emerged by 1860.

Two related aspects of that problem-constellation require special emphasis, for both are relevant to the way in which statistical considerations entered research on the black-body problem, and both changed rapidly soon after that entry had occurred. First, until after 1900 the research subject of the men who applied statistics and molecular mechanics to the study of heat was gas theory, not statistical mechanics. From Watson's *Treatise on the Kinetic Theory of Gases*

(1876), through Boltzmann's *Lectures on Gas Theory* (1896, 1898), to Jeans's *Dynamical Theory of Gases* (1904), the former phrase was regularly selected to describe their work.[43] Second, because the object of gas theory was to explain the observable behavior of gases (specific heats, viscosity, thermal conductivity, etc.), very few of its practitioners were significantly concerned with thermodynamics. Very occasionally a discussion of gas theory would point out that, when the Maxwell distribution obtains, dQ/T is an exact differential.[44] There were also, as will shortly appear, a few isolated discussions of the statistical basis of the second law. But only Boltzmann attempted to develop a statistical theory of entropy, and that aspect of his work was entirely ignored by other gas theorists until after Planck took it up at the end of 1900. In short, the set of concerns now covered by the phrase "statistical mechanics" or "statistical thermodynamics" scarcely existed during the nineteenth century. They appear first during 1902 in a famous book by J. Willard Gibbs (1839–1903) and in an almost forgotten article by Albert Einstein (1879–1955).[45†] With the rapid assimilation of those works, statistical physics became a different field, one unavailable to Planck when his attitude towards the relation between mechanics and the second law was formed.

Though he probably did not follow its technical development closely, Planck knew at least the main lines of gas theory. Clausius, whom he admired, continued a significant contributor to the field. After Clausius's death in 1888, Planck helped to prepare his *Kinetic Theory of Gases* for the press.[46] Shortly after that he was sole editor for the posthumous publication of Kirchhoff's *Lectures on the Theory of Heat*,[47] a task which involved him deeply enough with gas theory to produce a significant confrontation with Boltzmann. Together with other evidence to be examined below, that confrontation suggests that, until the last years of the century, Planck's knowledge of Boltzmann's own gas-theory papers was spotty, but he was surely aware of their existence and probably of their main lines. Though he had no great interest in gas theory, Planck did not find it repulsive. His attitude thus differed markedly from that of the energeticists and other anti-mechanists of the day.

Planck was, in the first place, a convinced if undogmatic believer in mechanics or the mechanical world view. It did not, he thought, yet display either the generality or the virtually incontestable empirical base that characterized thermodynamics. But, as he wrote in 1887, its evolution had been marked by a long series of striking successes, and it was therefore likely to continue "to point the direction in which research should move.... One ought to use all available means," he continued, "to develop the final [*letzten*] consequences of the mechanical viewpoint for all areas of physics, chemistry, and so on."[48] A letter to be examined below suggests that Planck began his black-body research in response to that directive.

Towards the atomic or molecular articulation of mechanics, Planck's attitude was more equivocal. For the most part, however, it is well represented by remarks he made at the start of his Habilitationsschrift, the second dissertation which, in 1880, licensed him to teach at Munich. To study equilibria between isotropic materials, Planck began, one needs a knowledge of the forces at work in the interior of bodies. Such forces depend, he pointed out, "not only on the relative position of the smallest particles of the body but also on their temperature." The special value of thermodynamics, he then emphasized, is that it provides a means to investigate the temperature dependence of these forces between particles "without its being necessary to introduce special assumptions about the molecular constitution of the body."[49] Here, as in many passages in his subsequent work, Planck appears simply to have taken atoms and molecules for granted.[50] He urged, however, that the physicist make minimal use of special hypotheses about their structure and about the forces between them.

Applied to research on gas theory, that viewpoint suggests, not that such research is wrong in principle, but rather that it is an error in strategy. One of Planck's rare published statements on the subject, read to the Physics and Chemistry Sections of the annual Naturforscherversammlung in 1891, makes the point explicit:

> Anyone who has studied the works of Maxwell and Boltzmann—the two scientists who have penetrated most deeply into the analysis of molecular motion—will scarcely be able to escape the impression that the remarkable physical insight and mathematical skill exhibited in conquering these problems is inadequately rewarded by the fruitfulness of the results gained. A remarkable situation has very recently reinforced this evaluation. The far-reaching analogy between the behavior of matter in dilute solutions and in perfect gases was not discovered by kinetic theory. On the contrary, kinetic theory, even after the event, has had to overcome great difficulties in accounting for this behavior, so that it cannot, at least in the near future, be expected to contribute to further progress with this problem.[51†]

Three years after those remarks were made, Boltzmann would open his criticism of Kirchhoff's *Lectures* by attributing to their editor, Planck, the view that efforts devoted to gas theory were a waste of time and talent.⁵² That paraphrase seems just (and the pique displayed in its expression understandable), but it has a converse. To the extent that Planck's skepticism about gas theory derived from strategic considerations, little more than demonstrated need or achievement would have been required to make him change his mind.

In the event, Planck's own work was to demonstrate a need, and he would then change his mind. In doing so, however, he was surrendering more than is suggested by any of his usual remarks about the status of molecular mechanics. Planck did have another sort of reason for distrusting kinetic theory, and it doubtless reinforced expressions of skepticism like those above. But, perhaps because it depended entirely upon his own by no means typical attitude towards the second law, he voiced it publicly on only one occasion in the years before the black-body research program it motivated was under way. The occasion was early, however, just two years after he had defended his thesis. Biographically as well as logically, Planck's reservations about gas theory are closely associated with his special relationship to the second law.

Planck had, as previously noted, provided an early restatement of his formulation of that law in the conclusion to his paper of 1882, "Vaporization, Melting and Sublimation." Emphasizing the special congruence of that formulation to equilibrium problems, he at once continued:

> In conclusion I should like to call explicit attention to a previously known fact. Consistently developed the second law of the mechanical theory of heat is incompatible with the assumption of finite atoms.* [For the accompanying note, see the discussion below.] It can therefore be foreseen that the further development of the theory will lead to a battle between these two hypotheses in which one of them will perish. An attempt to predict the conflict's outcome with precision at this time would be premature. Nevertheless, a variety of present signs seems to me to indicate that atomic theory, despite its great success, will ultimately have to be abandoned in favor of the assumption of continuous matter.⁵³

Fifteen years later Planck outlined virtually the same position in an important letter to be considered shortly. Not mechanics but its atomistic formulation conflicts with the second law; the resulting difficulties can presumably be eliminated by developing the mechanical world view for a continuum. Planck may not yet have had the ether in mind when that passage was written but that is what his continuous medium was to become.

The asterisk in the passage just quoted leads to a citation of the penultimate section of Maxwell's *Theory of Heat*. First published in

1871 in a series of "Textbooks of science adapted for the use of artisans and students in the public and science schools," the book appeared in German translations in both 1877 and 1878. Planck's reference is to the two-page section "Limitations of the Second Law of Thermodynamics," the first place where the statistical nature of the second law was pointed out. That law, Maxwell said,

> is undoubtedly true as long as we can deal with bodies only in mass, and have no power of perceiving or handling the separate molecules of which they are made up. But if we conceive a being whose faculties are so sharpened that he can follow every molecule in its course, such a being, whose attributes are still as essentially finite as our own, would be able to do what is at present impossible for us. For we have seen that the molecules in a vessel full of air at uniform temperature are moving with velocities by no means uniform, though the mean velocity of any great number of them, arbitrarily selected, is almost exactly uniform. Now let us suppose that such a vessel is divided into two portions, A and B, by a division in which there is a small hole, and that a being, who can see the individual molecules, opens and closes this hole, so as to allow only the swifter molecules to pass from A to B, and only the slower ones to pass from B to A. He will thus, without expenditure of work, raise the temperature of B and lower that of A, in contradiction to the second law of thermodynamics.[54]

The special "being" able to make heat flow from a colder to a warmer body is, of course, the one since known as "Maxwell's demon."[55]

In England, Maxwell's view of the second law was quickly accepted. Thomson developed it somewhat further in 1874; soon after that P. G. Tait (1831–1901) included it in the second edition of his widely used *Sketch of Thermodynamics*[56]; and Maxwell then gave it further currency in a long review of Tait's book published in *Nature* at the beginning of 1878. "The truth of the second law," Maxwell wrote, "is therefore a statistical, not a mathematical, truth, for it depends upon the fact that the bodies we deal with consist of millions of molecules.... Hence, the second law of thermodynamics is continuously being violated, and that to a considerable extent, in any small group of molecules belonging to a real body."[57] Thereafter the problem of the second law disappeared from British gas theory for almost twenty years.

On the Continent, attitudes were far more varied. At one extreme were the energeticists, who rejected atomism, mechanics, and the independence of the second law. At the other were Clausius and Boltzmann, who conceded that the second law was only macroscopic but who continued, as though dissatisfied by that conclusion, to display great interest in mechanical systems that showed non-statistical analogies to that law.⁵⁸ Neither of those positions was satisfying to Planck, who believed in mechanics and thought the second law absolute like the first. Recollecting the years before the turn of the century, he later wrote: "At that time, I thought the principle of entropy increase was valid without exception, like the principle of energy conservation, while for [Maxwell and] Boltzmann this principle was only a law of probability and thus subject to exceptions."⁵⁹ While he held that position, Planck could not have accepted the statistical approach.

But why should he have held it, at least for so long? By the 1880s no one else is known to have viewed the status of the second law as Planck did. Nor was there apparent reason to do so, for until 1896 no one even suggested that the "exceptions" to which that law was subject might conceivably be observable or have physical consequences.⁶⁰ Clausius dismissed Tait's claim that Maxwell had disproved his version of the second law with the words:

> The excess kinetic energy which may at a given moment pass from a colder to a warmer body by virtue of an accidental deviation from the [molecules'] average behavior is a magnitude of the same order as the mass of a single molecule compared with the masses directly accessible to our senses. Magnitudes of that order are, however, negligible for considerations relevant to the second law.⁶¹

Why could not Planck have taken a similar position? Is it possible that his autobiographical recollection did not quite catch the essentials of his objection to the probabilistic approach?

No conclusive answer is possible, but a plausible one relates directly to the reformulation of the second law that Planck worked out in his doctoral thesis and subsequent publications: all isolated physical systems move irreversibly from states of lower to states of higher entropy. In the thesis itself that strong a generalization is not yet fully explicit. There Planck introduces the term "natural" for processes to which the inequality sign in equation (5) applies; if the equality sign holds instead, the processes are called "neutral." The terminology itself suggests that neutral processes are an idealization, but in his thesis Planck never quite said so.⁶² Three years later, however, the terminological implication had become both explicit and central:

> There is in nature a function (for which we can retain Clausius's designation entropy) that is perpetually increasing, in the sense that the sum of the entropies of all bodies participating in any process is larger at the end than at the beginning. The entropy is thus to be recognized as completely determined by the state of a body (*i.e.*, *the totality of the positions and velocities of all its parts*).... Those processes for which the entropy is the same in the initial and final state, for example directly reversible

processes, form a limiting case. These sorts of processes do not really occur, however, but are to be regarded as an idealization.[63]

That passage makes irreversibility the core of the second law, and the italicized phrase suggests that the relevant irreversibility is microscopic. No one but Planck had taken such a position. The energeticists did not believe in atoms or recognize the distinction between reversible and irreversible change. Clausius, who did, had nevertheless developed thermodynamics in terms of reversible processes, introducing irreversible ones as a special case. In both versions of his *Mechanical Theory of Heat*, the second law first appears in the third chapter as $\oint dQ/T = 0$, and that form of the law is thereafter applied to a variety of physical problems. Only in the tenth chapter does Clausius turn to "Processes which are not reversible" and provide a "Completion of the mathematical expression of the second law" by writing $\oint dQ/T \leq 0$.[64] At that point only three more chapters remain in his book.

For Planck, then, a mechanical explanation of the second law had to be an explanation of irreversibility, preferably microscopic, and this he felt the statistical approach had failed to provide. Maxwell, though his treatment was suggestive, had not even touched on the general problem. Both Thomson and Boltzmann had gone farther, arguing that the final state of an irreversible process is more probable than its initial state.[65] Boltzmann had even given reasons to suppose that the entropy of a state is proportional to the logarithm of its probability. How and when Planck first heard of this approach is unclear (he may not have known Thomson's paper at all), but he was entirely dissatisfied with it when he did. That emerges clearly from a long letter that he wrote in the spring of 1897, at which time his black-body research was under way.

The letter requires a brief introduction. During 1895 Planck's young Berlin Assistant Ernst Zermelo (1871–1923) developed what has since been known as the recurrence paradox. Applying a mathematical theorem published by Poincaré five years before, Zermelo argued that no mechanical proof of the second law is possible, because any mechanical system left to itself must ultimately return to a configuration arbitrarily close to the one from which it began. "Hence," Zermelo wrote, "in such a system *irreversible processes are impossible* since (aside from singular initial states) no single-valued continuous function of the state variables, such as entropy, can continuously increase; if there is a finite increase, then there must be a corresponding decrease when the initial state recurs."[66]

Boltmann's reply emphasized that the second law was not simply mechanical but also statistical, and the argument continued to the end of 1896. Early in the following year Planck commented on it in a letter to his friend Leo Graetz (1856–1941). The problem of reconciling the second law with mechanics is, he wrote, "the most important with which theoretical physics is currently concerned." Then he continued:

> On the main point I side with Zermelo, in that I think it altogether hopeless to [attempt to] derive the speed of irreversible processes— e.g., viscosity or heat conduction in gases—in a really rigorous way from contemporary gas theory. Since Boltzmann himself admits that even the *direction* in which viscosity and heat conduction act can be derived only from considerations of probability, how can it happen that under all conditions the *magnitude* of these effects has an entirely determinate value. Probability calculus can serve, if nothing is known in advance, to determine the most probable state. But it cannot serve, if an improbable [initial] state is given, to compute the following [state]. That is determined not by probability but by mechanics. To maintain that change in nature always proceeds from [states of] lower to higher probability would be totally without foundation.

Here the question is not whether the second law admits exceptions, but a more fundamental one: can probability alone determine the direction in which a system develops? Planck's negative answer to that question appears to have provided his primary reason for rejecting statistical versions of the second law.

Shortly after challenging the proposition that nature must proceed towards more probable states, Planck's letter returned to the problem of irreversibility:

> Zermelo, however, goes farther [than I], and I think that incorrect. He believes that the second law, considered as a law of nature, is incompatible with any mechanical view of nature. The problem becomes essentially different, however, if one considers continuous matter instead of discrete mass-points like the molecules of gas theory. I believe and hope that a strict mechanical significance can be found for the second law along this path, but the problem is obviously extremely difficult and requires time.

Planck's hope is the one he had expressed in 1881, but its relation to his research had meanwhile changed. His effort to retrieve irreversibility from continuum mechanics had, in fact, begun more than two years before. Another passage in his letter suggests the spirit of the undertaking:

> I see only one way to reach a definitive conclusion about the question. One must embrace one of the two positions [mechanics or statistics] in advance and see how far one can proceed towards the light or towards

the absurd. That task will certainly be easier and more promising if one assumes the second law to be strictly valid (something which certainly cannot be shown from the kinetic theory of gases in its present form) rather than having recourse to hypotheses about the [improbable] initial state of the world simply to *save* gas theory.... That is to renounce any deeper insight.[67]

When Planck wrote those sentences, the first reports of his black-body research had already begun to appear. Their content will shortly concern us, but the first developed public statement of their objective belongs here. The passage that follows is the opening section of a paper Planck read to the Prussian Academy of Sciences in early February 1897. It introduces the first installment of a series, "On Irreversible Radiation Processes," which was to occupy him until the summer of 1899.

The principle of energy conservation requires that all natural occurrences be analyzable ultimately into so-called conservative effects like, for example, those which take place in the motion of a system of mutually attracting or repelling material points, or also in completely elastic media, or with electromagnetic waves in insulators. Non-conservative forces, on the other hand—like friction, inelastic impact, galvanic resistance—are not elementary in nature since they do not independently satisfy the energy principle. Therefore, in a complete system of physics, they must be reduced to the former [i.e., to conservative effects, all of them apparently reversible].

On the other hand, the principle of the increase of entropy teaches that all changes in nature proceed in one direction. When, for example, a system is isolated from its environment, then, after a longer or shorter time, an equilibrium or stationary state must set in, one from which the system does not depart under any circumstances while external conditions remain the same.

From this opposition arises the fundamental task of theoretical physics, the reduction of unidirectional change to conservative effects.[68]

That "fundamental task" was what had brought Planck to the black-body problem. After an intervening paragraph outlining the failure of earlier attempts to solve it, Planck allowed his audience a glimpse at the grounds for his hopes:

I believe I must recognize as a unidirectional process made up of entirely conservative effects the influence of a frictionless or resistance-less vibrating resonator on the wave which excites it .... Any such resonator is excited by absorbing energy from the vibration which falls upon it and is damped by radiating energy. The radiated energy is, however, generally not of the same sort as the absorbed, so that the vibrations of the resonator alter the character of the electromagnetic waves propagated in its vicinity. It can be shown that these alterations are unidirectional in various respects and that they have an equilibrating [*ausgleichende*] tendency.[69]

In the absence of one or more resonators, Planck pointed out, no such equilibration would occur. Electrical waves would be governed by Maxwell's equations alone, and they, like the equations of mechanics, are invariant when the direction of time is changed. But a resonator, Planck insisted, would alter the field, for example by absorbing energy from an incident plane wave and re-emitting it in the form of a spherical wave. In addition, because its amplitude changes only slowly with that of the field, a resonator will gradually eliminate fluctuations in the intensity of the radiation incident upon it. Finally, because it has a finite response range due to radiation damping, a resonator will interact through the field with resonators at neighboring frequencies, and that interaction will alter the distribution of radiation "color" or wavelength, as molecular collisions alter the distribution of molecular velocities. Analysis of the last of these equilibrating tendencies should, Planck pointed out, permit "important conclusions about the distribution of radiant energy in the stationary state,"[70] i.e., about the form of Kirchhoff's universal function.

Planck on the continuum and electromagnetism

The preceding pages have brought Planck step by step to the blackbody problem, but an obvious question remains. Why should Planck have supposed that resorting to the continuum would permit a mechanical or quasi-mechanical explanation of irreversibility and thus allow "the reduction of unidirectional processes to conservative effects"? That question, in turn, needs to be discussed twice, once for the years immediately following the completion of Planck's thesis, when he was apparently aware only that Maxwell's demon could upset the second law, and again for the 1890s, when he focused on irreversibility and turned his attention to the electromagnetic field. For neither period are complete answers possible, partly from lack of relevant information about Planck and partly because late nineteenth-century attitudes towards mechanics, the continuum, and the electromagnetic field have been too little studied.[71] But some relevant considerations can be isolated, and examining them will supply information also needed elsewhere about the state of electromagnetic theory at the time Planck's black-body work was done.

When Planck in 1882 announced his avowedly premature belief that atomic theory, despite its great success, would ultimately have to be abandoned in favor of the assumption of continuous matter, he is likely to have had something like the following in mind. In a continuum, which Planck may have conceived as studded with massy bits, motion in neighboring regions is strictly correlated. When the hole guarded by Maxwell's demon is open, regions on opposite sides of it are neighbors and move together. Energy may be transmitted through the hole nonetheless, but only from regions of more, to regions of less violent motion. Planck, who was a fine musician and often used acoustical analogies in his work, may from the start have thought of the resonant response of a stretched string coupled by a spring or other continuous medium to a driving source.[72]

Though it must remain speculative, that suggestion is indirectly reinforced by Helmholtz's views on the status of continuum mechanics. During 1877 Planck had heard Helmholtz lecture on the subject, and his student notes refer to an important and presently relevant distinction still discussed, many years later, in the published form of the lectures. There, Helmholtz began by reminding readers of his earlier treatment of particle mechanics. "We were then conscious," he wrote, "that that picture was only one of the observation-simplifying abstractions the consequences of which can represent the facts in certain areas of experience with adequate completeness and brevity." Elastic phenomena, Helmholtz went on, cannot be developed in their full generality by a theory that posits a particulate representation of matter. To treat elasticity, he concluded, "we will instead make use of the *opposite limiting conception* of the constitution of matter, namely the one in which matter continuously fills the occupied space."[73] Neither here nor elsewhere in his discussion is there a suggestion that one "picture" fits reality while the other is merely a convenient idealization. Rather the two are polar abstractions, useful in different realms of experience. Other cases, Helmholtz suggests, may require a compromise between the two, for example the theory of optical dispersion, a problem closely related to the one Planck would pose about resonators.[74]

That attitude is one that Planck, who believed in mechanics but was wary of models, would have found congenial. The distinction to which his notes on Helmholtz's lectures refer was intended to articulate and reinforce it. Particle mechanics, Helmholtz pointed out, recognizes two sorts of motion. One is the "ordered motion" in which all particles participate together (his example is the movement of a swarm of midges). The other is "disordered motion," the independent movements which the particles (or midges) carry out within the space their aggregate occupies. For Helmholtz the distinguishing feature of continuum mechanics was the prohibition of disordered motion.[75] With it vanished, however, the function for which Maxwell had conceived his demon.

Considerations of this sort were almost certainly still in Planck's mind when he took up black radiation in 1894, but their context had meanwhile changed. Now his problem was to demonstrate that irreversibility could characterize a system governed by conservative forces. In elaborating a solution, furthermore, he turned from the equations of mechanics to those governing the electromagnetic field. For Planck at this time, however, that second step was not a large one. He had chosen the black-body problem because his primary specialty, thermodynamics, was of unquestioned relevance to it. As a result, he was required to proceed from the field equations, which had been central to the thermodynamic arguments with which he was concerned. But in 1895 and again in 1897 he insisted that his method of demonstrating irreversibility could be applied equally well to the mechanical vibrations of acoustic resonators in a continuous medium of air.[76]

Quite possibly Planck thought the electromagnetic formulation the more fundamental: acoustic resonators cannot function in a vacuum[77]; for air, unlike the electromagnetic ether, there was a satisfactory particulate representation. But, during the early years of his research on the black-body problem, electromagnetic theory was for Planck a special sort of continuum mechanics, and his previous hopes for the continuum therefore applied to it unchanged. Only after 1897 did references to an equivalent acoustic proof of irreversibility disappear from his writings, and only at the end of 1900 did his papers begin to suggest that the transition from Newton's to Maxwell's equations might involve fundamentals. Both for Planck and for others, that was an important change of view, but it played no apparent role in Planck's thought until after the twentieth century had begun.

Though not the only viewpoint current, the close association of electromagnetism and mechanics was still widespread through the middle 1890s. Maxwell's own work provided clear precedent. In his *Treatise on Electricity and Magnetism*, published in 1873, the route to equations governing the field began with a series of Lagrangians in which the positions of circuits and the values of the currents in them served as generalized coordinates. Though the structure of the medium in which both currents and circuits moved remained unspecified, Maxwell continued to expect that it would be possible to work backwards from these mechanical equations governing the macroscopic variables to the hidden motions of unseen masses. Meanwhile, he conceived both the variables and the equations governing them as mechanical, and he repeatedly resorted to mechanical analogy in developing the equations and in clarifying their effects.[78] Very similar attitudes toward the mechanical basis of electromagnetism can be found in the relevant writings of Helmholtz, Boltzmann, and Hertz. Even H. A. Lorentz (1853–1928), soon to be known for his pioneering attempts to derive mechanics from electromagnetism, treated the field as a dynamical system in the first version of his electron theory, published in 1892.[79] The same conceptual continuum bridging mechanical and electromagnetic theories is a frequent feature of technical research papers in the 1880s and 1890s, the theory of optical dispersion supplying a particularly clear and apposite example. Until the late 1880s dispersion theory was primarily based on the laws of mechanics, thereafter increasingly on the laws of electromagnetic theory. But many of the same equations appeared in both modes of treatment; people engaged in one could borrow freely results achieved by the other; there was little sense of conceptual division between the two.[80]

Subject to further investigation, the preceding remarks on continuum mechanics must supply the context in which Planck's initially surprising references to that subject are to be viewed. In addition, they may explain Planck's early belief that continuum mechanics would defeat Maxwell's demon. But they provide no clue at all to his conviction that irreversibility could be derived from consideration of purely conservative effects. How could Planck, who recognized that the equations governing the field in an empty reflecting cavity were time-reversible, have supposed that the insertion of a resonator could generate unidirectional effects? Part of the answer has already been given. Planck's intuition plausibly suggested that the resonator's equilibrating effects—e.g., the conversion of a plane wave, traveling in a single direction, to a spherical wave, traveling in all—could not be reversed. But by 1896 Planck had another more concrete reason for faith in his research program. It depends on the state of electromagnetic theory in the 1890s, and considering it will bring us quickly to Planck's first black-body papers and to his earliest significant contribution to the theory of the electromagnetic field.

Until the development of Lorentz's electron theory during the 1890s, none of the main formulations of Maxwell's theory provided an account of the interaction between the field quantities (ether displacements) and matter.[81] Maxwell and many of his main successors, including Hertz, denied the existence of isolated electric charges and thus of charged material particles on which the field might have acted directly. Electric current did, of course, figure in the field equations, but ordinarily as an additional field quantity, another sort of ether-polarization, not as charge in motion. All three field quantities—electric, magnetic, and current—existed and satisfied Maxwell's equations both in free space and in the interior of bodies, their values in different media being related by appropriate proportionality constants and boundary conditions. While he conceived electromagnetism in this way, Planck's "resistanceless vibrating resonator" could not have been the charge-bearing mechanical vibrator which, as a bound electron, it was later to become. Instead, it was a tiny oscillating current, and thus a field quantity. The techniques appropriate to its study were a microscopic version of those Hertz had applied to the tuned circuits and dipoles with which, a few years earlier, he had generated and detected electromagnetic waves.

These differences are consequential. If Planck had analyzed Hertz's tuned circuits with twentieth-century field equations, in which isolated charges figure and in which current is represented as charge in motion, he would have recognized immediately that the equations relevant to the resonator problem are fully time reversible, like those for the field in free space. As it was, however, Planck followed Hertz in representing a physical resonator by means of boundary conditions on a surface exterior to it. Under those circumstances, reversibility was a characteristic to be investigated, not one to be discovered at a glance. Planck's first two black-body papers laid the groundwork for its exploration, and what emerged in the second of them was the first theoretical expression for what is now called radiation damping. When a small term representing that damping was introduced into the differential equation for a resonator, the equation ceased to be reversible in time, precisely the sort of effect on which Planck's hopes for the continuum depended. His treatment involved approximations, of course, and their elimination would have restored reversibility, not to his resonator equation, which is still in use, but to the equations for resonator plus field. But Planck was not to know that for a number of years, by which time it was inconsequential. He had long since abandoned hope that irreversible processes might be generated by conservative effects alone.

Detailed examination of the two papers in which Planck established the properties of his damped resonators would require a longer excursion into turn-of-the-century electromagnetic theory than is appropriate to the central concerns of this book. A brief sketch will, however, give body to what has just been said and simultaneously prepare for what is to come. Both papers open by underlining the conceptual novelty of the situation Planck is about to explore. For him its essential feature is the notion of "conservative damping."[82] Resonators are to be resistanceless if electric, frictionless if acoustic. They may, that is, lose energy only by radiation, so that total energy is retained in a directly mechanical form. In actual radiative processes, Planck concedes, there is also "consumptive damping," which converts vibratory energy to heat by friction or electrical resistance. But these consumptive effects may be minimized in practice, and they must be eliminated from theory to achieve "the reduction of unidirectional change to conservative effects."[83]

Planck's first paper, presented to the Prussian Academy in the spring of 1895, considers the case of a resonator in equilibrium with an incident field. Modeling his treatment explicitly on a paper Hertz had published six years before,[84] Planck first produced a free-space solution of Maxwell's equations in the form of a spherical wave train expanding with velocity c and wavelength λ from the origin of coordinates where the resonator is located. Assuming the resonator capable only of linear oscillations, Planck restricted attention to expanding waves symmetric about its axis. The field thus specified is the one emitted by the resonator (more accurately, the "resonator" is defined as the localized oscillating current that permits the field equations to be satisfied at the origin by the specified field), and it is fully determined by two constants, one fixing amplitude, the other phase. In his next paper and thereafter Planck would refer to it as the "secondary" field.

Planck then introduced a primary radiation from which the resonator might receive energy. For simplicity, it was restricted to a train of plane waves polarized parallel to the resonator's axis and with wavelength equal to that of the secondary radiation, and its nature was therefore also specified by a pair of constants fixing amplitude and phase. Adding the two fields together, Planck was able to compute the net energy flux through a small spherical surface about the origin. For equilibrium, that flux was required to be zero: the resonator at equilibrium emits as much energy as it absorbs. Assuming that the resonance was maximum, so that the two phase constants were equal, Planck was able finally to compute the ratio at equilibrium of the amplitude factors of the secondary and primary radiation. It proved to be dependent only on wavelength, not at all on the particular structure of the resonator. The latter would, Planck had shown, affect the field only very near the origin, and he had integrated over a surface farther out.

That is just the sort of result Planck had been looking for. After pointing out, near the end of his paper, that his methods apply equally to acoustic vibrations in air, he continued:

> Neither acoustical nor narrowly electrical problems have, however, motivated the preceding investigation. Rather, it was stimulated by the question of stationary radiation processes within a mechanical medium at rest at a constant temperature and surrounded by bodies with the same temperature.

Referring to his computation of the ratio of emission to absorption, he then went on:

> Kirchhoff's law for the proportionality of the emission and absorption capacities is an unmistakable consequence of this behavior. For the computation of vibration amplitudes one need not consider the detailed structure of the emitting particle; it suffices if the dimensions of the center of vibration are small compared with the wavelength.[85]

That result is Planck's first contribution to black-body theory, but its content is small and limited by numerous special hypotheses, particularly the restrictions to equilibrium and to a primary field at precisely the resonator's frequency. Within a year, however, Planck presented the Academy with a second paper in which the first restriction was dropped and the second relaxed. In that paper the primary and secondary fields were treated much as before except that their amplitudes were permitted to vary. In addition, the net energy flux through the small test sphere was allowed to depart slightly from zero so that the accumulated energy within the sphere might slowly change. Lacking information about the resonator's structure, Planck simply assumed that his computation would be unaffected if he extrapolated the dipole field outside of the sphere to its interior. The energy U of the dipole producing that field was known to be

$$U = \tfrac{1}{2}Kf^2 + \tfrac{1}{2}L\dot{f}^2, \tag{7}$$

where f is dipole moment, and K and L are resonator-dependent constants. Since any net flux across the sphere must appear within it as a change in dipole energy, Planck was able finally to combine his previous flux computation with equation (7) to produce a differential equation for the resonator itself. It proved to be

$$Kf + L\ddot{f} - \frac{2}{3c^3}\dddot{f} = E, \tag{8a}$$

where E is the electric field in the direction of the resonator axis due to the primary radiation.

Planck quickly pointed out that a resonator governed by equation (8a) has a natural period τ_0 and a logarithmic damping constant σ given by

$$\tau_0 = 2\pi\sqrt{\frac{L}{K}}, \quad \text{and} \quad \sigma = \frac{2\pi}{3c^3}\sqrt{\frac{K}{L^3}}. \tag{9}$$

Rewritten in terms of those parameters, the condition that resonator energy respond slowly to variations in the external field becomes $\sigma \ll 1$. For a resonator which satisfies that condition, Planck further showed that his first resonator equation is equivalent to

$$Kf + \frac{2K}{3c^3L}\dot{f} + L\ddot{f} = E, \tag{8b}$$

a form he subsequently used interchangeably with equation (8a). Both displayed the characteristic likely to have concerned him most. The radiation-damping term—third on the left in (8a), second on the left in (8b)—changes sign on time reversal, the other terms remaining the same. That result is a probable source of Planck's confidence that he was en route to a proof of irreversibility.

Planck did not make that point, however, presumably because he did not yet have such a proof. His resonators were imaginary entities of unspecified structure. Recourse to them in black-body theory was licensed only by Kirchhoff's law, which made the black-body field independent of cavity material. Both theoretically and experimentally, the black-body problem was about fields, not resonators, and it was therefore to fields that Planck had now to extend his work. The series of papers "On Irreversible Radiation Processes," of which the opening paragraphs were reproduced above, undertook that task, and its pursuit is the subject of Chapter III.

Before turning to Planck's work on irreversible changes in the electromagnetic field, however, a long detour is essential. Through most of the year 1897, Planck continued to believe that he could prove irreversibility directly, without the aid of any statistical or other special hypotheses. That proof had been his initial objective in taking up the black-body problem at all. But by the spring of 1898 he had recognized that that goal could not possibly be achieved, and the concepts deployed in his subsequent papers came more and more to resemble those developed by Boltzmann for gas theory. Some aspects of that resemblance may have been the result of independent discovery, but they cannot all be accounted for in that way. An understanding of selected aspects of Boltzmann's thought is therefore prerequisite to the study of Planck's development after 1897, especially because part of what Planck found in Boltzmann has been lost from view during the subsequent development of statistical mechanics.

II

PLANCK'S STATISTICAL HERITAGE: BOLTZMANN ON IRREVERSIBILITY

Electromagnetic radiation in a perfectly absorbing cavity is not the only physical system that moves irreversibly to equilibrium. If, for example, the velocity distribution of the molecules of a gas is initially different from Maxwell's, intermolecular collisions will produce that distribution with the passage of time and will thereafter maintain it. At least, that should be the case if the Maxwell distribution is, as its author claimed, the only one that remains stable as molecules continue to collide within the gas. Concern like Planck's with the second law of thermodynamics was not, therefore, the only motive for considering irreversibility. The natural way to show the uniqueness of the Maxwell distribution is to demonstrate that an arbitrary initial distribution of molecular velocities must, over time, evolve into and stabilize at Maxwell's form. Planck's electromagnetic problem, though not initially his motives, had had a precise parallel in gas theory.[1]

Until the end of the nineteenth century, however, only Boltzmann devoted significant attention to a kinetic theory of irreversibility. His first formulation appeared in 1872,[2] early in his career, and he repeatedly returned to the subject throughout his professional life, partly to generalize and extend his first result and partly to defend it against often cogent criticism. Those criticisms and Boltzmann's response to them are the sources of three interrelated sets of ideas which were to re-emerge in Planck's research as he approached the version of black-body theory that ultimately transformed physics.[3] The first set clusters about Boltzmann's so-called H-theorem, an apparent proof from mechanics that a gas must evolve to equilibrium from an arbitrarily chosen initial state. That proof and Boltzmann's initial interpretation of it are the subjects of the two sections that follow. The third section is devoted to a second set of ideas, derived from the very different probabilistic or combinatorial treatment of irreversibility which Boltzmann developed during 1877 to defend the H-theorem against paradox, but which also altered his understanding of what that theorem proved. Even with those alterations the H-theorem was not free from difficulties, and Boltzmann's wrestling with them produced a third conceptual cluster centering on the notion of "molecular disorder," to be discussed in the fourth and fifth sections below. Plank's simultaneous resort to the first and third of these sets of ideas will be treated in Chapter III. His first use of the second is a central theme of Chapter IV.

Boltzmann's H-theorem

Boltzmann's proof of irreversibility, the H-theorem, appeared in several versions, and Planck probably first encountered the somewhat simplified form developed for the opening chapter of Boltzmann's *Lectures on Gas Theory*, of which the first volume was published in 1896.[4] For present purposes, the accompanying discussion, which includes Boltzmann's only developed presentation of his recently conceived notion of molecular disorder, is more important than the proof itself. But unfortunately the discussion is incomprehensible without a prior knowledge both of the structure of Boltzmann's demonstration and of the debates it had previously provoked.

After a brief introduction, "A Mechanical Analogy for the Behavior of Gases," Boltzmann opens the first chapter of the *Gas Theory* by asking his readers to consider a container filled with a gas consisting of identical, perfectly elastic spherical molecules. Then he introduces a function f of velocity and time such that $f(u, v, w, t)\,du\,dv\,dw$ (usually abbreviated thereafter as $f\,d\omega$) is the number of molecules per unit volume that, at time t, have velocity components between the limits u and $u + du$, v and $v + dv$, and w and $w + dw$.[5] That function, $f(u, v, w, t)$, specifies the velocity distribution of molecules in the container at each instant of time. For simplicity, Boltzmann next introduces two additional assumptions which he later shows how to eliminate: the molecules are exposed to no external forces, and the walls of the container are perfectly smooth and elastic. On these assumptions, he writes:

> the gas is exposed to the same conditions everywhere within the container. It follows that, if at the beginning of some time interval the number of molecules per unit volume with velocity components between the [above] limits is on the average the same at each position in the gas ..., the same will hold at all future times.[6]

Elsewhere Boltzmann mentions a further condition, also required by the argument he is about to develop. His derivation requires the assumption that the volume $d\omega$ "is infinitesimal, but still contains a very large number of molecules."[7]

For simplicity, Boltzmann takes his molecules to be perfectly elastic spheres of diameter σ, and he then examines a collision between two of them, one initially contained in the velocity cell $d\omega$, the other in $d\omega_1$. To specify the collision further, he also posits that the line connecting the centers of the two molecules at the instant of collision lies in an infinitesimal cone of solid angle $d\lambda$ and given axial direction. All collisions between molecules originating in $d\omega$ and $d\omega_1$ and for which the line of centers is contained in $d\lambda$ will, Boltzmann writes, "for the sake of brevity be called 'collisions of the specified kind'."[8]

Boltzmann's next problem is to evaluate the number $d\nu$ of such collisions occurring in a unit volume during a time interval dt. For that purpose he elaborates a now standard technique first developed by Clausius to investigate the mean free path of gas molecules. Imagine a moving sphere of radius σ (twice the radius of an individual molecule) moving with velocity g, equal to the relative velocity of two molecules contained in $d\omega$ and $d\omega_1$, respectively. Any collision of the specified kind is equivalent to an encounter between the sphere and a fixed point provided that the fixed point lies in the surface element intercepted by the cone $d\lambda$ on the sphere. As the sphere moves during time dt, that surface element sweeps out a cylinder of volume $\sigma^2 \, d\lambda g \cos \theta \, dt$, where θ is the angle between the direction of the sphere's motion and the axis of the cone ($-\pi/2 < \theta < \pi/2$). If the sphere represents a molecule initially in $d\omega$, then the density of the points with which it can make a collision of the specified kind is $f_1 \, d\omega$, and the total number of such collisions during dt will be $f_1 \, d\omega_1 \sigma^2 \, d\lambda g \cos \theta \, dt$. (Here, f_1 is an abbreviation for $f(u_1, v_1, w_1, t)$; similar abbreviations will enter below without special comment.) Since, in addition, molecules in $d\omega$ have a density $f \, d\omega$, and since any one of them may enter into a collision of the specified kind, the total number of such collisions occurring in a unit volume during dt is given by[9]

$$d\nu = f \, d\omega f_1 \, d\omega_1 \sigma^2 \, d\lambda g \cos \theta \, dt. \tag{1}$$

Treating the same problem today, a physicist would describe that equation as yielding the average number of collisions. But Boltzmann, though he surely knew the difference, does not explicitly distinguish the average from the actual number of collisions, a fact that will shortly be important.

Having analyzed the rate at which collisions occur, Boltzmann calls upon the laws of mechanics to determine the velocity cells, $d\omega'$ and $d\omega_1'$, in which molecules initially in $d\omega$ and $d\omega_1$ will be found after a collision. These new cells, he shows, have two important properties. First, $d\omega' \, d\omega_1' = d\omega \, d\omega_1$. Second, a collision between molecules initially in the new cells $d\omega'$, $d\omega_1'$ will just return those molecules to the original cells $d\omega$, $d\omega_1$. Boltzmann contrasts these collisions, which he calls "collisions of the inverse kind,"[10] with the ones that originated in $d\omega$, $d\omega_1$. The latter, his collisions of the specified kind, remove molecules from $d\omega$ and $d\omega_1$, thus reducing the values of f and f_1. Inverse collisions, on the other hand, return molecules to $d\omega$ and $d\omega_1$, thus increasing the corresponding values of the distribution function. Both sorts of collisions are going on in the gas at all times, and the rates at which molecules enter and leave $d\omega$ and $d\omega_1$ can therefore readily be computed from equation (1). By combining previously derived results, Boltzmann can provide a formula for the rate at which the distribution function itself changes. It proves to be[11]

$$\frac{\partial f}{\partial t} = \int\int (f'f_1' - ff_1)\sigma^2 g \cos\theta \, d\omega_1 \, d\lambda. \qquad (2)$$

The preceding formula is a special case of the so-called Boltzmann equation, perhaps its author's most important contribution to science (its centenary was celebrated by a special meeting in Vienna in 1972). But it entered physics, not for its own sake, but simply as an intermediate step on the way to the H-theorem, and Boltzmann scarcely paused in his argument before putting it to work. Immediately after presenting equation (2) in the fourth numbered section of the first chapter of the *Gas Theory*, Boltzmann proceeds to its application in a new section called "Proof that the Maxwell Velocity Distribution Is the Only Possible One."[12] Its argument is mathematically complex but conceptually straightforward, so that only its result is required here. Boltzmann first defines a function $H(t)$ by the equation[13†]

$$H(t) = \int f \log f \, d\omega. \qquad (3)$$

Then he shows that

$$\frac{dH}{dt} = -\tfrac{1}{4} \int\int\int [\log(f'f_1') - \log(ff_1)][f'f_1' - ff_1]\sigma^2 g \cos\theta \, d\omega \, d\omega_1 \, d\lambda. \qquad (4)$$

In the integrand of that equation, the two expressions in brackets are either both positive, both negative, or both zero, so that $dH/dt \leq 0$. The equality sign, furthermore, obtains only if $f'f'_1 = ff_1$, a condition from which the Maxwell distribution can easily be shown to follow. Those results constitute Boltzmann's H-theorem. The function H, defined by equation (3) for an arbitrary initial distribution, can only decline to a minimum with the passage of time. As it reaches its minimum value, the distribution becomes Maxwell's, and that distribution is thereafter maintained. "We have shown," Boltzmann said, "that the quantity we have called H can only decrease, so that the velocity distribution must necessarily approach Maxwell's more and more closely."[14]

After an intervening section, "The Mathematical Meaning of the Quantity H" (to be considered in a later part of this chapter), Boltzmann puts the H-theorem to work. First he shows, by already standard techniques, that when $dH/dt = 0$ and the Maxwell distribution obtains, then Boyle's, Charles's, and Avogadro's laws all follow if temperature is equated with the mean translational energy of a molecule. Next, by equating heat content Q with total kinetic energy, Boltzmann derives an expression for the specific heat of a gas. Finally, he returns to the function H, now evaluated for the whole of a gas in a container rather than for a unit volume, and he proves by a lengthy but presently irrelevant argument a consequential theorem about it. When H has reached its minimum or equilibrium value, $H_{\min}$, it can differ from the negative of the entropy of the gas only by an arbitrary additive constant. Boltzmann shows, in short, that

$$\Delta H_{\min} = -\frac{\Delta Q}{T} = -\Delta S, \tag{5}$$

where the variation Δ is taken along a reversible path, the gas being maintained in equilibrium. At least for reversible processes, Boltzmann has derived the second law by combining mechanical and statistical premises.[15]

The first interpretation of the H-theorem

The preceding sketch of the H-theorem and its immediate consequences is modeled on Boltzmann's presentation in the first chapter of the *Gas Theory*, because Planck, who cites no other sources in which the theorem occurs, presumably found it there.[16†] But the parts of the *Gas Theory* so far selected for discussion differ only in mathematical detail from the corresponding portions of Boltzmann's original presentation of the theorem in 1872. That parallelism does not, however, extend to the paragraphs of the *Gas Theory* in which Boltzmann discusses the presuppositions required by his derivation and describes the significance of his results. When he wrote these interpretive passages, his views on the issues they raise were in a state of flux, inaugurated by discussions at the annual meeting of the British Association in 1894 and probably also shaped by a brief but sharp controversy with Planck at the very end of that year. As a result, Boltzmann's discussion of the H-theorem in Volume I of the *Gas Theory* includes elements which were very new as well as elements which would quickly disappear from his writings.

One of these elements, the concept of "molecular disorder," displays both characteristics. Boltzmann's only developed statement about it is the one in the *Gas Theory*, and he appears to have abandoned the notion very shortly after its publication in 1896. But despite that short life (for its subsequent reincarnation see the end of this chapter), "molecular disorder" played an important role in the development of physics, because Planck found it in the *Gas Theory* and, in 1898, applied to the black-body problem an electromagnetic equivalent: his concept of "natural radiation." Both are, by current standards, very strange notions, and they stand or fall together. Understanding Planck's position therefore requires an examination of Boltzmann's. Since the latter was short-lived and incompletely expressed, it can best be recaptured by an excursion through aspects of Boltzmann's developing thought of which Planck himself was only partly aware. Other Boltzmann contributions, equally vital to Planck, will be encountered along the way.

Look first at an apparent conflict or tension in the text of the *Gas Theory*, one which the concept of molecular disorder was intended to remove. When Boltzmann first published the H-theorem in 1872, he spoke of it in terms which, taken literally, imply that the values of f and H at an arbitrary time t are strictly determined by their values at an earlier time t_0. Shortly after deriving the Boltzmann equation, for example, he singles out equation (2) as showing that: "If the distribution function $[f(u, v, w, t)]$ is at any time determined by the formula [for the Maxwell distribution], then $\partial f/\partial t = 0$, i.e., *the distribution does not change further with the passage of time*." Later in the paper, after developing the H-theorem, he writes in a similar vein: "It can therefore be proved that, for the atomic motion of systems of arbitrarily many material points, there always exists a certain magnitude [the H-function] which cannot increase due to the atomic motion."[17] For statements like these the model is clearly a non-statistical mechanics, one in which the values of a system's coordinates are determined for all times by given equations of motion and specified initial conditions. But that model is not quite appropriate, a fact that any contemporary student of physics will recognize at once but one that was entirely missed by Boltzmann and his contemporaries throughout the nineteenth century. Statistics enters Boltzmann's derivation at the start in the computation of collision number, equation (1), and the statistical character of that parameter is carried over into the derivations of both the Boltzmann equation and the H-theorem. Given values of f and H at t_0, equations (2) and (4) permit the computation only of the most probable values of those functions at a later time. H, for example, *may* increase with time; it just very rarely does.

Conceivably, in 1872, Boltzmann thought his results were of the deterministic sort that his descriptive phraseology suggests. Maxwell's *Theory of Heat*, with its discussion of the sorting demon, had only just appeared. Boltzmann had sought a strictly mechanical proof of the second law during the 1860s, and he may not yet have realized the magnitude of the change introduced by his resort to Maxwell's statistical techniques.[18†] But he clearly came to do so during the decade after his first development of the H-theorem. During 1877, for example, in his first discussion of Loschmidt's paradox, to be considered below, Boltzmann identified a set of initial conditions from which H would necessarily increase for a time. That case is, he then wrote,

> extraordinarily improbable, and for practical purposes it may be regarded as impossible. Similarly, mixing a flask of hydrogen and nitrogen in such a way that after a month chemically pure hydrogen will have collected in the lower half, nitrogen in the upper, may be regarded as impossible, even though from the viewpoint of probability theory that outcome is only extremely improbable, not impossible.[19]

Passages of that sort, in which Boltzmann's understanding of statistical physics seems identical with that of Maxwell and his British followers, are a rarity in his writings before 1894. But they do occur, and they carry over into Volume I of the *Gas Theory* as well, for example, "The fact that H now increases does not contradict the laws of probability, for they imply only the improbability, not the impossibility, of an increase in H."[20] But in that volume (unlike Volume II) such statements are repeatedly juxtaposed with incompatible passages that retain the full non-statistical implications of Boltzmann's original

1872 paper. Two of them have already been quoted: "if at the beginning of some time interval [the value of the distribution function] is on the average the same at each position in the gas..., the same will hold true at all future times"; and "the quantity we have called H can only decrease." Others will be found in the discussion of molecular disorder, below. The juxtaposition is what creates the apparent conflict or tension previously referred to as inherent in Boltzmann's text. One wonders why it is there, why Boltzmann had not been able, twenty-four years after formulating the H-theorem, to abandon a deterministic phraseology when describing its results. That question is important because, as will soon become apparent, more than phraseology is involved.

The most plausible answer, for which additional evidence will appear here and there below, lies in the structure of Boltzmann's derivation. To carry it through he makes the distribution function f depend explicitly on time, develops a differential equation for its time dependence, and treats its form at t_0 as an initial condition which the appropriate solution of the equation must satisfy. That treatment is precisely the one appropriate to a function of a system's coordinates in classical, non-statistical mechanics, but it does not fit Boltzmann's problem. The form of $f(u, v, w, t_0)$ describes the actual distribution of molecular velocities at t_0, and it may thus seem, given an equation for $\partial f/\partial t$, to provide an initial condition capable of functioning like the initial conditions of classical mechanics. But, in fact, f is a coarse-grained distribution; an infinite number of different arrangements of the molecules within each cell $d\omega$ is compatible with the same form for f at t_0. Each of these arrangements corresponds to a different initial condition for a fully specified mechanical system, and each results in a different trajectory for that system over time. The actual form attributed to f at t_0 does not, therefore, determine the form of f at later times, excepting at times so close to t_0 that few intermolecular collisions have occurred. Like his formula for collision number, equation (1), Boltzmann's equations for the time rate of change of f and H determine only average or most probable values. Many other rates of change are compatible both with the given initial distribution and with mechanics.

Boltzmann ultimately recognized many of the problems that result from his initial misperception, some in 1877, others during and after 1894. That growing awareness of difficulty—one stage of which spawned the notion of molecular disorder—will be traced below. But Boltzmann never revised his proof of the H-theorem to eliminate the explicit dependence of f on t or to permit a family of f-curves to be generated from a given initial form for f. The concepts and vocabulary those revisions would have called forth remained undeveloped. That, I think, must be why, even when he knew better, his phraseology often continued to suggest that f and H were well-behaved, single-valued functions of time, functions whose values at time t were determined by their initial values at t_0. No wonder the essentially statistical premises of Boltzmann's derivation seem to vanish without a trace from its results.

Boltzmann's way of viewing the time dependence of f and H had another important consequence. Leaving no room for fluctuations, it inevitably obscured the existence of physical conditions under which the theorems at which he arrived would break down. With occasional exceptions, some of which will be encountered below, both the Boltzmann equation and the H-theorem appear in his writings as purely mathematical results the validity of which is independent of physical situation. But, as will appear below, both fail at gas densities sufficiently high that the molecular mean free path is of the same order of magnitude as the mean space between molecules. And it can already be seen, by consideration of collision number, that they also fail when gas density becomes too low. For a given unit of time and an appropriately chosen gas density, the collision rate will be high, and the actual and mean rates of collision will very nearly coincide. For much lower densities, however, the collision rate will be low and its value in successive intervals of time will fluctuate widely, causing corresponding variations in the time derivatives of both f and H. Only under appropriately chosen physical conditions and for appropriately limited periods of time does either the Boltzmann equation or the H-theorem provide a probable approximation to actual behavior. Boltzmann, however, remained insensitive to the need for specifying such physical conditions, and one example of that insensitivity is likely to have been significant to Planck.

Loschmidt's paradox and the combinatorial definition of entropy

The first stage in Boltzmann's transition towards a more fully probabilistic understanding of the H-theorem began early in 1876 when the Austrian physicist Josef Loschmidt (1821–95) presented to the Vienna Academy of Sciences a paper challenging Maxwell's conclusion that the temperature of a column of gas in equilibrium must, according to kinetic theory, be independent of height.[21] Close to its end, Loschmidt enunciated, almost in passing, a theorem intended to show the impossibility of deriving the second law from mechanics. If, he said, order or entropy is a specifiable function of the positions and velocities of the particles of a system and if that function increases during some particular motion of the system, then reversing the direction of time in the equations of motion will specify a trajectory through which the entropy must decrease. For every mechanically possible motion that leads towards equilibrium, there is another one, equally possible, that leads away and is thus incompatible with the second law. That statement presents what has since been called the reversibility paradox.

Though Loschmidt's challenge had not been addressed to Boltzmann in particular, the latter quickly took it up. His first response, a paper presented to the Academy at the beginning of 1877, suggests that he had thought deeply about the problem during the intervening year. Much in it, including the recently quoted passage about a hydrogen-nitrogen mixture, was new, especially the direction in which Boltzmann felt one must move to rescue the H-theorem from the difficulties posed by reversibility:

> Loschmidt's theorem merely teaches us to recognize initial states which lead, after the passage of a determinate time t_1, to a highly non-uniform state. It does not, however, prove that there are not infinitely more initial conditions which would lead, after the same interval t_1, to a uniform [distribution]. On the contrary, it follows from the theorem itself that, since *there are infinitely many more uniform than non-uniform states*, the number of states which lead to a uniform distribution after a time interval t_1 must be much larger than the number of those which lead to a non-uniform distribution. The latter, however, are just the ones which must, according to Loschmidt, be chosen as initial conditions if a non-uniform state is to occur after time t_1.[22]

Boltzmann provided no justification for the essential clause italicized above, but the sentence immediately following the quoted passage indicates that he already knew how such justification was to be sought. "One could even," he wrote, "calculate the probability of the various states from their relative numbers, [a procedure] which might lead to an interesting method of calculating thermal equilibrium." That computation of the probability of states is the topic to which Boltzmann proceeded in a major paper presented the following fall.[23] Planck followed it closely when deriving his distribution law in late 1900. For what he took from it, there is, as we shall see, no other source. Though introduced here for its contribution to the evolution of the concept of molecular disorder, the paper in which Boltzmann first computed the probability of states has another importance too, and must be examined in considerable detail.

After a brief introduction intended primarily to provide evidence that his probabilistic approach antedated that of Oskar Emil Meyer (1834–1909),[24] Boltzmann begins to develop for his readers the concepts and techniques required to compute the most probable state. Since both were then unfamiliar, he proceeds slowly, doubtless replicating to some extent the route he had followed himself. By the time he reaches the theorem he had sought, he has guided the reader through three different sets of computations, each slightly more complex and simultaneously more physical than the one before. Though modern readers require no guidance so detailed, Planck clearly did. In the event, all three of Boltzmann's derivations as well as the relations between them prove important to an understanding of Planck's thought.

Boltzmann's first model is both specially simple and also explicitly fictional. Imagine, he suggests, a collection of n molecules each restricted to an energy on the following finite list: $0, \varepsilon, 2\varepsilon, 3\varepsilon, \ldots, p\varepsilon$, where $p\varepsilon$ is the total available energy. If w_k is the number of molecules in the collection with energy $k\varepsilon$, then the set of numbers $w_0, w_1, \ldots, w_p$ is sufficient to define a particular state [*Zustandvertheilung*] of the gas.[25] Since individual molecules are, for Boltzmann, distinguishable, a state can be achieved in various ways, each of which Boltzmann calls a distinct complexion. There are, for example, n different ways (complexions) to achieve the state in which one molecule has energy $p\varepsilon$, all others zero. More generally, if a complexion is specified by a set of n numbers k_i, each fixing the energy ($k_i\varepsilon$) of the ith atom, a second complexion belonging to the same state can be achieved by any permutation of two molecules, i and j, which have different energies. The number of such permutations is, by known combinatorial techniques,

$$Z = \frac{n!}{(w_0!)(w_1!)\cdots(w_p!)}, \qquad (6)$$

where Z is the "permutability" (*not* the "probability").[26]

Boltzmann next imagines a large urn filled with well-mixed slips, each bearing a number from 0 to p inclusive, each number occurring equally often in the urn. The number on the first slip drawn determines the number (k_1) of energy elements to be attributed to the first atom, that on the second slip determines the energy of the second atom, and so on until n slips have been draw, thus determining a complexion. At that point the slips are returned to the urn and remixed. Repeating the drawing then determines a second complexion, and after many drawings, each possible complexion is represented approximately equally often. Finally, all complexions are discarded except those for which the total energy of the n atoms is just $p\varepsilon$. Within this subset all complexions compatible with the condition on total energy occur roughly equally often. But if these remaining complexions are sorted into piles representing not different complexions, but different states, then the states with high permutation numbers will contain more complexions than those with low numbers. The most probable state will be that for which Z in equation (6) is largest. It can be determined by maximizing Z subject to the two constraints:

$$\sum_{k=0}^{p} w_k = n, \quad \text{and} \quad \sum_{k=0}^{p} k w_k = p.$$

Since n is fixed, that problem is equivalent to minimizing the denominator of equation (6) or, for the sake of computational simplicity, of minimizing its logarithm.

Boltzmann therefore seeks the minimum of

$$M = \sum_{k=0}^{p} \log[(w_k)!],$$

a formula which can, for large values of w_k, be rewritten by a standard approximation now called Stirling's formula. The quantity to be minimized then becomes

$$M' = \sum_{k=0}^{p} w_k \log w_k,$$

where M' differs from M by an ε-dependent constant which has no effect on the location of the minimum. Standard variational techniques lead directly to the conclusion that, for $p \gg n$, Z will be maximum if the w_k's are given by

$$w_k = \frac{n\varepsilon}{\mu} e^{-k\varepsilon/\mu}, \tag{7}$$

where μ is the average energy of a molecule. That formula specifies the most probable energy distribution, which is what Boltzmann sought.

Having established his concepts and techniques by considering a fictional case, Boltzmann takes a first step back towards a physically realizable model by allowing the molecules to take on continuous values of energy.[27] Because his analysis again requires factorials, he begins by dividing the energy continuum into small finite ranges, 0 to

ε, ε to 2ε, 2ε to 3ε, etc. Then he writes the w_k's, which again define the state of the gas, as $w_k = \varepsilon f(k\varepsilon)$, where $\varepsilon f(k\varepsilon)$ is the number of molecules with energies between $k\varepsilon$ and $(k + 1)\varepsilon$. Thereafter the solution proceeds very much as before. The function to be minimized, again excepting an ε-dependent constant, is,

$$M' = \varepsilon \sum_0^\infty f(k\varepsilon) \log f(k\varepsilon),$$

and the minimization is subject to the constraints,

$$n = \sum_0^\infty \varepsilon f(k\varepsilon), \quad \text{and} \quad E_T = p\varepsilon = \sum_0^\infty k\varepsilon^2 f(k\varepsilon),$$

with E_T the total energy of the gas. For sufficiently small values of ε, these sums may be approximated by integrals, so that the final mathematical problem is to minimize the integral,

$$M' = \int_0^\infty f(E) \log f(E) \, dE,$$

subject to the constraints,

$$n = \int_0^\infty f(E) \, dE, \quad \text{and} \quad E_T = \int_0^\infty E f(E) \, dE.$$

Standard manipulations lead directly to the form of f for which M' is minimum: the number of molecules with energy between E and $E + dE$ must be given by,

$$f(E) \, dE = C \, e^{-hE} \, dE. \tag{8}$$

Presumably, this discussion of the continuous case was needed to prepare the way for Boltzmann's next step. It should nevertheless be noted that, since $E = k\varepsilon$ and $w_k = \varepsilon f(k\varepsilon)$, equation (8) follows directly from equation (7) when ε is small. Boltzmann's second derivation is in that sense redundant, and the redundancy will prove essential to the interpretation of Planck's early papers provided in Chapter V.

Boltzmann at once points out that equation (8) corresponds to the Maxwell distribution in two, rather than in three, dimensions. To treat the three-dimensional case, he continues, one must proceed to the actual physical situation, dividing up not the energy continuum but a three-dimensional velocity space.[28] If the permissible velocity components u, v, w are specified in small finite ranges, $a\varepsilon$ to $(a+1)\varepsilon$, $b\zeta$ to $(b+1)\zeta$, and $c\eta$ to $(c+1)\eta$, with a, b, and c integers running from $-\infty$ to $+\infty$, then the w_k's specifying a state become

$$w_{abc} = \varepsilon\zeta\eta f(a\varepsilon, b\zeta, c\eta),$$

and the corresponding permutation number is

$$Z = \frac{n!}{\prod\prod\prod_{-\infty}^{+\infty} w_{abc}!}.$$

Boltzmann now describes a new way of numbering slips for the urn, so that Z becomes the relative frequency of occurrence of the corresponding distribution in velocity space, and he then seeks, as before, to maximize Z by minimizing the product in the denominator of the expression for Z. To that end he first writes as sums the logarithm of the expression to be minimized and also the relevant constraints, discards a constant dependent on the values of ε, ζ, and η, and finally replaces the sums by integrals. His problem is thus reduced to maximizing the expression,

$$\Omega = -\iiint_{-\infty}^{+\infty} f(u, v, w) \log f(u, v, w) \, du \, dv \, dw,$$

subject to the constraints,

$$n = \iiint_{-\infty}^{+\infty} f(u, v, w) \, du \, dv \, dw,$$

and

$$E_T = \frac{m}{2} \iiint_{-\infty}^{+\infty} (u^2 + v^2 + w^2) f(u, v, w) \, du \, dv \, dw.$$

In this form, Boltzmann points out, the problem of a maximum has already been solved. The two constraints are simply expressions for the conservation of matter and of energy, and the quantity Ω is just the negative of the H-function he had introduced five years before. At that time Boltzmann had proved that H reaches a minimum when f corresponds to the Maxwell distribution, and he need not now repeat the demonstration.[29] Reference to it completes his proof that the case of thermal equilibrium, previously shown to correspond to minimum H, also corresponds to the most probable state of the gas.

Interestingly enough, Boltzmann claims little more than this here or elsewhere in the paper. In particular, he nowhere claims to have fulfilled the promise which led him to combinatorials in the first place. The initial states identified by Loschmidt, those from which H must increase, are not referred to, much less said to be extremely improbable. Boltzmann may already have become aware that no theorem quite like that follows from the notion of probability he has been developing, a point to be examined in the next section. But Boltzmann does, immediately after the argument just described, take the first of a series of steps that, when briefly reformulated in Volume I of the *Gas Theory*, make the logarithm of probability proportional to a generalized entropy function. Referring back to the first of the most recent set of formulas, Boltzmann writes:

> We shall call the magnitude Ω the permutability measure. It differs from the logarithm of the permutability only by an additive constant, and it has a special importance for the material to follow. I also note an advantage of suppressing the constant: the total permutability measure of the union of a pair of bodies is then equal to the sum of their individual permutability measures.[30]

By "the material to follow" Boltzmann means the discussion with which, after much intervening generalization of the theorems already presented, his paper closes. Apparently he regards it as demonstrating an important thesis, but the demonstration clearly fails in ways that will return us to one of the main criticisms of the probabilistic approach which Planck enunciated in his letter to Graetz.

Boltzmann's closing section is called "The Relation of the Entropy to the Quantity I Have Called the Probability of a Distribution,"[31] and it opens by showing that $\int dQ/T = 2\Omega/3 + C$ for a perfect monatomic gas at equilibrium. If the integration constant C is ignored, the entropy of the gas must thus be just two-thirds of the permutability measure Ω. From the second law, Boltzmann continues, it is known that the total entropy of all bodies involved in a given change of state must increase or, for reversible processes, remain constant, and the same must therefore be true of the permutability measure for systems in equilibrium. That characteristic of Ω Boltzmann immediately extends to transitions of a gas between non-equilibrium states, i.e., between states that do not obey the Maxwell distribution and for which entropy and temperature had not previously been defined. Unlike the entropy of a gas, he points out, "the quantity I have called the permutability measure can always be computed; and its value surely will be necessarily greater after the change of state than before."[32] These results are the ones gathered together in a special italicized paragraph close to the end of Boltzmann's paper:

> Consider an arbitrary system of bodies that undergo an arbitrary change between states that need not be characterized by equilibrium; then the total permutability measure of all the bodies will increase continuously during the change of state, and it can at most remain constant if all the bodies throughout the transformation approximate thermal equilibrium infinitely closely (a reversible change of state).[33]

Substitute "entropy" for "permutability measure" and Boltzmann's statement becomes the one which Planck, two years later, would present in his thesis as the most fundamental formulation of the second law. By the same token, it is a statement of just the sort that Planck hoped to derive from electromagnetic theory when he took up the black-body problem in 1894. At that time he felt that Boltzmann's attempts at a derivation had failed, and in the present instance, at least, he was clearly right. Even in the case where the initial and final states are equilibrium states, Boltzmann has had to call upon the second law to show that the permutability measure must increase. For the non-equilibrium case, he has provided no argument at all.

Though Boltzmann's concluding section nowhere makes the point explicit, there is a likely reason why he thought his argument stronger than it was. The title of the section had referred to the "relation of the entropy to... the probability of a distribution," and the permutability measure of a state is, by definition, proportional to the logarithm of its probability. Boltzmann's concluding generalization about permutability measure therefore transforms directly into the plausible assertion that all natural changes proceed from states of lower to states of higher probability. Almost certainly he had that formulation in mind in 1877, when the preceding passage was written: early in the same paper he had written, "from it [a highly improbable state], the system will hasten to ever more probable states until it finally reaches the most probable of all." In any case, the formulation is the one he introduced when summarizing his combinatorial derivation in an early section of the *Gas Theory* under the title "Mathematical Meaning of the Quantity H."[34] There, Boltzmann dropped the term "permutability measure" entirely and simply showed that $-H$ is proportional to the logarithm of the probability W and that both are proportional to the entropy for equilibrium states. The generalization to non-equilibrium states was then introduced with the assertion that, "In nature, transformations always tend to proceed from less probable to more probable states." That assertion, in turn, permitted him to conclude that, "In one respect we have even generalized the entropy principle by showing how entropy is to be defined for a non-equilibrium state of gas."[35] These conclusions are, of course, immensely plausible, but they remain, in the *Gas Theory*, entirely without proof. Planck must have had that gap in mind when, early in 1897, he wrote to Graetz:

> Probability calculus can serve, if nothing is known in advance, to determine the most probable state. But it cannot serve, if an improbable [initial] state is given, to compute the following [state]. That is determined not by probability but by mechanics. To maintain that change in nature always proceeds from [states of] lower to higher probability would be totally without foundation.[36]

In matters involving irreversibility and entropy, the burden of proof still, in 1897, lay with mechanics and the H-theorem, not with the combinatorial approach. And the H-theorem was threatened by Loschmidt's reversibility and Zermelo's recurrence paradox.

The conflation of "molecular" and "molar"

Immediately after providing a definition of the permutability measure, Boltzmann looks once more at the ground he has already covered and briefly analyzes two issues that will prove relevant to an understanding of Planck's work. The first concerns the appropriateness of the techniques used when numbering the slips to be drawn from the urn. Boltzmann's brief analysis discloses another important basis, in this case surprising, for his retention of both a deterministic phraseology for the H-theorem and also of the apparently fully probabilistic combinatorial approach, the one with which all the probabilistic passages previously quoted are textually associated.

Boltzmann had previously noted the importance of the method used to populate the urn. If slips are numbered to distribute molecules at random in velocity space, the Maxwell distribution results. But distributing molecules at random over the energy continuum yields a quite different distribution, one incompatible with thermal equilibrium. Boltzmann's justification for the former is worth quoting in full, for the conclusions to be drawn from it may prove controversial.

> The reason why only the one distribution of slips leads to the correct [final] state will not escape those who have studied problems of this sort

with penetration. It is as follows: consider all those molecules the coordinates of which are contained at a given time in the interval between

x and $x + dx$, $\qquad y$ and $y + dy$, $\qquad z$ and $z + dz$

and the velocity components of which lie in the interval between

u and $u + du$, $\qquad v$ and $v + dv$, $\qquad w$ and $w + dw$.

Next allow all these molecules to collide with some given molecule under specified circumstances [*gegebenen Verhältnisse*], so that after a given time their coordinates lie in the interval between the limits

X and $X + dX$, $\qquad Y$ and $Y + dY$, $\qquad Z$ and $Z + dZ$

and their velocity components in the range between

U and $U + dU$, $\qquad V$ and $V + dV$, $\qquad W$ and $W + dW$.

Then it is always the case that

$$dx\, dy\, dz\, du\, dv\, dw = dX\, dY\, dZ\, dU\, dV\, dW.$$

This theorem holds still more generally. If at time zero the coordinate and velocity components of arbitrarily chosen molecules (material points) lie within the [first pair of] limits and if arbitrary forces act on these particles, and if, after the passage of one and the same time t, the coordinates and velocity components of all these molecules lie within the [second pair of] limits, then [the last] equation is always satisfied.[37]

Boltzmann concludes by citing Watson's *Kinetic Theory of Gases* (p. 12), for a full and still more general version of this theorem and by pointing out that if energy and the two angles that determine the direction of motion replace the three velocity components in the equations of motion, then the invariant equation in the quotation above contains the square root of energy as well as its differential. These considerations provide, he says, the basis for a proof that only the last of his methods of distributing slips should yield a final most-probable state corresponding to the Maxwell distribution.

Boltzmann's intuition is sound, and good physics is spawned by sound intuition. Watson's result, which its author had attributed to Boltzmann and Maxwell,[38] is a special case of a general proposition known in this century as Liouville's theorem. Its subject is the trajectory of an isolated mechanical system described in a phase space which, for present purposes, may be taken to have $6N$ dimensions (i.e., the phase space appropriate to a gas consisting of N monatomic molecules interacting under central forces). If the point that specifies the position of such a system is contained in a $6N$-dimensional volume $d\omega$ at time t_0, then Liouville's theorem states that it will be contained in a phase-space volume of the same size at all future times. Together with some form of ergodic hypothesis or its equivalent, that theorem can be used to prove that equal volumes of phase space (or of position-velocity space) are equiprobable.

Clearly, Boltzmann knew the theorem and glimpsed its relevance. Nevertheless, in the passage just quoted he rewrites it as a theorem in six-dimensional rather than $6N$-dimensional space. An individual molecule, he says, will at all times be contained in an infinitesimal N-dimensional cell of the same size. If there happen to be N molecules in that cell initially, then they will all move together in that cell with the passage of time. That statement can be correct, however, only if the individual molecules are restricted to interactions with fixed scattering centers and exert no forces at all on other molecules, a conception actually suggested by Boltzmann's phrase, "collide with some given molecule under specified circumstances." But in a gas all individual molecules can interact; none may be treated as fixed during a collision. Liouville's theorem then applies only to the point specifying the position of the entire system in $6N$-dimensional space; individual molecules originally in the same N-dimensional cell will be widely dispersed as the gas develops over time. Unless considerably extended in ways not hinted at in Boltzmann's discussion, his argument, though highly suggestive, permits no conclusions about the relative probability of the different possible phase-space locations of individual molecules.

To a contemporary physicist, accustomed to conceiving even the neighboring particles of a gas as rapidly dispersed, the physical difficulties presented by Boltzmann's resort to molecules that move together are apparent at once. Boltzmann himself would surely have acknowledged so obvious a point if anyone had called it to his attention. But no one did so, and it is nowhere to my knowledge corrected in his later work. On the contrary, the only other passage in which Boltzmann even touches upon the justification for his method of numbering slips raises precisely the same difficulty. Deriving the H-theorem in Volume I of the *Gas Theory*, he considers, as we have seen, two separate molecules initially contained in volume elements $d\omega$ and $d\omega_1$, and he proves without difficulty that the product $d\omega \, d\omega_1$ is conserved through a collision. Later, when discussing the numbering of slips, he refers to that proof, saying, "as we have seen, if the velocity point of one of the colliding molecules lies in an infinitesimal volume before the collision, then it will afterwards lie in a volume of precisely the same magnitude, the other variables characterizing the collision remaining constant."[39] The closing phrase makes the statement literally correct, but simultaneously deprives it of relevance. If the collision is between molecules, both must be affected by it; the "other variables" characterizing it cannot be held constant; and the cell containing an individual molecule must therefore ordinarily change size. Thus, as late as 1896, Boltzmann still sometimes attributes to individual velocity or phase-space cells the mechanical, but non-statistical, behavior that he knows cannot be attributed to individual molecules.

What these discussions of cell behavior suggest, of course, is that Boltzmann's error was in some part an essential constituent of his viewpoint. It helped, that is, to preserve something very like his initial deterministically phrased interpretation of the H-theorem. Loschmidt had persuaded him, if persuasion were needed, that that theorem could not be through and through deterministic, that there were some initial configurations from which H would for a time increase. But Boltzmann at once reintroduced an almost equivalent interpretation by changing the focus of concern from the individual molecules to the cells that surround them. Cells pass through phase space in accordance with Newton's laws almost as the planets and stars move through the heavens, or as the sphere of radius σ had moved among fixed scattering centers in Boltzmann's treatment of the collision problem.

As a statement of Boltzmann's position, the last sentence is, of course, excessively explicit. He would doubtless have rejected it at once. But the sentence does capture something about his intuition of his subject, and what it captures can be made more precise by reference to a distinction introduced in our earlier examination of the H-theorem. Until 1896, when he first developed the concept of molecular disorder, Boltzmann repeatedly said things about cell coordinates that apply literally only to molecular coordinates, and vice versa. He conflated, or conceived as one, two very different notions of state, order, and probability. One set of concepts, to which in 1896 he would apply the term "molar," applies to the function f, i.e., to the distribution of molecules over cells, nothing being specified about molecular position within cells. The other set, for which Boltzmann simultaneously supplied the term "molecular," is determined by position within cells, i.e., by the precise specification, required by non-statistical initial conditions, of the position and velocity of each molecule. Only by moving subconsciously back and forth between these two ultimately independent conceptions was Boltzmann able to preserve for so long a predominantly deterministic way of discussing his H-theorem in uneasy equilibrium with his largely probabilistic combinatorial formulation.

Two passages in the 1877 papers just discussed will clarify what is involved and in the process provide some conceptual requisites for a discussion of molecular disorder. The first is from Boltzmann's initial response to Loschmidt and was quoted at the beginning of the preceding section:

> Loschmidt's theorem merely teaches us to recognize the initial states which lead, after the passage of a determinate time t_1, to a highly nonuniform state. It does not, however, prove that there are not infinitely many more initial conditions which would lead, after the same interval t_1, to a uniform [distribution].

In the first of these two sentences the phrase "initial states" must be read in the molecular sense. Loschmidt's paradox depends on tracing backwards the *precise* trajectories of molecules which have previously moved from an ordered to a disordered state. If terms like "state" and "initial conditions" are read in this way, as Loschmidt's paradox requires, then for every state that leads from disorder to order there is another that leads back from order to disorder. The theorem which the second sentence of the quotation identifies as the resolution of the paradox is then simply wrong. But if, in the second sentence, the phrase "initial conditions" is read in the molar sense, as Boltzmann uses it when actually computing the probability of states, then the theorem is right, though no longer an answer to Loschmidt. Used in that sense, "state" and "initial conditions" depend only on the number, not at all on the arrangements, of the molecules within each cell. There are then many more disordered than ordered states. For any given size of cell, furthermore, the two senses are independent. Having specified a distribution which is highly improbable in the molar sense, one may still arrange the molecular initial conditions within cells so that the gas will move to a still less probable state. Boltzmann's combinatorial techniques have not resolved Loschmidt's paradox.

An even more revealing passage occurs in the introduction to the 1877 paper in which Boltzmann first worked with combinatorials. After citing the passage just discussed (and quoting its closing remark: "One could even calculate the probability of the various states from their relative numbers"), Boltzmann proceeds to paraphrase what he had said and to indicate what the paper he is introducing will show.

> The passage thus makes explicit that one can compute the state of thermal equilibrium by investigating the probability of the various possible states of the system. The initial state will in most cases be a highly improbable one; from it the system will hasten to ever more probable states until it finally reaches the most probable of all, i.e., thermal equilibrium.[40]

Note that Boltzmann has reversed the terminology used in the passage he cites. There, improbable initial states led to violations of the H-theorem; here, they are the starting points for processes which exemplify it. That the passage nevertheless makes sense is due to Boltzmann's having simultaneously shifted his use of "state" from molecular to molar. That shift, however does more than save the sentence; it also makes the results obtained in the paper irrelevant to the deterministic theorem announced by Boltzmann's closing words. Troubles of this sort could not forever have remained hidden.

Turn now to the second analytic comment in Boltzmann's combinatorial paper.[41] It deals with the process by which Boltzmann, in the second and third of his derivations, passes from sums to integrals, and it will recall a second distinction introduced above in the discussion of the H-theorem. The resort to combinatorials demands, Boltzmann states, that one begin by dividing the energy or the velocity continuum into finite cells of size ε or ε, ζ, η. These cells must, furthermore, be large enough so that each contains many molecules. The quantities w_k or w_{abc} must, in short, be very large. While cells retain their initially given size, one may discard small quantities as well as size-dependent constants which cannot affect the form of the minimum. In this way one arrives at a sum the value of which is to be maximized, for example at:

$$\Omega = - \sum_{-\infty}^{+\infty} \sum \sum \varepsilon \zeta \eta f(a\varepsilon, b\zeta, c\eta) \log f(a\varepsilon, b\zeta, c\eta).$$

Only after reaching this point, Boltzmann says, may the quantities ε, ζ, η be further reduced and the transition to an integral made. Furthermore, one must even then remember that the symbols dE or $du\,dv\,dw$, though they appear as mathematical differentials, still represent cells large enough to contain many molecules. "This may at first glance seem strange," Boltzmann writes, "since the [total] number of gas molecules, though large, is still finite, while du, dv, dw are mathematical differentials. Yet on closer inspection the assumption must be regarded as self-explanatory, for all applications of the differential calculus to gas theory depend upon it."[42]

The last of those sentences is, of course, a non sequitur, and it suggests that there is something about the argument which Boltzmann is not seeing. The transition to an integral form depends for its legitimacy, not on some mathematical limiting procedure but on the validity of a physical hypothesis, and the latter is by no means self-evident. For plausible values of gas density and thus of n, it must be possible to choose cell size large enough so that each cell contains many molecules, yet small enough so that the variation of f is small as one moves from the center of one cell to that of the next. Boltzmann sees the first of these conditions as merely mathematical and never alludes to the second one at all, an oversight presumably facilitated by his little studied views about the relation between the continuous and the discrete.[43] Planck's subsequent insensitivity to an important difference between his distribution of energy over resonators and Boltzmann's distribution over gas molecules may be traceable in part to this aspect of Boltzmann's thought. Until sometime after 1906, Planck did not notice, or at least did not note the consequences of, the fact that under quite usual physical circumstances his distribution function varies markedly from cell to cell.

Molecular disorder

Boltzmann separated the concepts molecular and molar during 1895 or 1896, and the event is traceable to a complex series of encounters starting during the second half of 1894. Though not well known in Germany, Boltzmann's papers on gas theory were closely followed in England, where a number of Maxwell's successors worked actively on the subject. Striking evidence of both the extent and level of interest is provided by a long report on the field's current state presented by G. H. Bryan (1864–1928) to the meeting of the British Association for the Advancement of Science held at Oxford in August 1894.[44] Boltzmann was present and participated actively in the lively exchange that followed. Recapitulated and extended in a series of letters to *Nature* during the following twelve months, the discussion had a substantial impact on Boltzmann's understanding of what he had done.[45]

Few details of the debate need to be considered here, for the relevant parts were directed to problems already discussed.[46] E. P. Culverwell (1855–1931), describing himself as a relative outsider to gas theory, re-invented Loschmidt's paradox in order to ask how anything like the H-theorem could be derived from mechanics alone. His letter closed with the plaintive request that "someone say exactly what the H-theorem proves."[47] The principal answer was quickly supplied by S. H. Burbury (1831–1911) in a letter that provided a brief proof of the H-theorem and highlighted the essential role played in that proof by a special assumption. "If the collision coordinates be taken at random," Burbury wrote, "then the following condition holds, viz.:—For any given direction of R [the relative velocity of the colliding molecules] before collision, all directions after collision are equally probable. Call that Condition A." When he wrote again the following month, Condition A was explicitly the hypothesis that all the molecular coordinates are independent or uncorrelated. That hypothesis might not, Burbury emphasized, be a unique basis for a proof of the H-theorem, but it was certainly sufficient as well as plausible.[48]

These and other relevant letters persuaded Boltzmann to explain his views once more, and the explanation was of a sort he had not previously provided. In a long letter to *Nature* in February 1895, he discussed the shape of the H-curve and conceded for the first time that it might very occasionally rise to peaks greater than its established minimum value. Then he continued:

> What I proved in my papers is as follows: It is extremely probable that H is very near to its minimum value; if it is greater, it may increase or decrease, but the probability that it decreases is always greater.... The theory of probability itself shows that the probability of such cases [in which H increases] is not mathematically zero, only extremely small.[49]

For the first time Boltzmann had tied an apparently fully probabilistic statement directly to a discussion not of the combinatorial but of the mechanical H-theorem.

That important step was followed four months later by an even more decisive one. But the latter, though reported in a letter to *Nature* commenting on a communication from Burbury, had probably been prompted by a set of events independent of and apparently unrelated to the discussions that grew out of the British Association meeting. Ironically, those events (they amount to a confrontation) deeply involved Max Planck, whose posthumous edition of Kirchhoff's *Lectures on the Theory of Heat*[50] had been published early in 1894. Of the eighteen lectures that constitute that volume, the last six deal with gas theory, and preparing them for publication was Planck's first public involvement with Boltzmann's subject. Boltzmann responded at a meeting of the Munich Academy in May, opening his remarks with praise for the volume and with the suggestion that its very merits made it an appropriate target for critical comment.[51] Criticism was then introduced with a remark which deeply offended Planck: "Even those who—like the editor of the [volume]...now under discussion—think gas theory unworthy of the acumen expended on it would not wish those who do write on the subject to expend less." Turning next to matters of substance, Boltzmann concentrated on the Kirchhoff-Planck derivation of the Maxwell distribution, emphasizing that it involved an assumption for which no justification was given and which was, in any case, implausible.

The derivation in the lectures employs a distribution function $f(x, y, z, u, v, w)$ such that $f \, dx \, dy \, dz \, du \, dv \, dw$ is the probability of finding a molecule with position coordinates between x and $x + dx$, etc., and with velocity in the range u to $u + du$, etc. If two ranges of coordinates are chosen so that any pair of molecules contained, one in each, will collide, then the probability of such a collision's actually occurring is given by the product $f(x_1, \ldots, w_1) f(x_2, \ldots, w_2)$, where the subscripts 1 and 2 refer to the first and second molecule, respectively, and where the molecules are assumed to be independent. To this much Boltzmann has no objections, but he rejects out-of-hand the step that follows. To arrive at the Maxwell distribution by the route in the lectures requires the evaluation of the probability of inverse collisions, and it is taken to be $f(x_1', \ldots, w_1') f(x_2', \ldots, w_2')$, where the primed quantities are the coordinates of the molecules after collision. Justification for the use of the product form again rests upon the assumption that the two molecules are independent, and it is to this assumption that Boltzmann objects. Though it is legitimate, he points out, to treat as independent, molecules which have not yet collided, one may not do the same for a pair specially selected because they are receding from a recent collision.[52]

Planck's reply was delivered at a November meeting of the Academy.[53] In it he first defends himself against what he took to be a charge of editorial irresponsibility. He had, he insists, been entirely faithful to Kirchhoff's manuscript texts; if editorial responsibility demands the preparation of a new book, who would be willing to see a posthumous work through the press? As to Boltzmann's substantive criticism, Planck continues, he had noticed the point himself and will now suggest the following remedy. Though one may not generally equate the probability of an inverse collision with the product of the distribution functions of two independent molecules, one may do so when seeking the form of the equilibrium, or Maxwell, distribution. Equilibrium is, by definition, the state in which the distribution function is unaffected by collisions. Though the particular molecules receding from collision may not be independent, the probability of an inverse collision can, since the state of the gas is unchanged, be computed by the same formula used to compute that of the corresponding direct collision.

That response interested Boltzmann. At a meeting of the Academy in January 1895 he treated it with considerable respect, insisting only that Planck's argument would not suffice to prove that the Maxwell distribution was unique.[54] But more was to come, for Boltzmann had not yet realized either the full force of his own criticism or the consequence for it of Planck's reply. In deriving his H-theorem, Boltzmann had, like Kirchhoff, used the product of distribution functions to compute the effects of both direct and inverse collisions. Since his function had, however, represented an actual rather than a most probable distribution, the question of statistical independence was at least disguised. But within the more probabilistic framework of his 1895 letters to *Nature*—letters written after his response to Planck—the problem of inverse collisions was more likely to stand out. Boltzmann had learned from Burbury that the proof of the H-theorem demanded recourse to some explicit assumption, Condition A or an equivalent, not derivable from the premises on which his theory had previously been based. With that in mind, his criticism of Kirchhoff could be rephrased as follows. Though molecules engaged in a direct collision may be independent and thus obey Condition A, the correlation of their subsequent motions bars the application of Condition A to inverse collisions.[55] Therefore, if Boltzmann's criticism of Kirchhoff had been altogether correct, which it was not, then it applied equally to his own proof of the H-theorem. In that application, furthermore, Planck's defense of Kirchhoff's argument could be of no use, since Boltzmann's theorem, unlike Kirchhoff's, extended to the non-equilibrium case.

Nevertheless, analysis of Planck's argument may well have provided the clue required to save the H-theorem. By invoking the stationary character of the equilibrium distribution in order to show the independence of molecules undergoing inverse collisions, the argument isolates a puzzle. If collisions do introduce correlations and thus order, how can the distribution function possibly remain the same? Once the question is put in that form, the answer is straightforward. Imagine that two molecules receding from collision are contained in infinitesimal elements $d\omega'$ and $d\omega'_1$ of velocity space. Their coordinates are then correlated, but those of the other molecules *in the same cell* need not be. If, furthermore, the number of other molecules is large, the effect of the correlation due to the first pair will be negligible. The presence of many molecules per cell and the absence of special interrelationships between them are thus necessary if the distribution function is to be stationary and Planck's argument preserved.

Boltzmann's assimilation of these possibilities occurred in stages, of which the first is recorded in a much revised version of his January reply to Planck, submitted to the *Annalen* before the issue in which it appeared was closed at the beginning of May. The revisions are noteworthy for the first appearance of the phrase "molecularly disordered," though the notion to which Boltzmann there attaches the phrase is very nearly Burbury's Condition A, not the concept developed in the *Gas Theory*. "For the proof [of the H-theorem]," he writes, "it is necessary to assume that the state of the gas is and remains molecularly disordered, i.e., that the molecules of a given kind [contained, that is, in a given cell or pair of cells] do not always or even preponderantly collide in a certain way, but that the frequency of every sort of collision can be calculated from the laws of probability."[56]

Writing to *Nature* in July, Boltzmann develops the point further in ways that make its relation both to Planck's argument and to Condition A more apparent:

> Mr. Burbury points out, indeed, the weakest point of the demonstration of the H-theorem. If condition (A) is fulfilled at $t = 0$, it is not a mechanical necessity that it should be fulfilled at all subsequent times. But let the mean path of a molecule be very long in comparison with the average distance of two neighbouring molecules; then the absolute position in space of the place where one impact of a given molecule occurs will be far removed from the place where the next impact of the same molecule occurs. For this reason, the distribution of the molecules surrounding the place of the second impact will be independent of the conditions in the neighbourhood of the place where the first impact occurred, and therefore independent of the motion of the molecule itself. Then the probability that a second molecule moving with given velocity should fall within the space traversed by the first molecule, can be found by multiplying the volume of this space by the function f. This is condition (A).
>
> Only under the condition that all the molecules were arranged intentionally in a particular manner, would it be possible that the frequency (number in unit volume) of molecules with a given velocity, should depend on whether these molecules were about to encounter other molecules or not. Condition (A) is simply this, that the laws of probability are applicable for finding the number of collisions.[57†]

As here rephrased by Boltzmann, Condition A is a stipulation that molecules are situated at random both within and between cells. Burbury's version of Condition A had dealt only with the latter. This letter of Boltzmann's and the revised version of his reply to Planck are the places where the distinction between molecular and molar order begins to emerge. The position that results is very close to that of contemporary physics. If Boltzmann could have stopped with it, the passages on molecular disorder in the *Gas Theory* would not have been written, and the subsequent history of both kinetic theory and the black-body problem would have been at least slightly different. But there is a fundamental difficulty in reconciling Condition A with Boltzmann's derivation of the H-theorem, and he may well have discovered it when attempting to juxtapose the two for the first chapter of his lectures.[58] If the distribution function f is to be an explicit function of time, and if the collision number, equation (1), is to be used to derive a formula for $\partial f/\partial t$, then the "laws of probability" utilized in deriving the collision number must yield an exact result, not simply a most probable one. The difficulty is the one referred to in the first section of this chapter as the probable reason for Boltzmann's delay in relinquishing a deterministically phrased interpretation of the H-theorem.

The context of the difficulty is now different, however, and it leads us at long last back to the interpretive passages accompanying the mathematical presentation of the H-theorem with which this chapter began. They provide Boltzmann's only extended discussion of molecular disorder, a notion that enters immediately after the first introduction of a formula, equivalent to equation (1), for the number of collisions between molecules contained in cells $d\omega$ and $d\omega_1$. "Behind this formula," Boltzmann says at once, "lies a special assumption, as Burbury in particular has clearly shown."[59] To retrieve the required assumption he next explains what he means, not yet by molecular, but rather by molar disorder. It is the assumption that the distribution function is independent of position within the container, and it has been introduced solely to simplify the first demonstration of the H-theorem to be given in the book. The same demonstration can, as he will later show, be carried through for a gas possessing molar order, one in which temperature, density, and velocity distribution vary from place to place. But even with molar disorder presupposed, Boltzmann continues, in a passage later to be cited by Planck,[60]

> pairs of neighboring molecules (or groups of several molecules contained in infinitesimal space) can exhibit determinate regularities. A distribution which displays regularities of this sort will be called molecularly ordered. A molecularly ordered distribution (to consider only two examples from the infinity of possible cases) would be one in which the velocity of each molecule was directed towards its nearest neighbor or in which each molecule with velocity less than some limiting value had 10 specially slow molecules in its immediate vicinity....[61]

These examples do not, of course, define molecular disorder or even make the concept very clear. But a definition follows quickly. If one of these special molecular arrangements were to occur throughout the gas, then the condition for molar disorder would be satisfied but that for molecular disorder would not. Probability theory would not then apply, and the derivation of collision number would be endangered. To save it, Boltzmann suggests, one must stipulate the validity of equation (1), a step which in passing supplies the missing definition. "The validity of [equation (1)] can...," Boltzmann's paragraph concludes, "therefore be regarded as a definition of the expression: the distribution is molecularly disordered." That device—defining a concept as the condition required for the validity of a previously derived equation—is precisely the one Planck would use two years later to define his own related concept, natural radiation. In that case, as in this, its effect is to guarantee that the most probable distribution will be actualized.

After defining molecular disorder, Boltzmann at once proceeds to indicate why the concept is needed.

> If in a gas the length of the mean free path is large compared with the distance between neighboring molecules, then the particular molecules which are neighbors will quickly change with time. Under those circumstances a molecularly ordered but molar-disordered distribution would with high probability quickly become molecularly disordered as well.... But if, after computing in advance the trajectory of each individual molecule, we deliberately choose a suitable initial distribution, thus deliberately violating the laws of probability, then we can produce long-enduring [non-equilibrium] regularities or can construct an almost molecularly disordered distribution which will become molecularly ordered after a time. Kirchhoff, too, places the assumption that the state is molecularly disordered within the definition of the concept of probability.[62]

Since initial distributions chosen by "violating the laws of probability [!]" can result in violations of the H-theorem, they must be banned in advance. "We shall therefore," Boltzmann concludes, "now explicitly make the assumption that the motion is...molecularly disordered *and remains so throughout all future time.*"[63]

These are strange passages. For some years they have ordinarily been read as advancing an hypothesis of randomness, more or less equivalent to Burbury's Condition A.[64] Boltzmann's reference to Burbury and his cryptic description of the basis of Kirchhoff's derivation provide support for that reading, suggesting that it is not simply wrong. But, as will be seen, neither Planck nor Boltzmann's other contemporaries equated molecular disorder with randomness, and it is easy to see why. Throughout these passages Boltzmann's distribution function continues to specify an actual rather than a most probable distribution. Molecular order itself is an actual arrangement of nearby molecules, as Boltzmann's examples show. His "special assumption" does not demand simply that such arrangements be improbable, but rather that they never occur at all, either initially or as the motion proceeds. Finally, even when the special assumption was introduced in 1896, the H-theorem remained, for all Boltzmann's equivocations, the deterministically phrased theorem he had first developed in 1872.[65]

Because the equivocations in the *Gas Theory* are both substantive and new, dating principally from 1894, they are important to an understanding of Boltzmann's continuing development. No univocal reading of the preceding passages will quite catch the nature of their author's thought in the mid-1890s. Nevertheless, one of the ways in which the hypothesis of molecular disorder must be read is as a prohibition of certain actual configurations of the molecules within individual cells, configurations which the laws of mechanics, taken alone, would otherwise allow. On that reading molecular disorder is a physical hypothesis to be tested by experiment. That, in any case, is the way Planck read it, beginning shortly after Volume I of the *Gas Theory* appeared and continuing until sometime after 1906.

Epilogue: Molecular disorder and the combinatorial definition after 1896

Excepting one minor point, shortly to be supplied, the preceding pages complete the required account of Planck's statistical-mechanical heritage. To close the chapter at this point would, however, be to leave the statistical problem of irreversibility in a strange and, in the event, unstable state. Planck, who dealt with radiation rather than with gases, was for some years content with what he had learned from Boltzmann in the late 1890s, but gas theorists were not. Before we turn from their subject to the pursuit of black-body theory and the quantum, a few necessarily tentative words about the future of two formulations considered in this discussion may prevent frustrations which could result from a prematurely terminated narrative.

As presented in Volume I of the *Gas Theory*, molecular disorder was clearly a problematic concept. It could have been clarified and developed, but for many years it was not. Though Planck continued to apply an equivalent notion to radiation until at least 1906, molecular disorder had by then been dropped from gas theory. Boltzmann himself does not use it again after 1896. In particular, it is absent from Volume II of his *Gas Theory*, published in 1898. When Boltzmann rederives the H-theorem there, his recourse is exclusively to the independence of molecular coordinates outside of the limited sphere of molecular interaction. The increase in H with time is consistently treated as only highly improbable, not impossible. Speaking of the equilibrium distribution, Boltzmann emphasizes that: "For a finite number of molecules in a rigid container with completely smooth walls, the Maxwell distribution cannot hold exactly and throughout all times."[66] Abandonment of the concept of molecular disorder is thus only one indication of the extent to which Boltzmann's thought had continued to develop in the two years since 1896.

Elsewhere the notion of molecular disorder was discussed, but ordinarily for the problems it presented rather than as a useful premise. In 1899, for example, Burbury published a book intended to explore the limits of Condition A, which, he concluded, were severe. He does not there use molecular disorder, but recognizes it as Boltzmann's premise, and he occasionally asks whether a particular motion he examines is or is not "molekular-ungeordnet" in Boltzmann's sense. Significantly, he always leaves the phrase in German and often in quotation marks. Investigating a motion that satisfies Condition A when time flows in one direction but violates it when time is reversed, Burbury writes:

> We must now consider Boltzmann's own assumption that the motion is "molecular-ungeordnet." It being assumed that in the direct course the motion is molecular-ungeordnet, is it molecular-ungeordnet in the reverse course, or not? I think Boltzmann's answer to this question would be in the negative.... If this be so, then "molecular-ungeordnet" has, as applied to this theorem, precisely the same properties as my Condition A.[67]

Clearly, Burbury was puzzled about what "molekular-ungeordnet" could mean if, as Boltzmann's text suggests, it meant more than Condition A, and that puzzlement burst out publicly four years later. J. H. Jeans (1877–1946), in a paper published in the November 1902 issue of the *Philosophical Magazine*, had introduced "Boltzmann's assumption that the gas is in a 'molekular-ungeordnet' state" in the course of a proof of the equipartition theorem. Two months later Burbury responded, first quoting that clause and then continuing:

> I do not see the use of making it [the assumption of molecular disorder] unless we can reason from it, and that we cannot do till we know what it

means. It is possible to believe what we do not understand. It is not possible to reason about what we do not understand. Now Boltzmann gives us no adequate explanation of "molekular-ungeordnet." Nor does Mr. Jeans. Boltzmann makes no use of the assumption in argument. Nor does Mr. Jeans. Unless, indeed, the very definite assumption which they both (as I think) make, and which is stated below, is to be taken as the interpretation of "molekular-ungeordnet."... The condition thus assumed I call Condition A.[68]

Jeans more than conceded the point, though without reference to Burbury's criticism, in a long paper published the following June. The notion of "a molekular-ungeordnet state" was at best vague, he pointed out. As used by Boltzmann, he continued, it had the consequence that H could never increase, whereas from mechanics it must increase in the reverse motion, a contradiction. Finally, developing his own approach, Jeans derived an expression for the probability of finding a pair of molecules, one in each of two specified regions of position-velocity space. About it, he went on:

The probability given by [this] expression is exactly that which would be found in a homogeneous gas, with the help of the "molekular-ungeordnet" assumption. The whole supposed point of this assumption is, however, to exclude a certain class of systems, whereas it has just been seen that the result arrived at is only true upon the understanding that all conceivable systems—*geordnet* as well as *ungeordnet*—are included. It would therefore appear that the effect of this assumption is simply to defeat its own ends.[69]

Boltzmann's concept of molecular disorder as a prohibition of physically permissible states was rapidly disappearing from gas theory when that passage appeared in mid-1903. Indeed, outside of Volume I of the *Gas Theory*, it had never had a significant function.

The difficulties which had led Boltzmann to the concept remained, however. They are presented more or less acutely by any problem that requires resort to equation (2), the Boltzmann equation—e.g., viscosity, thermal conductivity, and, recently, plasma turbulence—and the term "molecular disorder" (or "molecular chaos") has figured repeatedly in their discussion. Sometimes it has been taken to be a stipulation of randomness, but that usage by itself solves little, for it is unclear what it can mean to say that a distribution function $f(u, v, w, t)$ describes a random distribution of molecules at time t_0. On other occasions the term has been applied to one or another mathematical condition that the distribution function must satisfy at t_0 if the Boltzmann equation is to apply during a subsequent interval $t_0 + \tau$. In this case the difficulty is to specify conditions on f at t_0 which will, under circumstances of physical interest, ensure the applicability of the Boltzmann equation for a significant length of time. The literature on these problems is immense, and the existing solutions both incomplete and controversial.[70] However imperfect and short-lived Boltzmann's original concept of molecular disorder may have been, it was addressed to a deep and enduring problem in statistical physics.

In marked contrast to molecular disorder, a second set of notions which Planck took from Boltzmann had scarcely come into its own. The probability calculus and the combinatorial definition of entropy did not attain their currently central position in statistical mechanics until sometime after 1910. Through the nineteenth century, the main approaches to gas theory began with the laws of mechanics, which were then applied to tracing molecules through collisions, or more complex systems through phase space. In this respect, though not in all, they resemble the approach Boltzmann had employed to derive the H-theorem, and they thus often obscured the point at which and the manner in which probabilistic assumptions entered and affected derivations. The elements that distinguish Boltzmann's 1877 paper—the identification of equiprobable states, reference to drawing slips from an urn or to rolling dice, and explicit recourse to the probability calculus—were notably lacking. Only after Planck applied them to radiation theory did they begin to assume their now standard place. That change in their status must be due in part to persistent efforts, dating from around 1906, to understand Planck's work, which was, unlike gas theory, apparently entirely dependent on them.[71†]

There were, of course, some nineteenth-century exceptions, of whom Boltzmann himself is the most prominent. But his example serves primarily to prove the rule, for the combinatorial approach presented in his 1877 paper was never a prominent ingredient in his work. He had developed it in an attempt to resolve Loschmidt's paradox. On the only other occasion he even sketched its content—the section "Mathematical Meaning of the Quantity H" in Volume I of the *Gas Theory*—his motive was the same. I am aware of only five other papers, all written in the aftermath of his initial response to Loschmidt, in which Boltzmann uses the calculus of probabilities at all, and in only the last of these is he applying it to a physical problem. Of the others, two are brief replies to O. E. Meyer and two are short elaborations of the mathematical probability theory he had developed in 1877.[72] For Boltzmann, in short, the probability calculus was primarily a technique for evading paradox; the mechanical approach to gas theory, here exemplified by the H-theorem, was always his fundamental tool, the one to which he returned again and again. Naturally enough, therefore, though his H-theorem was much discussed elsewhere, his resort to the probability calculus remained very nearly unknown. To the best of my present knowledge, Bryan's outline of it in his 1894 report to the British Association is the only published discussion of this aspect of Boltzmann's work before Planck took it up in December 1900. And Bryan says nothing of the relation between probability and entropy.[73]

Few other authors applied the probability calculus to gas theory at all. In a considerable literature I have as yet found only four. William Thomson (1824–1907), in a semi-popular lecture delivered in 1874, used probability theory (and, in an appendix, combinatorials) to explain the nature of order-disorder transitions.[74] O. E. Meyer, as previously noted, derived the Maxwell distribution from probability theory in 1877, and Kirchhoff, before turning to the more standard treatment which Boltzmann criticized, did the same in the thirteenth of the lectures published under Planck's editorship in 1894. Also in that year, Burbury applied error theory to a proof of the equilibrium distribution, noting in closing how one might represent "Boltzmann's Minimum Function" with the techniques developed in his proof.[75] But these are all scattered examples, apparently little known. When Jeans in 1903 redeveloped the Maxwell distribution and other results from the probability calculus, he felt justified in describing his approach as "New."[76]

Planck was, therefore, turning from the main developmental path previously taken by statistical physics when, in December 1900, he briefly abandoned the electromagnetic H-theorem he had recently developed in favor of combinatorials. In one respect, furthermore, that change was drastic, for only briefly in Volume I of the *Gas Theory* and at more length in Boltzmann's 1877 paper is a quantitative relation between entropy and probability even mentioned. Planck was apparently the first person other than Boltzmann to use it. Under those circumstances it is not surprising that his first work with combinatorials was very closely modeled on Boltzmann's article of 1877, including some of its idiosyncratic details. Nor is it odd that others found Planck's early quantum papers hard to follow, though for that more than an unfamiliarity with Boltzmann's combinatorial definition is responsible.

III

PLANCK AND THE ELECTROMAGNETIC
H-THEOREM, 1897–1899

Turn now to Planck's work on irreversible change in the radiation field. During 1895 and 1896 his research had, as previously noted, dealt primarily with the response of resonators to a field, and it had reached a first culmination with the derivation of a time-asymmetric differential equation for a radiation-damped resonator. Then, early in 1897, Planck produced the first installment of a five-part series entitled "On Irreversible Radiation Processes,"[1] a series that until mid-1899 constituted his entire published research. In all five papers, as well as in the major article, which recapitulated their results for the *Annalen* in early 1900, his concern was with the behavior of an entire system consisting of one or more resonators interacting with a field. Such systems provided his model for a black-body cavity, and, by analyzing them, he hoped both to prove irreversibility and also to reach "important conclusions" about the form of Kirchhoff's universal function.[2]

Even before introducing his black-body model in the first installment of the series, Planck announced his belief that the area in which he hoped to succeed was one in which kinetic theory had failed.

> To be sure, the kinetic theory of gases has undertaken to explain the approach to thermal-mechanical equilibrium... in terms of conservative effects, namely as the end result of all collisions between the numerous molecules which, conceived as points, interact through conservative forces. But a closer investigation [due to Zermelo] shows that the molecular motions assumed by the kinetic theory of gases are in no sense unidirectional, that quite generally any state which has once existed will, in the course of time, recur arbitrarily often to any desired degree of approximation. No rigorous theory of viscosity will be provided from the standpoint of the kinetic theory of gases without resort to some additional [non-mechanical] hypothesis.[3]

One "additional hypothesis" of the sort Planck hoped to avoid was molecular disorder, and he at once indicated the basis for that hope in a series of passages discussed in Chapter I. Opening with the sentence,

"I believe I must recognize as a unidirectional process made up of entirely conservative effects the influence of a frictionless or resistance-less vibrating resonator on the wave which excites it," Planck continues with a qualitative description of the "equilibrating tendency" of a resonator.[4] By averaging out inhomogeneities in the directions of propagation, in the phases, and in the amplitudes of the various components of an initially arbitrary field, resonators will ensure that that field proceeds towards equilibrium. Developing those ideas was Planck's object through the first three installments of his series. In the fourth, he executed an abrupt about-face and thereafter his black-body theory came more and more to resemble Boltzmann's theory of irreversibility in gases.

Cavity radiation without statistics

The initial form of Planck's black-body theory is developed in his first and third installments (the second, a brief response to criticism, will be discussed below). Mathematically, the discussion is elaborate, but only one aspect of it need concern us here.[5] For simplicity, Planck takes his cavity to be a conducting sphere at the center of which a resonator may be placed. He then specifies a particular resonator-free field (the "primary field" of his earlier papers) in terms of a potential function $\phi(t - r/c)$, developed in a Fourier series of long base period T. At the sphere's center ϕ is given by

$$\phi = \sum_n D_n \cos\left(\frac{2\pi nt}{T} - \theta_n\right), \qquad (1)$$

and Planck shows that the electric field component parallel to the z-axis is $4\ddot{\phi}/3c^3$. This is the field which will, at a later stage of his derivation, excite the resonator.

Planck is not yet ready, however, to introduce a resonator into his cavity. Instead, he continues to explore the resonator-free case, in the process developing concepts and techniques that will remain basic to his black-body theory until after 1906. Rewriting equation (1) with $t - r/c$ for t, Planck easily shows that the rate at which energy is radiated outward through a sphere of radius r about the origin is

$$\frac{dE}{dt} = \sum_n \left\{C_n \cos\left(\frac{2\pi n}{T}\left(t - \frac{r}{c}\right) - \theta_n\right)\right\}^2,$$

with $C_n = (2/3c^3)^{1/2}(2n\pi/T)^2 D_n$. That equation, in turn, he rewrites in the more cumbersome but physically more revealing form,

$$\frac{dE}{dt} = \frac{1}{2} \sum_n C_n^2 \left\{ 1 + \cos\left(\frac{4\pi n}{T}\left(t - \frac{r}{c}\right) - 2\theta_n\right)\right\}$$
$$+ \sum_\alpha \sum_n C_{n+\alpha} C_n \left\{ \cos\left(\frac{2\pi\alpha}{T}\left(t - \frac{r}{c}\right) - \theta_{n+\alpha} - \theta_n\right)\right.$$
$$+ \left.\cos\left(\frac{2\pi(2n+\alpha)}{T}\left(t - \frac{r}{c}\right) - \theta_{n+\alpha} - \theta_n\right)\right\}. \quad (2)$$

This is the equation Planck proceeds to analyze for its physical significance.

From the start he has assumed that the coefficients D_n in equation (1) contribute significantly to the field only for large values of n, those corresponding to wavelengths small compared with cavity diameter and thus, physically, to frequencies in the infrared and above. That assumption applies even more forcefully to the coefficients C_n in equation (2), so that all terms with small n may there be neglected. The terms that remain divide into two sorts. Within each of the sums in equation (2) the first term varies slowly with time (frequency 0 or α/T), the second rapidly (frequency $2n/T$ or $(2n + \alpha)/T$). But any measurement of radiant energy, Planck points out, determines, not the instantaneous value of dE/dt, but rather an average radiation rate over an interval long compared with the period of the corresponding radiation but short compared with the base period T of the series. To that average, the high frequency terms in equation (2) should make no contribution.

By carrying out the appropriate averaging procedure, Planck shows that the rapidly varying terms do not contribute. What he is then left with is a formula for what he calls "the 'radiation intensity J at time t.'"[6] Thereafter, J will be his fundamental measure of field intensity, the quantity he follows over time to demonstrate irreversibility. But he includes "intensity" within quotation marks when J is first introduced, partly because J represents the rate at which energy is radiated through an entire sphere rather than through a unit surface and partly because it is an average for an interval about t rather than the value at time t itself.

Straightforward manipulation of equation (2) yields the formula,

$$J = \frac{1}{2}\sum_n C_n^2 + \sum_\alpha \left\{ A_\alpha \sin\frac{2\pi\alpha}{T}\left(t - \frac{r}{c}\right) + B_\alpha \cos\frac{2\pi\alpha}{T}\left(t - \frac{r}{c}\right)\right\}, \quad (3)$$

with
$$A_\alpha = \sum_n C_{n+\alpha} C_n \sin(\theta_{n+\alpha} - \theta_n),$$
$$B_\alpha = \sum_n C_{n+\alpha} C_n \cos(\theta_{n+\alpha} - \theta_n). \tag{4a}$$

Since the coefficients A_α and B_α, unlike C_n and D_n, contribute most significantly for small values of the index of summation, J is a slowly varying quantity and in that respect differs from ϕ and dE/dt. Until 1906 much of the further development of Planck's black-body theory depends upon his production of other Fourier series which, like equation (3), represent the slowly varying intensity of a rapidly varying field. The techniques of manipulation and approximation developed for the derivation of equations (3) and (4a) are fundamental to Planck's entire classical radiation theory as are the concepts developed with them. In particular, though Planck may not yet have recognized it, the distinction between slowly and rapidly varying quantities parallels the distinction in gas theory between macroscopic and microscopic quantities. The first are measurable the second are not, and only the first are the subject of proofs of irreversibility.

Having developed these techniques for the resonator-free case, Planck at once applies them to the behavior of radiation in a cavity with a resonator at its center. From equation (1) he derives the exciting primary field at the origin, inserts it in the resonator equation (I-8a), and finds a Fourier series for resonator moment f as a function of time. From it he computes the secondary field radiated outward by the resonator, and adds it to the outward-moving component of the primary field. Applying to the resultant total field the manipulations developed for the field-free case, he obtains a new pair of formulas for the intensity of the field. The formula for J, equation (3), retains its form, but coefficients A_α and B_α are redefined as

$$A_\alpha = \sum_n C_{n+\alpha} C_n \sin\left(\frac{2\pi \Delta_\alpha}{T}\left(t - \frac{r}{c}\right) + \eta_{n+\alpha} - \eta_n\right)$$
and
$$B_\alpha = \sum_n C_{n+\alpha} C_n \cos\left(\frac{2\pi \Delta_\alpha}{T}\left(t - \frac{r}{c}\right) + \eta_{n+\alpha} - \eta_n\right). \tag{4b}$$

In these equations the constants Δ_α and η_n depend upon the resonator parameters K and L. Since it can be shown, in addition, that $\Delta_\alpha \ll \alpha$, the variation with time of A_α and B_α is again very slow.

Equations (3) and (4b) for the intensity J in the presence of a resonator are the ones Planck had hoped to use in demonstrating irreversibility without recourse to special assumptions. But, when he presented them at the end of 1897, he was almost surely aware that they would not fulfill his purpose. He was not yet prepared to admit defeat, however, and his discussion indicates how he had hoped to achieve his goal. The two terms in equation (3) behave, he says, very differently with the passage of time. To show that the radiation intensity approaches equilibrium irreversibly, Planck therefore argues that the variable term either is always negligible compared with the constant term (the case of equilibrium) or will become and remain small after some time interval within the base period T.

To carry through the argument, Planck points out, two plausible assumptions about the nature of the radiation are required. First, a large number of the amplitudes C_n must contribute significantly to the intensity or, in his words, the radiation field must not be "'tuned to the system.'"[7] Second, the phase constants η_n in equation (4b) must not vary systematically with n, must not be "'ordered.'"[8] Subject to these conditions there will be many terms in the summations determining A_α and B_α, and they will be no larger individually than the significant terms in the series $\frac{1}{2}\sum C_n^2$. Since their signs will vary unsystematically while the signs in the series $\frac{1}{2}\sum C_n^2$ remain positive, the variable portion of the series for intensity will generally be negligible compared with the constant. A special choice of initial conditions, i.e., of the phases η_n, may yield a significant variable term at $t = 0$, but it will thereafter decay irreversibly towards zero, the behavior Planck wishes to demonstrate. He is, however, forced to admit that:

> Probably cases are also possible in which the phase constants η_n possess values such that the radiation process is disordered at the beginning but at later times appears ordered. Under those circumstances, the radiation intensity would be constant at the start, and later experience noticeable variations. Whether such a process actually occurs in nature or not depends upon the nature of the initial [radiation] state.[9]

Those words are, excepting a parenthetical back reference, the closing statements of Planck's third installment. They mark his recognition of the crucial importance of suitable initial conditions in non-statistical proofs of irreversibility. Whether or not he yet entirely realized it, Planck's initial resort to electromagnetic theory had failed.

The entry of statistics and natural radiation

Planck's remarks about the role of initial conditions were a concession, and he must have felt their irony himself. Midway through 1897, four months after he had read his first installment to the Academy, a brief critique by Boltzmann was presented to the same audience. In it Boltzmann argued that, though Planck's new formulas for resonator absorption and emission were of great value, the program for which they were designed must necessarily fail. Whether or not resonators are present, both Maxwell's equations and the boundary conditions on their solution are invariant under time reversal. All processes that satisfy them can run in either direction and are thus reversible. "Any unidirectionality which Hr. Planck finds in the effect of resonators must," Boltzmann continued, "therefore derive from his choice of unidirectional initial conditions."[10] A mechanical analogue is, he concluded, provided by the case of numerous small spherical balls impinging from the same direction on a large fixed sphere. After reflection from the sphere, their direction of motion is more disordered than before, but that is only because they were specially arranged in the first place. The opposite case, a transition from disorder to order, could equally well be arranged. One would simply reverse the direction of motion of the balls receding from collision. In his second installment, a brief response to Boltzmann presented in July, Planck brushed these remarks aside as a product of "misunderstanding."[11] Five months later, closing his third installment, Planck tacitly conceded an essential part of Boltzmann's point.

After that concession, Planck's program for cavity radiation entered a new phase. The fourth installment of his series, presented to the Academy in mid-1898, opens with the announcement that the explanation of irreversible radiation processes requires a special hypothesis, precisely the step his program had initially been undertaken to avoid. The hypothesis itself, natural radiation, resembles an electromagnetic version of Boltzmann's molecular disorder, and its development demands explicit recourse to averages over resonator bandwidth. Finally, with the hypothesis in hand, Planck proceeds at once to seek a function that, like Boltzmann's H, can vary only monotonically, approaching a stationary value with time. These parallels, to be further developed below, are too close to be plausibly attributed to independent discovery, though that possibility cannot be categorically barred. Instead, they strongly suggest that by mid-winter 1897–98, at the latest, Planck was studying Boltzmann's version of the second law with care, was exploiting suggestions he found there, and had abandoned or all but abandoned his resistance to Boltzmann's approach. Unfortunately for historians, he did not explicitly acknowledge his change of mind for almost two years, a delay that has reinforced the almost universal impression that his conversion to a statistical viewpoint was intimately associated with his introduction of a quantum hypothesis at the end of 1900.[12]

Turn now to Planck's fourth installment, the one in which his revised program takes shape. After the briefest of references to previously established results, it begins:

> If the theory here developed is to be made useful for the general explanation of irreversible processes..., it is above all necessary to bar once and for all by a positive stipulation in advance, all radiation processes which do not display the characteristic of irreversibility. After carrying out this task mathematically [i.e., finding a mathematical expression of the necessary and sufficient conditions for irreversibility], it is then necessary to introduce the physical hypothesis that all irreversible processes in nature actually satisfy the stipulation under all circumstances.
>
> That step will be completed in the following paper by the introduction of the concept of natural radiation.... It will, that is, be shown that all radiation processes which possess the characteristic of "natural" radiation are necessarily irreversible in that the intensity of the waves which pass over the resonator always show smaller fluctuations afterwards than before.[13]

Planck is here looking ahead. Elaborate mathematical manipulations will be required before he can define natural radiation and discuss its consequences. But he quickly points to a more immediate difference between his present treatment and his past attempts. In earlier papers he has taken the intensity of radiation of period τ to be one-half the square of the amplitude of the corresponding term in the Fourier series for the total radiation. Now he recognizes that:

> A single term of the Fourier series has no independent physical meaning since there is no way in which it can be physically isolated and measured. To determine the intensity of a particular color one must rather, as is also the case in acoustics, determine the energy absorbed from the total radiation by a resonator of period τ and suitable damping.[14]

It is in this passage that the basis for parallel treatments of radiation and gas theory begins to emerge. The individual amplitudes and the corresponding phase angles, which together specify the Fourier series for an actual field or for its intensity, are like the actual coordinates of the individual molecules of a gas. Both are microscopic coordinates, and many different specifications of microstructure correspond, in both cases, to the same values of all measurable physical quantities. To determine the variation of the latter with time, one must therefore average over all microstates compatible with the same set of physical quantities. For radiation, as well as for gases, forbidden microstates must be postulated to ensure that change proceeds irreversibly.

Planck's mathematical argument opens by specifying, in terms of a Fourier integral, the axial component near the resonator of the net electric field,[15]

$$E = \int_0^\infty d\nu\, C_\nu \cos(2\pi\nu t - \theta_\nu). \tag{5}$$

Since E is now the net field, rather than a resonator-free field as in previous installments, the intensity J is given by $\overline{E^2}$, where the average is again taken over an interval long enough to eliminate effects with a period of the same order as that of the resonator. Straightforward manipulation yields for the intensity

$$J = \int d\mu (A_\mu \sin 2\pi\mu t + B_\mu \cos 2\pi\mu t), \tag{6}$$

with

$$A_\mu = \int d\nu\, C_{\nu+\mu} C_\nu \sin(\theta_{\nu+\mu} - \theta_\nu)$$

and
$$\tag{7}$$
$$B_\nu = \int d\nu\, C_{\nu+\mu} C_\nu \cos(\theta_{\nu+\mu} - \theta_\nu).$$

These are the equations that determine the energy relations in the field.

Inserting the field of equation (5) in the resonator equation (I-8b) and using also the equation for resonator energy (I-7), Planck next investigates the changing energy U_0 of a resonator with period ν_0 and damping constant σ. The result is, after considerable manipulation,

$$U_0 = \int d\mu (a_\mu \sin 2\pi\mu t + b_\mu \cos 2\pi\mu t), \tag{8}$$

where

$$a_\mu = \frac{3c^3}{16\pi^2 \sigma \nu_0^3} \int d\nu\, C_{\nu+\mu} C_\nu \sin \gamma_{\nu+\mu} \sin \gamma_\nu \sin(\theta_{\nu+\mu} - \theta_\nu)$$

and
$$\tag{9}$$
$$b_\mu = \frac{3c^3}{16\pi^2 \sigma \nu_0^3} \int d\nu\, C_{\nu+\mu} C_\nu \sin \gamma_{\nu+\mu} \sin \gamma_\nu \cos(\theta_{\nu+\mu} - \theta_\nu).$$

In these equations, which hold only after sufficient time has elapsed to eliminate the effect of initial conditions, γ_ν is defined by the equation

$$\cot \gamma_\nu = \frac{\nu_0^2 - \nu^2}{\sigma \nu_0 \nu}, \tag{10}$$

so that the value of $\sin \gamma_\nu$ is appreciable only for ν near ν_0. Thus, $\sin \gamma_\nu$ is like a response curve for the cavity resonator.

The integrals in equations (6) through (9) are to be taken from zero to infinity, but the behavior of γ_ν and the condition that physical intensity be well behaved[16] restrict significant values of the integrands to a region where μ is near zero and, at least for equation (9), ν near ν_0. If the integrals were transformed to sums, they would yield a constant term for $\mu = 0$ and a series of variable terms for μ near zero. The behavior of equations (6) through (9) is therefore like that of equations (3) and (4a) with which they may usefully be compared. To this point in the mathematical development of his revised theory, Planck has simply refined the version he had developed between 1895 and 1897. If that theory had given the desired result, he would have only to apply the argument developed for the pre-statistical equation (3) to the new equations (6) and (8). As the most recent quotation indicates, however, Planck now knows that that argument will not work, and he believes he also knows why. Equations (7) and (9) contain products of the form $C_{\mu+\nu} C_\nu$, which corresponds to nothing physical and which must be replaced by quantities that do before a usable result can be expected. In the course of that replacement Planck's derivation becomes, first, statistical and, then, dependent upon a special hypothesis which closely resembles molecular disorder.

To determine a physically measurable field intensity, Planck introduces what he calls an "analyzing resonator." Mathematically, its behavior is described by the same equations that govern the original cavity resonators, but it has a much larger damping constant ρ, set in the range $1 \gg \rho \gg \sigma$.[17†] An analyzing resonator thus behaves like a tuned laboratory probe. Planck emphasizes that, though it can respond quickly to changes in field intensity, it will leave the field it measures essentially unchanged. Defined in terms of this new resonator, the physical intensity at frequency ν_0 is a function of time governed by equations of a by now familiar form:

$$J_0 = \int d\mu (\mathbf{A}_\mu^0 \sin 2\pi\mu t + \mathbf{B}_\mu^0 \cos 2\pi\mu t), \tag{11}$$

with

$$\mathbf{A}_\mu^0 = \frac{2}{\rho\nu_0}\int d\nu\, C_{\nu+\mu}C_\nu \sin^2\delta_\nu \sin(\theta_{\nu+\mu} - \theta_\nu)$$

and (12)

$$\mathbf{B}_\mu^0 = \frac{2}{\rho\nu_0}\int d\nu\, C_{\nu+\mu}C_\nu \sin^2\delta_\nu \cos(\theta_{\nu+\mu} - \theta_\nu).$$

In those equations, δ_ν is defined by $\cot\delta_\nu = \pi(\nu_0^2 - \nu^2)/\rho\nu_0\nu$, so that $\sin\delta_\nu$ resembles a response curve for the *analyzing* resonator, takes its maximum value for $\nu = \nu_0$, and is very small except near that value.[18]

Unlike the quantities J, A_μ, and B_μ, which are determined by the form of some actual solution of the electromagnetic and resonator equations within the cavity, the quantities J_0, $\mathbf{A}_\mu^0$, and $\mathbf{B}_\mu^0$ are physically measurable. It is to them, not the microstate variables determining a particular field, that the desired proof of irreversibility should apply, and they ought therefore to be used to eliminate the original C_ν and θ_ν from equation (9), which, together with equation (8), applies to the interaction between a cavity resonator and the field. But there is, Planck notes, a serious difficulty about performing the elimination. A resonator responds to some actual field, not to the grosser quantities measured by an analyzing resonator, and the latter are not sufficient to determine the former. More precisely, if ε and η are rapidly varying functions of μ and ν such that

$$C_{\nu+\mu}C_\nu \sin(\theta_{\nu+\mu} - \theta_\nu) = \mathbf{A}_\mu^0 + \varepsilon$$

and (13)

$$C_{\nu+\mu}C_\nu \cos(\theta_{\nu+\mu} - \theta_\nu) = \mathbf{B}_\mu^0 + \eta,$$

then any actual field, determined by amplitudes C_ν and phases θ_ν, will yield the same values of $\mathbf{A}_\mu^0$ and $\mathbf{B}_\mu^0$ provided only that the corresponding ε and η satisfy

$$\int \varepsilon \sin^2\delta_\nu\, d\nu = \int \eta \sin^2\delta_\nu\, d\nu = 0. \qquad (14)$$

In these last equations, Planck points out, the effects of the rapid variation with frequency of the amplitudes and phases, C_ν and θ_ν, of a typical actual field are contained in the rapidly varying functions ε and η. The amplitudes $\mathbf{A}_\mu^0$ and $\mathbf{B}_\mu^0$ of the physically measurable intensity are, on the other hand, slowly varying quantities obtained by averaging, with appropriate weights, over neighboring Fourier components of the actual field. If these amplitudes are to be used in computing the response of a cavity resonator to the field, then one must assume that the actual field that stimulates the resonator behaves in the way represented by these averaged quantities. One must, that is, introduce an hypothesis of the same form as Boltzmann's stipulation that the actual rate of collisions in a gas is the same as the average rate.

That conclusion is of great consequence, and Planck's presentation of it is worth quoting in full.

> If we now return to the investigation of the [cavity] resonator with resonant frequency ν_0 and damping constant σ, then it is illuminating [to note] that the average values A_μ^0 and B_μ^0 are in general insufficient for the computation of the influence which the exciting radiation E [equation (5)] exerts on the resonator. The quantities C_ν and θ_ν must themselves be known....
>
> Under those circumstances the only alternatives are: either to abandon the attempt to find a general relation between the quantity U_0 and J_0—which all experience speaks against—or else to bridge the gap by introducing a new hypothesis. The physical facts decisively favor the second alternative.
>
> We shall now introduce and use throughout the following [discussion] the hypothesis which lies nearest to hand and is probably the only one possible. It is the assumption that, in computing U_0 from equations [(8) and (9)]..., the rapidly varying magnitudes $C_{\nu+\mu}C_\nu \sin(\theta_{\nu+\mu} - \theta_\nu)$ and $C_{\nu+\mu}C_\nu \cos(\theta_{\nu+\mu} - \theta_\nu)$ can, without significant error, be replaced by their slowly varying average values A_μ^0 and B_μ^0. That step gives the task of computing U_0 from J_0 a determinate solution to be tested by experiment. In order to make explicit, however, that the law to be derived below holds, not for all sorts of radiation but only after the exclusion of certain special cases, we shall designate those sorts of radiation which fit the hypothesis as "natural" radiation....
>
> A more intuitive but less direct grasp of the concept of natural radiation is provided by the following statement [of its defining characteristic]: the deviation of the rapidly varying magnitudes, such as $C_{\nu+\mu}C_\nu \sin(\theta_{\nu+\mu} - \theta_\nu)$, from their slowly varying average values A_μ^0, etc., shall be small and irregular.[19]

In effect Planck is defining natural radiation as any actual field which permits the use of equations (13) with $\varepsilon = \eta = 0$. That device is precisely the one Boltzmann had employed in defining molecular disorder as any molecular distribution that satisfied equation (II-1).

Planck's "fundamental equation"

Having developed the concept of natural radiation, Planck returned to a line of argument he had earlier elaborated in its absence. What he had then sought was an equation relating resonator energy U_0 to the actual field intensity J. Now he looked instead for a relation between U_0 and the physical intensity J_0. Both in a section heading and at the moment of its introduction, he described the formula that resulted as "the fundamental equation" of his theory.[20]

Planck had begun his earlier argument by considering the rate of change of resonator energy with time. As in the case of the field, fluctuations during intervals of the order of the resonator period were irrelevant, and he therefore began by averaging the resonator equations (I-7) and (I-8b) over a time interval containing a number of oscillations. Noting also that, for small damping, $\overline{Kf^2} = \overline{Lf^2} = U_0$, he readily showed that

$$\overline{Ef} = \frac{dU_0}{dt} + 2\nu_0 \sigma U_0. \tag{15}$$

In that equation the term on the left represents the rate at which energy is absorbed by the resonator; the second term on the right yields the rate of emission. This part of his argument Planck had largely completed before introducing even the expansion of the actual field E in a Fourier series.

Once that expansion was introduced he developed his result further, first writing the right-hand side of equation (15), with the aid of equations (8) and (9), in the form,

$$\overline{Ef} = \int d\mu (a'_\mu \sin 2\pi\mu t + b'_\mu \cos 2\pi\mu t), \tag{16}$$

with

and

$$a'_\mu = 2\sigma\nu_0 a_\mu - 2\pi\mu b_\mu$$

$$b'_\mu = 2\sigma\nu_0 b_\mu - 2\pi\mu a_\mu. \tag{17}$$

These equations Planck then restricted to the case of natural radiation, defined by equations (13) with $\varepsilon = \eta = 0$. After laborious manipulation, which involved several of the preceding equations, he was able finally to rewrite equation (16) in the form,

$$\overline{Ef} = \frac{3c^3\sigma}{16\pi^2\nu_0} \int d\mu (\mathbf{A}^0_\mu \sin 2\pi\mu t + \mathbf{B}^0_\mu \cos 2\pi\mu t).$$

The integral on the right is the quantity defined in equation (11) as J_0, so that comparison with equation (15) enables Planck to write

$$\frac{dU_0}{dt} + 2\nu_0 \sigma U_0 = \frac{3c^3 \sigma}{16\pi^2 \nu_0} J_0. \tag{18}$$

He has at last produced a simple differential equation relating the two physical quantities U_0 and J_0 of his theory.[21] That is what he calls his "fundamental equation."

One short last step permits Planck to rewrite equation (18) as a relation between resonator energy and density of radiant energy in the field. The latter is determined by the standard relation $u = (\overline{E_x^2} + \overline{E_y^2} + \overline{E_z^2})/4\pi$, where E_x, E_y, and E_z are the three components of the electric field vector. At equilibrium, $\overline{E_x^2} = \overline{E_y^2} = \overline{E_z^2} = \overline{E^2}$, the last term being the mean square field parallel to the resonator axis. Summed over all frequencies in the immediate vicinity of ν_0, $\overline{E^2}$ is just the intensity J_0, so that the energy density in the field is given by $u_0 = 3J_0/4\pi$. Equation (18) can therefore be rewritten in the form

$$\frac{dU_0}{dt} + 2\nu_0 \sigma U_0 = \frac{c^3 \sigma}{4\pi \nu_0} u_0.$$

Since at equilibrium, $dU_0/dt = 0$, the condition for equilibrium between radiation and field energy at frequency ν becomes

$$u_\nu = \frac{8\pi \nu^2}{c^3} U_\nu, \tag{19}$$

a form that will appear again and again in the pages to follow.

Entropy and irreversibility in the field

Planck's fundamental equation governs the slow secular variation of resonator energy with time. Given U_0 and J_0 at time t_0, he can use equation (18) to compute U_0 at some slightly later time $t_0 + \Delta t$, just as Boltzmann, after averaging over all possible collisions, could compute the value of his distribution function f at $t_0 + \Delta t$ when he knew its value at t_0. Planck is, therefore, also in a position to follow the remaining steps of Boltzmann's program for the second law, first providing a proof of irreversibility and then exhibiting an electromagnetic entropy function and deducing from it an equilibrium distribution. The first of these steps Planck took in both his fourth and fifth installments, though in somewhat different ways. The second step was introduced only in installment five, which apparently completed the program of research on black-body theory he had begun almost five years before.

Planck's fourth installment is still restricted to a special case, the single resonator at the center of a spherical cavity. Turning to the problem of irreversibility, Planck writes:

> The most direct symptom of the irreversibility of a process is the exhibition of a function which is completely determined by the instantaneous state of the system and which possesses the characteristic that it changes in only one direction, perhaps increasing, throughout the entire process. For the radiation process now being considered, there exists, by virtue of its extremely special character, not just one but a large number of functions which have this property. Because it is sufficient, for the proof of irreversibility, to know a single such function, we here select an especially simple one. In analogy with Clausius' thermodynamic function, we shall call it the entropy of the system consisting of a spherical cavity and resonator without thereby imputing to it any significance with respect to more general radiation processes.[22]

Planck then sets the entropy of the resonator equal to log U_0, where U_0 is the resonator energy. The entropy of radiation is equated, again without explanation or justification, to an expression proportional to $\int (\log J_0' + \log J_0'') \, dr$, where J_0' and J_0'' are the intensities of the incoming and outgoing waves, respectively,[23] and where the integral extends from the center to the periphery of the spherical cavity. Finally, Planck adds the two expressions and easily shows, using an earlier version of equation (18), that the time derivative of the total entropy can be expressed as the square of a real quantity and is thus necessarily positive or zero for natural radiation. His proof of irreversibility is then complete and his fourth installment virtually complete.

Planck's concluding installment deals with a far more general case, and in it he makes a very different claim. The cavity is now of arbitrary shape, and it contains infinitely many resonators tuned to all radiation frequencies. In this case the total entropy S_t is given by

$$S_t = \sum S_i + \int s \, d\tau.$$

The sum is over all resonators, their individual entropy being S_i, and the integral extends over the cavity volume, which contains radiation characterized by an entropy density s. Planck then, with no preparatory argument whatsoever, simply "defines" the entropy of a resonator of frequency ν and energy U by the equation

$$S = -\frac{U}{a\nu} \log \frac{U}{eb\nu}, \tag{20}$$

where e is the base of natural logarithms and "a and b are two universal positive constants the numerical values of which in the absolute c.g.s. system will be developed from thermodynamics in the next section."[24]

The entropy density s is also introduced by definition, but, since entropy is conserved in any reversible exchange between resonators and field, its form is in fact determined by equation (20) together with other previously elaborated relationships between field intensity and resonator energy at equilibrium. If K_ν is the intensity of a linearly polarized monochromatic wave of specified direction and with frequency ν, then, Planck says, the rate L_ν at which that wave transports entropy across unit surface perpendicular to its direction of propagation is given by

$$L_\nu = -\frac{K_\nu}{a\nu} \log \frac{c^2 K_\nu}{eb\nu^3},$$

where c is the velocity of light, and the other symbols have their previous meanings. For unpolarized radiation (Planck also treats the more general case), the total entropy crossing the same unit area is $L = 2\int d\nu L_\nu$, and the total entropy density is given by

$$s = \frac{1}{c}\int L \, d\Omega = \frac{4\pi L}{c}, \tag{21}$$

where Ω is solid angle. As in the previous installment, Planck next finds the total entropy from the preceding equations, and he then proves, laboriously but straightforwardly, with the aid of equation (18), that its derivative with respect to time takes a form that can only be greater than or equal to zero.

To this point, Planck's argument, though more general, is that of his fourth installment. Immediately, however, he goes farther. The quantity he has defined as entropy will, he points out, achieve a maximum value only when the total system of resonators plus field has reached equilibrium. In equilibrium, furthermore, the total entropy must be constant if the system undergoes some virtual displacement consistent with energy considerations. Planck considers the case in which, the other variables remaining constant, a small amount of energy is transferred from a resonator with frequency ν to another with frequency ν_1. The corresponding entropy and energy change will be

$$\delta S_t = \delta S + \delta S_1 = 0$$

and

$$\delta U + \delta U_1 = 0.$$

If these equalities are applied to the definition of resonator entropy in equation (20), it follows that

$$\frac{\delta S}{\delta U} = \frac{\delta S_1}{\delta U_1} = -\frac{1}{a\nu}\log\frac{U}{b\nu} = -\frac{1}{a\nu_1}\log\frac{U_1}{b\nu_1} = \frac{1}{\theta}, \qquad (22)$$

where $1/\theta$ is simply a constant to which $\delta S/\delta U$ for any single oscillator may be equated. Rewriting equation (22) yields

$$U = b\nu\, e^{-a\nu/\theta} \qquad (23)$$

for the distribution function governing oscillator energy as a function of frequency. The corresponding expression for the distribution of the density of radiant energy follows at once with the aid of equation (19),

$$u = \frac{8\pi b\nu^3}{c^3} e^{-a\nu/\theta}. \qquad (24)$$

Equation (24) is, of course, a form of the famous Wien distribution law, announced in 1896 after Planck's work had begun, and apparently well confirmed by experiment, if only in a restricted frequency range. It is therefore important to recognize that Planck is still two steps away from a derivation of that law. The first is trivial: Planck must rewrite equation (24) as a distribution function K_λ, for the intensity of radiation as a function of wavelength. That step he promptly takes to yield

$$K_\lambda = \frac{2c^2 b}{\lambda^5} e^{-ac/\lambda\theta}, \qquad (25)$$

the form he identifies as Wien's law. That identification depends, however, upon his taking the constant θ to be the absolute temperature, a crucially important step, which must now be examined at greater length. Much that will follow depends upon Planck's difficulties in justifying it.

If it could be assumed that the function S_t, which Planck has defined in equations (20) and (21), were the actual thermodynamic entropy, the step would present no problems. Planck could apply the standard thermodynamic equation $\partial S/\partial U = 1/T$ to equation (22) and conclude $\theta = T$. But all Planck knows about the function he has defined as entropy is that it approaches a maximum monotonically with the passage of time. In his fourth installment, dealing with a less general case, he had pointed out that there were a number of such functions and had explicitly refused to choose among them. Unlike the derivation of a distribution law, the proof of irreversibility did not depend upon such a choice.

A comparison with Boltzmann's derivation is instructive. Like Planck's S_t, Boltzmann's H could be shown to approach a limiting value monotonically with the passage of time. But that property it shared with many other functions—for example with H^2, log H, and so on. Among these alternates, only H and the functions that differed from it by an additive constant could serve as (negative) entropy, for only they would possess the additional property, $\partial H/\partial U = -1/T$. Boltzmann knew all this, and he had the means to show that he had chosen the appropriate function from the infinite set of possible candidates. Since classical thermodynamics attributes neither entropy nor temperature to a gas not in thermal equilibrium, he first let H proceed to its limit, $H_{\min}$. Then he showed that, at least for a perfect monatomic gas, $\delta H_{\min} = -\delta Q/T$, the change in thermodynamic entropy. The latter step depended, however, on his identifying T as the mean translational energy of a gas molecule, and he justified that identification by recourse to the empirically based laws of Boyle and Charles which govern the behavior of ideal gases. Without a detour through the gas laws, he would have had no basis for describing H as even an extension of the entropy function.

Planck can have no similar recourse in identifying his S_t with thermodynamic entropy. Temperature does not appear in Maxwell's equations or in Newton's laws. Planck's resonators are at rest, not subject to thermal motion. Under these circumstances he can only attempt to assure himself that his function S_t is uniquely determined by the conditions of his problem. Having done so, he sets $\partial S/\partial U = 1/\theta$, where θ represents the only possible "electromagnetic *definition* of temperature."[25] This "temperature" may, he then assumes without argument or discussion, be attributed to the cavity in which the radiation is contained. His derivation of the Wien law therefore rests on whatever reasons he can produce for the uniqueness of S_t, together with his tacit assumption that if S_t is unique, then its limiting or equilibrium form must be the same as that of thermodynamic entropy.[26]

Planck considers only the problem of uniqueness, and his remarks about it are brief:

> In the theory developed here...[Wien's] law appears as a necessary consequence of the definition of the electromagnetic entropy of radiation introduced in [equations (20) and (21)]; the question of the necessity of that law therefore coincides with the question of the necessity of that definition.... I have repeatedly tried to alter or generalize equation [(20)], which in turn determines [(21)],...in a way which would satisfy all well-grounded electromagnetic and thermodynamic laws. But I have

not succeeded. For example, the entropy of a resonator could be defined, not by equation [(20)] but by

$$S = -\frac{U}{f(\nu)} \log \frac{U}{\phi(\nu)}.$$

[The form of the distribution function that results has, however, been considered by Wien, who has shown that it is compatible with the displacement law, $U = \nu F(\lambda T)$, only if f and ϕ are both proportional to ν, precisely the form I have here introduced.]...

If one tries, conversely, to start from some distribution law other than [Wien's] and compute from it backwards an expression for the entropy, one always encounters contradictions with the theorem developed [above] concerning increase of entropy.

I believe I must therefore conclude that...the limits of validity of the [Wien] law, if they exist, are coextensive with the limits of the second law of thermodynamics. That conclusion of course increases even more the interest in further experimental tests of that law.[27]

Planck cannot have thought that *argument* very strong, and during the following year he was to find it mistaken. But until after the end of 1899 he seems to have been confident that his *conclusions*, the formulas for the entropy of resonators and field, were the only ones possible. Though his argument for uniqueness was weak, his confidence was well based.

Some reasons for confidence were direct. Planck could show that equations (20) and (21) had a number of non-trivial characteristics that any other entropy formula must share. Thermodynamics demands that the total entropy S_t be separable into two additive terms, one dependent only on the field variables, the other only on the state of the resonators. That condition bars such obvious choices of a monotonically increasing function as the square of Planck's S_t. The requirement that entropy be conserved in reversible processes prohibits such other alternatives as the sum of the squares of the functions specified by equations (20) and (21). Other obvious choices (for example the one Planck had developed for spherical cavities in his fourth installment) were presumably rejected for their failure to produce the required monotonically increasing function in the general case. Planck could show, in addition, as he did in print the following year,[28] that resonator entropy must be a function of the single variable U_ν/ν to satisfy the displacement law. Though he lacked a proof of uniqueness, he had many of its parts.

Strong indirect arguments must have suggested that the missing parts would be found. As Planck's reference to working "backwards" suggests, he had presumably found equation (20) by starting from Wien's law, conceived as an empirical regularity, and computing the corresponding entropy. That is, in any case, the route he would follow the next year after inventing his own distribution law. The very ease of the process and the simplicity of the formula that resulted from the use of an apparently well-confirmed law were in themselves suggestive. A further source of encouragement would have been the obvious parallels between Planck's $(U_\nu/\nu)\log(U_\nu/\nu)$ and Boltzmann's $f \log f$. To these considerations Planck added another, which was to assume still greater importance at the end of the following year. The constants a and b of the Wien law are, he noted, absolute natural constants, independent of time, place, or special convention. Two other well-known constants with the same characteristics are the velocity of light c and the gravitational constant G. Because of their dimensional relations, these four together permit the first definition of a set of units of mass, space, time, and temperature that share their absolute character and are thus in marked contrast to the normal c.g.s. units, which are products of "accident" and of the "special exigencies of our earthly culture."[29] Planck, who so greatly valued the absolute in science, clearly took great pleasure in this result: it provided still further evidence of the depth and validity of his approach.[30†]

Planck's program for research on cavity radiation was thus all but complete when he presented the fifth installment of his series to the Berlin Academy in May 1899. Though his initial hope of avoiding special hypotheses had been defeated and though a uniqueness proof remained to be supplied, the program had been largely successful. Planck had produced a new demonstration of irreversibility and, in the process, extended the depth to which thermodynamic arguments could penetrate radiation phenomena. One evidence of the increased power of such arguments was their ability to educe from theory the empirically based Wien distribution law. Another was the glimpse they provided of a new system of absolute units. Planck could be well satisfied with what he had achieved during the preceding five years. Only one step remained before he could consider turning to another topic.

Unlike his two preliminary papers on radiation theory, Planck's five-part series had appeared only in the Academy's *Sitzungsberichte*. To reach a wider audience, he would need to summarize his results systematically in the more widely read *Annalen der Physik*. Most of the paper he submitted to it in late 1899 was copied, nearly verbatim, from the fifth installment he had read to the Academy six months before. It opened, however, with a long new prefatory section in which Planck sketched his developed program in ways that displayed its close parallels to Boltzmann's, indicated the problems, especially of reversibility, that both programs shared, and outlined the similar additional hypothesis through which these could be bypassed. Boltzmann's need for such an hypothesis has, he wrote,

> led to objections to gas theory, on the one hand, and to doubts about the validity of the second law, on the other. But in reality there can be no question of alternatives like these.... Nothing stands in the way of the general development of the hypothesis of molecular disorder. The possibility of developing the second law in all directions on the basis of the kinetic theory of gases is thereby assured.[31]

Though somewhat belated, that is a not ungenerous capitulation.[32†] On the road to it, furthermore, Planck had retraced all but one of the steps taken by Boltzmann. Both men had initially sought a deterministic demonstration of irreversibility; both had been forced to settle for a statistical proof; and both had finally recognized that even that method of derivation required recourse to a special hypothesis about nature. By the beginning of 1900, only one aspect of Boltzmann's treatment of irreversibility was still absent from Planck's approach, the use of combinatorials, and by the end of the year, Planck had embraced that aspect, too. But what led him to do so was no longer the problem of irreversibility. It was rather the search for a radiation law that could pass the test of new, more refined experiments. Ultimately the change of focus was to reveal a new sort of physics.

IV

PLANCK'S DISTRIBUTION LAW AND ITS DERIVATIONS, 1900–1901

If only natural phenomena had been slightly different, Planck would now be remembered primarily for his considerable nineteenth-century contributions to the thermodynamic analysis of radiation and for his proof of the Wien distribution law. His name would not, in that case, be attached to the natural constant that has made it so nearly a household word. But he might well have received the Nobel Prize with or shortly after Wien, and their names would probably figure together in modern physics texts. Instead, he is remembered for the vastly more consequential, though not obviously more proficient, research he conducted during the first thirteen months of this century, work that rendered his earlier achievement not simply obsolete but mistaken. What accounts for his unexpected apotheosis is his response to the improved experiments which, just at the turn of the century, extended precise black-body measurements farther into the infrared than they had penetrated before. For that response both his earlier research and a set of events still to be considered had especially prepared him.[1]

Planck's uniqueness theorem and the new distribution law

The single obvious imperfection of the derivation of the Wien distribution law that Planck submitted to the *Annalen der Physik* in November 1899 was the lack of a uniqueness proof for the function he had defined as oscillator entropy. That difficulty he eliminated, or so he thought, in a paper he mentioned during discussions at the German Physical Society in early February 1900 and submitted for publication seven weeks later.[2] In that paper he claimed to have derived, rather than defined, oscillator entropy for the first time. Furthermore, since his derivation resulted in the same entropy function he had previously introduced by definition, the putative outcome of his paper was the first full proof of the Wien law. By the following fall Planck recognized that one of the assumptions invoked by his proof must be wrong. His triumph was therefore short-lived. But its importance is not to be measured by his proof's brief life in physics, for among the techniques Planck developed with it was the one that enabled him to propose, in October 1900, his own new distribution law, the law that has since withstood all experiment challenges.

The sources and motives for Planck's uniqueness proof are obscure and likely to remain so. He may simply have been trying to bridge the recognized gap in the argument that he had developed through November 1899. Or, more likely, he may have been responding to the criticisms O. Lummer and E. Pringsheim had included in a report on experimental findings that raised doubts about the validity of the Wien law. A third possibility, though it depends upon Planck's having known an as yet unpublished result, seems to me more plausible still. The subject of the discussion in which Planck first mentioned his new derivation of oscillator entropy was a paper in which Max Thiesen (1849–1936) suggested an alternative to the Wien law. Since Thiesen was a professor at the Physikalisch-Technische Reichsanstalt in Charlottenburg and a member of the German Physical Society, which met fortnightly in Berlin, Planck may well have known of his proposal a week or so before its formal presentation.[3] No more time would have been needed, for an investigation of Thiesen's distribution formula by techniques Planck had already developed would lead straightforwardly to all but the final step of his own forthcoming derivation. In addition, Thiesen's distribution formula presented a problem Planck is not likely to have ignored. It did differ from the Wien law, but it nevertheless satisfied just those thermodynamic criteria Planck had deployed to argue that his definition of resonator entropy, from which the Wien law followed, was unique.

Begin with Lummer and Pringsheim, who, like Thiesen, worked at Charlottenburg. Early in February 1899 they delivered the first of a series of reports on the frequency distribution of radiation from a new piece of laboratory apparatus, the first experimental black cavity. In the wavelength range from 0.7 μ to 6 μ, to which their first experiments were restricted, their results were generally favorable to the Wien law, but they did notice an apparently systematic dependence of the two constants in that law on temperature and wavelength. Further experiments would, they concluded, be required to determine "whether the variation is due to the nature of black-body radiation or to systematic, hard-to-control errors of observation."[4] By November, in a second installment, they were able to report that more careful experiments, extending to somewhat longer wavelengths (8.4 μ) and higher temperatures displayed the same systematic pattern of deviations. Though they concluded only that these could not be due to "accidental errors of observation," they did feel obligated to comment on the status of existing derivations of the Wien law. Concerning Planck's derivation, they remarked that it "would be conclusive only in the presence of a proof that any form differing from that [law] leads to an expression for entropy which violates the second law."[5] That remark, backed by their experimental findings, may itself have started Planck on his search for a uniqueness proof. A pressing challenge was, however, shortly to follow, and to show what it could have supplied, I shall present the next stage in the story as though it had been required.

In a paper read to the Physical Society on 2 February 1900, Lummer and Pringsheim's colleague Max Thiesen suggested that, writing x for the single variable λT, the well-known displacement law was compatible with a family of solutions

$$K_\lambda = \lambda^{-5} f(x) = T^5 \Psi_m \left[\frac{x_m}{x} e^{(1 - x_m/x)} \right]^a. \qquad (1)$$

In the right-hand side of that equation, x_m is the value of λT for which the radiation intensity K_λ reaches its maximum, Ψ_m is that maximum value divided by T^5, and a is a disposable parameter. For $a = 5$, the equation reduces to the Wien law, but Thiesen reported that the most recent data obtained by Lummer and Pringsheim, data they had kindly made available to him, were well matched only with $a = 4.5$.[6]

Planck's first response to that report, if he had not already taken equivalent steps, would surely have been to investigate the entropy function corresponding to equation (1) in an effort to see whether it was compatible with the second law. He would quickly have found that, for $a \neq 5$, equation (1) yields no closed form for $1/T$ ($= \partial S/\partial U$) and, therefore, no explicit form for $S(U)$. Planck's usual route to the proof that $dS_t/dt \geq 0$ would therefore have been barred. The very equation that disclosed that difficulty could easily, however, have suggested an alternate route. It and its derivatives with respect to U do yield relations between U and ν, on the one hand, and $(\partial S/\partial U)_0$, $(\partial^2 S/\partial U^2)_0$, etc., on the other, where the subscript indicates that the corresponding values are for the equilibrium case to which alone the putative distribution function applies. A power series in U is therefore at hand with which to investigate the behavior of entropy near that equilibrium. The result, as Planck mentioned in the paper he submitted in March,

PLANCK'S DISTRIBUTION LAW AND ITS DERIVATIONS   95

is a proof that the Thiesen distribution, like the Wien law, determines a local maximum of the entropy function and thus satisfies the standard thermodynamic criterion Planck had used before.[7]

Planck's article treats a more general case, asking simply what characteristics the function $S(U)$ must have if it is to possess local maxima. For this purpose he imagines a system initially in equilibrium and then displaced from it by adding a small extra energy ΔU to one of the resonators. The system will thereafter return by itself to equilibrium, the transition being governed, Planck shows, by a trivially altered form of equation (III-18), his "fundamental equation,"

$$\frac{dU}{dt} + 2\sigma\nu\,\Delta U = 0. \tag{2}$$

Further manipulation, using previously developed relationships between energy and field strength, leads to an equation for the change in total entropy dS_t, in a time interval dt during the system's return to equilibrium:

$$dS_t = dU\,\Delta U\,\frac{3}{5}\left(\frac{\partial^2 S}{\partial U^2}\right)_0, \tag{3}$$

where dU is the change in resonator energy during dt.[8†] Since, by equation (2), dU and ΔU must have opposite signs, dS_t will be positive and the equilibrium position must be a local maximum provided that

$$\frac{3}{5}\frac{\partial^2 S}{\partial U^2} = -f(U), \tag{4}$$

where $f(U)$ is any positive function of U. It is this last condition that, Planck states, is satisfied by the entropy function corresponding to the Thiesen distribution. Other forms, which correspond to different choices of $f(U)$, are available as well.

Planck could well have come this far simply by analyzing the Thiesen distribution. When publishing the result, however, he continues at once to an additional step. Suppose that instead of displacing a single resonator from equilibrium by an amount ΔU, one displaces n of them by that amount, a total displacement of the energy of the system by an amount $\Delta U' = n\,\Delta U$. Since entropy is additive, the corresponding change in dS_t should be $dS'_t = n\,dS_t$. On the plausible assumption that the total resonator entropy S' depends on the total resonator energy

U' in the same way that the entropy S of a single resonator depends on its energy U, we must have

$$dS'_t = -dU' \, \Delta U' f(U') = -n^2 \, dU \, \Delta U f(nU) = n \, dS_t = -n \, dU \, \Delta U f(U)$$

or

$$f(nU) = \frac{1}{n} f(U).$$

The last functional equation is, however, satisfied only when $f(U)$ is proportional to $1/U$ or, from equation (4), when

$$\frac{\partial^2 S}{\partial U^2} = -\frac{\alpha}{U}. \tag{5}$$

From that equation the entropy can, for the first time, be derived. Two integrations, the standard relationship $\partial S/\partial U = 1/T$, and the displacement law together yield both the Wien distribution and the expression for resonator entropy which Planck had previously introduced by definition. "Since this [direct] computation [of entropy] yields," Planck says, "the same expression for entropy as before, my view of the formula's significance is still further strengthened even though its basis has been in part somewhat displaced."[9]

By March 1900, then, Planck's black-body program had for the second time reached what might well have been its final stage. But experiment again proved stubborn, a fact that Planck must have known by early October at the latest. In a paper, the essential contents of which were reported to the Physics Section of the Naturforscherversammlung on 18 September 1900, Lummer and Pringsheim concluded that "the Wien–Planck distribution law does not represent our measurement on black radiation in the region from $12 \, \mu$ to $18 \, \mu$."[10] In this range, achieved only with the aid of recently developed techniques, the discrepancies between experiment and theory ranged systematically from 40 percent to 50 percent and could not conceivably be due to experimental errors.

Their evidence was entirely convincing, but Planck, however disappointed he may have been, was by now well prepared. In a paper presented to the Physical Society on 19 October, he referred to the proof of the Wien law he had submitted to the *Annalen* in March, and at once pointed out its shortcoming.[11] The entropy of n oscillators must, he said, depend not simply on their total energy U', as he had supposed, but also on the energy U of a single oscillator. A more complex form

PLANCK'S DISTRIBUTION LAW AND ITS DERIVATIONS 97

must therefore replace $-\alpha/U$ in equation (5), and Planck had found, he reported, a form that "is the simplest by far of all the expressions which yield S as a logarithmic function of U (*a condition which probability theory suggests*) and which besides coincides with the Wien law for small values of U."[12†] If Planck's original form, equation (5), is regarded as the first term $(-U/\alpha)$ in a power series expansion of $(\partial^2 S/\partial U^2)^{-1}$, his new form follows directly by addition of a term proportional to U^2. With

$$\frac{\partial^2 S}{\partial U^2} = -\frac{\alpha}{U(\beta + U)},$$

two integrations, the standard condition $\partial S/\partial U = 1/T$, and an application of the displacement law yield a new distribution law

$$K_\lambda = \frac{C\lambda^{-5}}{e^{c/\lambda T} - 1}. \tag{6}$$

This radiation formula, Planck stated, "so far as I can see by quick inspection, represents the hitherto published observational data just as satisfactorily as the best previously proposed distribution function.... I therefore feel justified in directing attention to this new formula, which, from the standpoint of electromagnetic radiation theory, I take to be the simplest [possible] excepting Wien's."

In the event, new measurements quickly showed equation (6) to be superior to all other existing distribution laws, and after the resolution of a few small experimental anomalies, it has continued to agree with observation ever since. But Planck can scarcely have been confident of that outcome, particularly after his recent experience with apparently better founded claims. The modesty with which he justified "directing attention" to the new formula was almost certainly genuine. Before he could say more, he would need not only new experiments but also a route to the formula that was less ad hoc. The former he could leave to colleagues, but the latter was his responsibility. It led to what he later described as "a few weeks of the most strenuous work of my life."[13]

Recourse to combinatorials

The retrospective passage just quoted also indicates to what "the most strenuous work of my life" was devoted. "On the very day," Planck says, "when I first formulated this [new distribution] law, I began to devote myself to the task of investing it with a real physical meaning, and that issue led me of itself to the consideration of the relationship between entropy and probability, and thus to Boltzmann's line of thought." Those remarks have regularly been read as recording Planck's initial conversion from a phenomenological to a statistical approach to thermodynamics, but that turn-about had, as we have seen, occurred at least a year, and more probably three years, before. When Planck refers to "the relationship between entropy and probability" he does not have in mind the statistical approach in general but rather, as his words suggest, Boltzmann's combinatorial definition of entropy. That definition was to be found only in Boltzmann's writings and there in only two places, in neither of which it was developed fully. Its status was thus entirely different from that of the widely discussed H-theorem. Planck, who must have discovered the combinatorial definition in Sections 6 and 8 of Boltzmann's *Gas Theory*, appears to have been the first man other than its author to acknowledge even its existence.

Hints in Planck's early papers on his new law permit a highly plausible reconstruction of the path that led him to Boltzmann's combinatorial approach, and that path did much to determine the form of his derivation. When presenting the new law to the Physical Society in October 1900, he had already located what he took to be the error in his earlier work. It lay neither in the strictly electromagnetic arguments he had developed before 1900 nor in the new local-maximum argument that had led to equation (4). These could and did remain the basis of Planck's approach through 1906. Instead, the difficulty to which Planck pointed was in the proof that $f(U)$ must be inversely proportional to U. Some other means of determining that function, and with it the relationship between entropy and resonator energy, would have to be found. In his second paper on the derivation of his new law Planck repeated that analysis and added, "A look at the breakdown [*Unhaltbarkeit*] of the previously introduced hypothesis provides a clue to the direction of the conceptual path to be followed."[14] More about the nature of that clue is to be found in Planck's October discussion of his error.

Just before presenting his new law, Planck recorded the relationship he had derived earlier in the year: $dU_n \Delta U_n f(U_n) = n \, dU \, \Delta U f(U)$, with U_n the total energy of a collection of n resonators. Then he continued:

> In that functional equation the expression on the right certainly yields the entropy variation under discussion, for n identical processes take

place independently of each other, and their entropy variations must therefore simply be added. On the other hand, I think it possible, though not easy to conceptualize and in any case difficult to prove, that the expression on the left does not generally possess the significance previously ascribed to it. In other words, the values of U_n, dU_n, and ΔU_n may not suffice to determine the entropy change in question. U itself may also need to be known.[15]

That obscure passage gains in meaning when read with the criticisms directed to the same argument by Lummer and Wien at the International Physics Congress held in Paris the preceding August, a meeting Planck attended. Lummer had singled out the preceding equation and remarked: "One may well ask whether the increase of entropy for any number n of resonators is indeed the same as it is for a single resonator displaced from equilibrium by an amount equal to the sum of the displacements of the n individual resonators."[16] Wien, a more sophisticated theorist, carried the same line of criticism further:

[Planck's] expression for the entropy can be established only by positing the existence of several (at least two) resonators. On the other hand, nothing in the earlier part of [Planck's] argument suggests that irreversibility and the [final] stationary state cannot be reached with a single resonator. If the resulting expression for entropy is a necessary consequence only when two resonators are at work, then these resonators are not mutually independent, a result which would contradict one of [Planck's] hypotheses. Though the expression for entropy probably holds also for a single resonator, it is desirable that it be demonstrated also for this case.[17]

Though Planck may still have believed in the Wien law when those criticisms were read, he doubtless saw their force. Past failures, however, gave him reason to doubt that his expression for entropy could be derived, as Wien suggested, from considerations involving a single resonator. He is more likely instead to have felt that his recent use of multiple resonators, like Boltzmann's of multiple molecules, was essential. That insight would not have been affected by his abandoning the Wien law. His first multi-resonator argument would need to be improved, not given up, and Wien's remarks could well have suggested the nature of the required improvement. Planck's initial derivation contained an internal contradiction: the n resonators he considered were required to be independent, but his argument depended on supposing that their total energy U_n was distributed equally among them. An improved argument would consider the various ways in which that energy might be divided between resonators as Boltzmann, in his combinatorial arguments, had divided the total energy of a gas among its component molecules. Presumably Planck was already groping towards that position in October when he wrote that "the values of U_n, dU_n, and ΔU_n may not suffice to determine the entropy change in question. U itself may need to be known." Surely some such considerations were on his mind when he later wrote that the attempt to provide a basis for his new distribution law "led me *of itself* to the relationship between entropy and probability, and thus to Boltzmann's line of thought." That transition was, in any case, one for which Planck was especially well prepared. Though he had not previously used Boltzmann's combinatorial definition of entropy, he had carefully studied the use of combinatorials in deriving the Maxwell distribution. Such proofs were a great rarity in the nineteenth century, but one of them constituted the thirteenth lecture of Kirchhoff's *Theory of Heat*, which Planck had prepared for publication in 1894, following Kirchhoff's death.

One last clue, first noted by Rosenfeld, was available to Planck, and he is almost certain to have exploited at least part of it from the beginning of his search.[18] His own distribution law, equation (6), is easily rewritten to yield average resonator energy as a function of frequency and temperature,

$$U = \frac{b\nu}{e^{a\nu/T} - 1}.$$

That equation, in turn, can be manipulated to yield $1/T$ as a function of U and ν, and $1/T$ is just $\partial S/\partial U$. One integration thus yields

$$S = \frac{b}{a} \log \left[\frac{\left(1 + \frac{U}{b\nu}\right)^{1 + U/b\nu}}{\left(\frac{U}{b\nu}\right)^{U/b\nu}} \right] + \text{Constant}. \tag{7}$$

That formula for entropy is the one for which Planck must discover a basis, and he probably found the formula itself easily and early. If so, he is likely also to have noted and been encouraged by its clear resemblance to Boltzmann's expression for the logarithmic relation between entropy and probability or permutation number. Equation (7), however, applies only to a single resonator with average energy U in equilibrium with a radiation field. It is, therefore, not yet suitable for interpretation in probabilistic terms. Planck's attempt to reformulate it may well have constituted an early stage of "the most strenuous work" of his life. If, as seems virtually certain, Planck traced the path outlined below, false starts and experimentation must repeatedly have led him from it.

Imagine N independent resonators of frequency ν in equilibrium with their radiation field. Their total entropy must be N times that given by equation (7), and their total energy must, since all have the same average energy over time, be NU. If combinatorials are to be introduced, that total energy must be subdivided into P elements of size ε, so that $P\varepsilon = NU$. Multiplying equation (7) by N and substituting $P\varepsilon/N$ for U yields

$$S_N = \frac{b}{a} \log \left[\frac{\left(N + \frac{P\varepsilon}{b\nu}\right)^{N+(P\varepsilon/b\nu)}}{N^N \left(\frac{P\varepsilon}{b\nu}\right)^{P\varepsilon/b\nu}} \right] + \text{Constant}.$$

That equation, unlike the preceding one, does yield the entropy due to a number of resonators, so that, if the parallel to Boltzmann is trustworthy, the probability corresponding to the equilibrium case should be proportional to the expression in square brackets on its right-hand side.

The bracketed expression is not yet, of course, a combinatorial form, but the steps required to transform it are very nearly as obvious as they are unprecedented. To obtain an expression involving only integers, the size of the energy element ε must be set equal to $b\nu$. The quantity in square brackets then reduces to $(N + P)^{N+P}/N^N P^P$, an expression for which Planck in his published papers adopts the special symbol R.[19†] Stirling's formula and an acquaintance with established combinatorial forms place the final step near at hand. For large N and P, the preceding expression for R may be written

$$R = \frac{(N + P - 1)!}{(N - 1)! \, P!}, \tag{8}$$

and that equation, in turn, is the standard expression for the number of ways in which P indistinguishable elements can be distributed over N distinguishable boxes.[20†] Except for a proportionality constant, which Planck, like Boltzmann, regularly absorbs in the additive entropy constant, it must yield the probability of some yet to be discovered physical situation. The first steps in Planck's utilization of Boltzmann's relation between entropy and probability have been completed.

Deriving the distribution law

Planck's discovery of equation (8) is likely to have encouraged him greatly, for it gave the problem of deriving his distribution law a far more concrete structure than it had previously possessed. But the problem itself remained. The combinatorial expression discovered by working backwards from his new distribution law is very different from the one Boltzmann had developed in deriving the equilibrium distribution of gas molecules. Planck must, therefore, still show that it is proportional to the probability appropriate to equilibrium radiation. Presumably that task called forth additional "strenuous work," but it was soon successfully concluded. By 14 December 1900, when Planck first described to the members of the German Physical Society the theoretical basis of the law he had presented to them two months before, he had in fact found two derivations, historically closely related but logically independent. In his December lecture, Planck outlined one and mentioned the existence of the other[21]; it was shortly made available in a paper received by the editors of the *Annalen der Physik* in early January 1901.[22]

As published, Planck's first derivation explicitly omitted one essential, though conceptually straightforward step. His second was complete, but it was presented in an extremely condensed form, one especially difficult to follow because, after introducing Boltzmann's relation between entropy and probability, Planck's derivations dealt with a distribution problem very different from Boltzmann's. As a result, until the appearance of his *Lectures on the Theory of Thermal Radiation* in 1906, many of Planck's contemporaries found his derivations extremely obscure, especially the second version, better known because published in the *Annalen*. Later historians have inherited their difficulties together with one still more severe. In 1910, H. A. Lorentz (1853–1928) derived Planck's law in a way that closely parallels Boltzmann's derivation of the distribution law for gases.[23] Planck adopted a similar method in the second edition of his *Lectures*, published in 1913, and it has been standard ever since. Under those circumstances the temptation to assimilate Planck's very different early derivations to the subsequently canonical Boltzmann-like form has proven irresistible. Since no such assimilation is possible, the few analysts of Planck's first quantum papers have concluded that he did not have at his command the probabilistic techniques on which his early derivations were based, and they have therefore dismissed his argument as hand-waving. More typical accounts simply paraphrase

Planck's second proof, further condensing it in the process until even the possibility of comprehension is lost. Both approaches block, though in different ways, an understanding of the process by which the quantum entered physics. The one that treats Planck's argument as unproblematic, inevitably concludes that resistance to it was due exclusively to his introduction of the energy element $h\nu$. But the alternate approach, which dismisses Planck's derivation as incompetent, fails to identify the respects in which it departs not simply from Boltzmann's argument but from his approach. As a result, it joins the standard alternate approach in misrepresenting both the nature and the function of the energy element, Planck's central innovation. The integrity of Planck's combinatorial argument must therefore be restored before the nature of that innovation can be understood.

Recall, to begin with, the structure of the now standard derivation with which Planck's first formulations are regularly confounded: N resonators, all with the same frequency ν, are imagined, and the various ways in which a given total energy E may be distributed among them are examined.[24†] For that purpose the energy is imagined subdivided into P elements of size ε, so that $P\varepsilon = E$. A given distribution or state is then defined by a set of integers w_k, with $k = 0, 1, 2, \ldots, P$, and with w_k the number of resonators possessing k energy elements. Two distributions are distinct if they are described by different sets w_k. Any individual distribution can, however, be achieved in Z different ways, with

$$Z = \frac{N!}{w_0!\, w_1! \cdots w_p!}. \tag{9}$$

If one can show that all the ways of distributing the P indistinguishable energy elements over the N distinguishable resonators are equally probable (a problem to be considered in the next chapter), then Z is proportional to the probability W of the distribution specified by the w_k's. The proportionality factor can, furthermore, be neglected, for it appears only as an additive constant in the entropy, which is itself proportional to $\log W$. The equilibrium distribution is therefore specified by the set of w_k's which maximizes $\log Z$ subject to the constraints

$$\sum_{k=0}^{P} w_k = N$$

and

$$\sum_{k=0}^{P} k w_k = P. \tag{10}$$

Clearly, this part of the standard proof of Planck's law is identical, both conceptually and mathematically, with the combinatorial proof developed by Boltzmann for gases.

In his two early derivation papers Planck's problem has a different structure, though it is described explicitly only in the first, his December lecture to the Physical Society. There, after a brief introduction explaining his reasons for using Boltzmann's relation between entropy and probability, Planck asks his audience to consider a reflecting enclosure that contains N resonators at frequency ν, N' at frequency ν', N'', at frequency ν'', and so on. The total energy of all these resonators is E_0, and it is distributed among them so that the set of N resonators at frequency ν has energy E, the set at ν' has energy E', and so on. Thereafter, Planck's problem is to compute the entropy of this particular distribution of the total energy E_0 over the $N + N' + N'' + \cdots$ resonators and, then, to discover its maximum with respect to the variation of the distribution of the total energy over frequency. That problem differs in two respects from the one considered in the now standard derivations that stem from Lorentz. First, sets of resonators at *different* frequencies are considered from the start. More important, the quantities to be varied in maximizing entropy or probability are simply the energies E, E', E'', etc., attributed to each frequency; the manner in which each of these energies is distributed over resonators at the corresponding frequency does not enter the argument; Planck has no need for parameters that correspond to the Boltzmann-Lorentz w_k's.

To compute the entropy of an arbitrary distribution Planck must introduce combinatorials, and for this purpose he follows Boltzmann in subdividing the energy continuum into elements of finite size. It is at this point that he introduces the further novelty that was soon to prove the most consequential of all. For his purpose, unlike Boltzmann's, the size of the energy elements ε, ε', ε'', etc., must be fixed and proportional to frequency.[25] Consideration of that vital step is the subject of the next chapter, but the passage in which Planck introduces it must be noted here, for it illustrates an aspect of his lecture that helped mislead readers about his intent.

> The distribution of energy over each type of resonator must now be considered, first the distribution of the energy E over the N resonators with frequency ν. If E is regarded as infinitely divisible, an infinite number of different distributions is possible. We, however, consider —and this is the essential point—E to be composed of a determinate number of equal finite parts and employ in their determination the

natural constant $h = 6.55 \times 10^{-27}$ (erg × sec). This constant multiplied by the frequency, ν, of the resonator yields the energy element ε in ergs, and, dividing E by ε, we obtain the number, P, of energy elements to be distributed over the N resonators.[26]

Because Planck, here and for some time after, considers only the single set of resonators with frequency ν and because he later omits the computation of a maximum, which would have demanded explicit recourse to resonators at other frequencies, the difference between his argument and that of Lorentz is obscured.

Planck next defines a "complexion" (an expression, he points out, "used by Boltzmann for a *similar* concept"[27]) as a particular specification of the set of numbers k_i, which fixes the number of elements ε attributed to the various resonators in the set of N. (No other term was available, but Planck might better have reserved "complexion" for the distribution determined by the full set of numbers k_i, k'_i, k''_i, etc.) The total number of complexions compatible with a distribution in which the N resonators of frequency ν possess energy $E(=P\varepsilon)$ is just $(N + P - 1)!/(N - 1)!P!$, i.e., the combinatorial expression first discovered by working backwards to equation (8). It is also, as it should be, the expression derivable by summing Lorentz's equation (9) over all values of the w_k's compatible with the constraints, equations (10). For Planck's problem, unlike Boltzmann's and Lorentz's, any set of the w_k's that satisfy these constraints corresponds to *the same distribution* of the total energy E_0.

From this point, Planck's path is straightforward. Having found the number of ways in which the energy E can be distributed over the N resonators at frequency ν, one must find the corresponding numbers for the N' resonators with energy E' and frequency ν', the N'' resonators with energy E'' and frequency ν'', etc. These numbers multiplied together yield "the total number, R, of possible complexions compatible with the provisionally selected [*versuchsweise vorgenomennen*] distribution of energy over all resonators."[28] To find the equilibrium distribution one then simply maximizes R or $\log R$ by varying the energies at the various frequencies subject to the constraint on total energy. Having found the equilibrium distribution R_0, one may, ignoring the additive entropy constant, write the equation for equilibrium resonator entropy as $S_0 = k \log R_0$, with k "a second natural constant" of value 1.346×10^{-16} erg/deg.[29] Temperature is then determined from the standard thermodynamic relation $\partial S_0/\partial E_0 = 1/T$, and the result manipulated to yield the distribution law.

Planck did not carry out these mathematical manipulations. Early in the sketch of his argument he had spoken of "seeking the [equilibrium] distribution, if need be with the aid of trial and error [*eventuell durch Probieren*]." After concluding his outline, he described the computations it would require as "obviously very roundabout [*freilich sehr umständlich*]." Rather than involve himself with any procedure so cumbersome, Planck mentioned the existence of "a more general, entirely straightforward means of computing the *normal* distribution which would result from the preceding steps and which follows immediately from their description."[30] The outcome of that computation he simply wrote down, reserving the description of the alternate method for the paper he submitted to the *Annalen* three weeks later.

Except for trial-and-error steps required to transform non-integral multiples of ε to integral multiples, the mathematical manipulations missing from Planck's paper can, in fact, be carried through straightforwardly. If Planck did not discover a quick way to do so, that is partly because, given an alternate, he had no occasion to work on the problem and probably also because he was following Boltzmann extremely closely. In the one place where Boltzmann had dealt with a mathematically closely related problem (the most probable distribution of molecules restricted to energies $0, \varepsilon, 2\varepsilon, \ldots$), he had employed an elaborate ad hoc method, which, including the last trial-and-error stages, occupied ten dense pages.[31] Adapting that argument to his own problem is presumably what Planck, with reason, wished to avoid. But the argument Planck omitted is nevertheless worth examining here, for doing so will both clarify his derivation-sketch and supply the background needed to understand his quite different alternate form. For that purpose, it is convenient to drop Planck's prime notation and write E_ν, N_ν, P_ν, and ε_ν for the energy, number of resonators, number of energy elements, and size of the energy element at each frequency. When Stirling's formula is then applied to equation (8), the entropy of any given distribution can be written, for large N_ν and P_ν,

$$S_{E_0} = k \sum_\nu \{(N_\nu + P_\nu) \log(N_\nu + P_\nu) - N_\nu \log N_\nu - P_\nu \log P_\nu\}. \quad (11)$$

This is the basic formula for resonator entropy. Planck's sketch envisages maximizing it subject to the constraint

$$\sum_\nu P_\nu \varepsilon_\nu = \sum_\nu h\nu P_\nu = E_0.$$

That task can be carried out somewhat more straightforwardly than Planck may have realized, but it ordinarily yields non-integral values of the P_ν's, and these must be adjusted by trial and error. Here we may avoid that problem with the aid of a substitution that Planck uses for other reasons. If U_ν is the *average* energy of the N_ν resonators at frequency ν, then $N_\nu U_\nu = P_\nu \varepsilon_\nu$. For sufficiently large N_ν, the variations of U_ν are therefore effectively continuous as P_ν runs through successive integral values. Equation (11), rewritten as a function of the U_ν's and the arbitrary integral parameters N_ν, thus yields, after brief manipulation, an equation for entropy as a continuous function of the mean resonator energies,

$$S_{E_0} = k \sum_\nu N_\nu \left\{ \left(1 + \frac{U_\nu}{\varepsilon_\nu}\right) \log\left(1 + \frac{U_\nu}{\varepsilon_\nu}\right) - \frac{U_\nu}{\varepsilon_\nu} \log \frac{U_\nu}{\varepsilon_\nu} \right\}. \quad (12a)$$

With the insertion of Planck's special hypothesis, $\varepsilon = h\nu$, that equation becomes

$$S_{E_0} = k \sum_\nu N_\nu \left\{ \left(1 + \frac{U_\nu}{h\nu}\right) \log\left(1 + \frac{U_\nu}{h\nu}\right) - \frac{U_\nu}{h\nu} \log \frac{U_\nu}{h\nu} \right\}. \quad (12b)$$

These are now the formulas to be maximized, subject to the constraint on total energy,

$$E_0 = \sum_\nu N_\nu U_\nu. \quad (13)$$

To find a maximum one sets $\delta(S_{E_0} - \mu E_0) = 0$, with μ a multiplier to be determined. Straightforward manipulation shows that the entropy will be maximum and the constraint satisfied only if the U_ν's are governed by

$$U_\nu = \frac{h\nu}{e^{\mu h\nu} - 1}.$$

The insertion of that result in equations (12b) and (13) yields formulas for the entropy and total energy at equilibrium as functions of μ. From those expressions, μ can be evaluated by applying the standard relation $\partial S_{E_0}/\partial E_0 = (\partial S_{E_0}/\partial \mu)/(\partial E_0/\partial \mu) = 1/T$. Straightforward manipulation yields $\mu = 1/kT$, and the equilibrium distribution becomes

$$U_\nu = \frac{h\nu}{e^{h\nu/kT} - 1}, \quad (14)$$

just the form Planck seeks. One of its significant characteristics, he quickly notes, is that the corresponding distribution for the field, $u_\nu = (8\pi\nu^2/c^3)U_\nu$, satisfies the Wien displacement law.

Other aspects of the paper in which Planck presented his first derivation will concern us later, both at the end of this chapter and in the next, but the relation of the preceding argument to the better known derivation he prepared for the *Annalen* must be considered first. In the latter, Planck does not introduce the ad hoc, and correspondingly implausible, relation $\varepsilon = h\nu$ as an hypothesis. Instead, he supposes from the start that he is dealing with resonators already in equilibrium with the radiation field, and he enforces that condition at the appropriate point in his argument by introducing the displacement law, now a precondition rather than a consequence of his derivation. The counting of states and the justification of recourse to the combinatorial expression, equation (8), proceed exactly as before, since the relevant arguments apply to equilibrium as well as to more general distributions. Now, however, since equilibrium is presupposed, there is no place for further maximization. Instead, Planck calls upon the displacement law to specify the still missing elements in his expression for entropy. Both his distribution law and the mysterious formula $\varepsilon = h\nu$ emerge at once, the latter now as a consequence of the derivation.

In Planck's new proof, equation (11) continues to express the total entropy of all resonators at all frequencies. Since he is now dealing with the equilibrium case, however, he can ignore exchanges between resonators at different frequencies and consider only the expression for the equilibrium entropy S_{N_ν} of N_ν resonators at any frequency ν. By eq. (11) or by a direct count of complexions, it is given by

$$S_{N_\nu} = k\{(N_\nu + P_\nu)\log(N_\nu + P_\nu) - N_\nu \log N_\nu - P_\nu \log P_\nu\}$$
$$= kN_\nu\left\{\left(1 + \frac{U_\nu}{\varepsilon_\nu}\right)\log\left(1 + \frac{U_\nu}{\varepsilon_\nu}\right) - \frac{U_\nu}{\varepsilon_\nu}\log\frac{U_\nu}{\varepsilon_\nu}\right\}.$$

Since, as Planck shows next,[32] the displacement law applied to resonators demands that $S = \phi(U/\nu)$, the preceding expression is compatible with an equilibrium distribution only if ε is proportional to ν. Imposing that condition in the form $\varepsilon = h\nu$, dropping the subscript ν, and dividing the preceding equation by N yields the entropy S of a single resonator at equilibrium:

$$S = k\left\{\left(1 + \frac{U}{h\nu}\right)\log\left(1 + \frac{U}{h\nu}\right) - \frac{U}{h\nu}\log\frac{U}{h\nu}\right\}. \tag{15}$$

A final application of the thermodynamic relation $\partial S/\partial U = 1/T$ gives

PLANCK'S DISTRIBUTION LAW AND ITS DERIVATIONS 109

the distribution law in the form of equation (14). Rewritten for the density of radiant energy in the field, it becomes

$$u_\nu = \frac{8\pi h \nu^3}{c^3} \frac{1}{e^{h\nu/kT} - 1}. \tag{16}$$

Except for the problems raised by the introduction of the relation, $\varepsilon = h\nu$, Planck's argument is, I think, unexceptionable. Unfortunately, however, his presentation of it was extremely condensed and his derivation correspondingly obscure. Writing for the *Annalen* in January 1901, Planck failed to describe the general problem—distributing given total energy E_0 over resonators at various frequencies—which he had considered in his December lecture and which provided the conceptual basis for his alternate proof. Instead, having postulated equilibrium, he immediately took up the problem of distributing energy $E(=NU = P\varepsilon)$ over N resonators at a single frequency, pointing out that there are just $(N + P - 1)!/(N - 1)!P!$ ways in which that can be done, from which point the argument continued as above. Under those circumstances it is not surprising that his contemporaries, especially those unfamiliar with his December lecture, found his presentation hard to follow. Nor is it difficult to understand why recent commentators, noticing that Planck deals explicitly only with resonators at a single frequency, have emphasized his apparent failure to introduce the Boltzmann–Lorentz count of complexions, equation (9), and to maximize the result by varying the w_k's. Planck's *Annalen* paper makes clear conceptual sense only when systematically juxtaposed with his December lecture, itself easily dismissed by virtue of its incompleteness.

During 1901 Planck published several more papers on his black-body theory, but none repeats, except by brief reference, either derivation of his radiation law. Thereafter, he published nothing further on the black-body problem until 1906, when the first edition of his *Lectures* appeared. In the *Lectures*, for reasons to be discussed in Chapter V, the first derivation is not even mentioned. Instead, Planck presents again the elements of his *Annalen* argument, but in a new order and with the previously missing explanatory comments supplied.[33] The displacement law is introduced before combinatorials, and Planck emphasizes both that its introduction restricts his treatment to the equilibrium case and that the restriction distinguishes his problem from Boltzmann's. Nevertheless, he describes how Boltzmann's method of counting complexions could be applied to his problem and points out why one would then have to sum over all possible Boltzmann distributions to obtain the number of complexions relevant to the problem he has in mind. Rather than perform the summation, he produces his own combinatorial form, equation (8), as its result. These and other additions to his argument suggest that, by the time he wrote his *Lectures*, Planck had recognized the problems his first formulation might present to readers. But they do not suggest that the conceptions that underlay the derivation had changed. A coherent derivation of Planck's law does not demand recourse to the Boltzmann–Lorentz count of complexions nor to explicit maximization.

The new status of the radiation constants

Planck closed his December lecture to the Physical Society by looking once more at the radiation-law constants and calling attention "to an important consequence of the theory [just] developed, one which makes possible a further test of its admissibility."[34] Nothing in his own earlier work or in that of his contemporaries had suggested that any such test might exist. Planck took its emergence especially seriously, as is indicated by the pattern of his publications relating to the new theory during 1901 and 1902. When he prepared an account of his second derivation for the *Annalen*, he reserved for a brief separate paper his remarks on the important new consequence of his theory, presumably hoping in this way to call special attention to it.[35] Later in the year, he described the new result at greater length in a contribution to a volume of papers honoring the Dutch physicist Johannes Bosscha. Though it added nothing of substance to the remarks he had reported in the *Annalen* earlier in the year, Planck republished it in that widely read journal during 1902.[36] With one exception, to be discussed briefly in the next chapter, these articles are the only ones on his new theory that Planck published anywhere between the beginning of 1901 and the appearance of his *Lectures* in 1906. Whether for its intrinsic importance or for its special evidential appeal, Planck emphasized the new consequence of his theory more strongly than the theory itself.

To discover the reason for Planck's special emphasis, return briefly to the Wien distribution law. When Planck first announced his success in deriving it, he also reported up-to-date values for the two constants it contained: $a = 4.818 \times 10^{-11}$ deg·sec; $b = 6.885 \times 10^{-27}$ erg·sec.[37] Simultaneously, he grew as nearly ecstatic as he ever could over the glimpse their determination offered of a natural system of units. Max Thiesen, in the paper discussed early in this chapter, extended the point. The appearance of two natural constants was not, he pointed out, characteristic of Wien's distribution alone. The displacement law itself necessitated their presence in any satisfactory radiation distribution law. Such a law must, Thiesen emphasized, be expressible by a known form involving an arbitrary function Ψ of a single argument λT. One constant would be required to determine the amplitude of Ψ and another—"since Ψ cannot, as is easily seen, be a simple power [of its argument]"[38]—to render dimensionless the argument of the exponential term. Under these circumstances, there was nothing surprising either about the reappearance of two constants in the distribution law Planck announced in October or about the values he attributed to them when deriving the law in December: $k = 1.346 \times 10^{-16}$ erg/deg; $h = 6.55 \times 10^{-27}$ erg·sec. If Planck's law were to coincide with Wien's in the high-frequency limit, the value of h would have to be very nearly that of b, and h/k of a. That the reported values were only close, not identical, was due partly to Planck's use of Lummer and Pringsheim's latest data and partly to the difference in the behavior of the two distribution laws in the region to which the data applied.[39]

What was extremly surprising, however, was the new significance these constants gained by virtue of the role of k in Boltzmann's combinatorial definition of entropy. Boltzmann had not himself introduced any similar constant, but he had, for a perfect monatomic gas in equilibrium, compared the value of his permutability measure Ω with that of the thermodynamic entropy $\int dQ/T$.[40] Planck rewrote the relationships thus obtained in the form $S = \omega R \log Z_0$, where Z_0 is the equilibrium permutation number for the gas, R the universal gas constant, and ω the ratio of the weight of a molecule to that of a mole of the corresponding gas, the reciprocal of Avogadro's number. If that gas were in equilibrium with radiation, the total entropy of the system could, he then showed, be the sum of the entropy of its parts only if the radiation constant k were equal to ωR. Since R was well known, that relationship enabled him to compute ω and from it both Loschmidt's number and the electronic charge. The values he obtained were comparable with existing estimates and, if his method were sound, considerably more precisely determined. For Loschmidt's number he found 2.76×10^{19} molecules/cm^3, which he compared with a standard previous estimate of 2.1×10^{19}; its modern value is 2.69×10^{19}. For the electronic charge he found 4.69×10^{-10} esu, which he compared with recent estimates of 1.29×10^{-10} and 6.5×10^{-10}; its modern value is 4.803×10^{-10}.

Planck concluded his lecture by urging that his new values be tested by more direct means, but the experiments he sought were slow in appearing. Ernest Rutherford (1871–1937) is the only scientist known to have been drawn to "the general idea of a quantum of action" by the special accuracy of Planck's computations. Though his interest proved consequential (it enabled him "to view with equanimity and even to encourage Professor Bohr's bold application of the quantum theory to explain the origins of spectra"), it was apparently unique.[41†] By the time additional measurements of the electronic charge unequivocally demonstrated the accuracy of Planck's prediction, his theory had won widespread acceptance by other means.

Accuracy was not, however, what made the new consequence of Planck's theory especially impressive. Rather it was that, in this area, he had obtained any results at all. Without apparently having intended to do so, Planck had produced a concrete quantitative link between electromagnetic theory, on the one hand, and the properties of electrons and atoms, on the other. At the turn of the century the search for such links was central to perhaps the most active, exciting, and troublesome area of physics research, for the relationship between electrodynamics and mechanics had, for a generation, been growing increasingly problematic. Maxwell's theory, which was almost universally accepted at the close of the nineteenth century, offered no obvious place for the introduction of either matter or discrete charge. Many physicists still expected that the resulting gap—widened and deepened by the discovery of the electron in the 1890s—would be bridged by the design of an appropriate mechanical model of the ether. A mechanical ether would, by its nature, interact with ordinary molecules, and its displacements would constitute the electromagnetic field. But the discouraging outcome of strenuous efforts to design ether models had led other physicists to doubt that a mechanical theory would ever succeed. Led by H. A. Lorentz, a number of them hoped ultimately to reduce matter and mechanics to electrodynamics within what they increasingly referred to as "the electromagnetic view of nature."[42]

Though he published nothing about these issues until he became involved with the special theory of relativity after 1905, Planck was very much aware of them. From the beginning of his career, mechanics had been for him a model science. Since 1894 he had been concerned with electromagnetic theory as well, and by late 1898 he was exchanging long letters on that subject with Lorentz. Their main topic is the standard one of ether drag considered in relation to the Fizeau and Michelson–Morley experiments. Questions about the interaction between ether and matter emerge repeatedly, and Planck proves willing to contemplate such possibilities as the ether's being subject to gravitational attraction. His conviction, however, is that there is "no basis for attributing the properties of ponderable matter to the optical ether, for the latter differs from the former in its most essential characteristics."[43] How, then, account for their interaction?

Remarks like this one are by no means unique to Planck. His letters to Lorentz suggest the context within which his response to the unexpected discovery that experiments on radiation could yield constants relating to matter and charge must be viewed. Though the joint entry of the constant k into the divergent realms of mechanics and radiation provided no conceptual bridge between the two, it was a striking, concrete clue to the direction in which such a bridge might be sought. Since, furthermore, that clue involved the universal natural constants which Planck had so emphasized the year before, the special pleasure and conviction its discovery generated is not surprising. It suggested that Planck had found something more important and fundamental than a derivation of his own distribution law.

Planck had a son Erwin who was seven years old in late 1900 and who, late in life, at least twice reported a memorable walk he had taken with his father at about the turn of the century. On that occasion, the younger Planck said, his father had told him that he had just made the greatest discovery in physics since Newton. The details of that story may well be retrospective—Planck need only have said, for example, that he was on the track of such a discovery—but it is likely that some such conversation occurred.[44†] If it did, however, then the discovery to which Planck referred almost surely involved his unveiling the special characteristics of the constant k. No other event in his scientific career provides so firm a basis for the claim that his research had provided or might provide a previously inaccessible glimpse into nature's innermost workings. His attempt to explain irreversibility without special assumptions had been abandoned. His radiation law had yet to be severely tested. His derivation of that law remained firmly within the classical tradition, a point to be further explored in the next chapter. Until others intervened during 1906 (by which time Erwin was thirteen, a developmental interval he would likely have remembered), neither the law nor its derivation provided a basis for a claim to fundamental innovation. The joint role of the constant k at least promised such a result, and I am aware of no other aspect of Planck's work that did so.

V

THE FOUNDATIONS OF PLANCK'S RADIATION THEORY, 1901–1906

Planck's achievements in the scant four months from late September 1900 to early January 1901 presage a turning point in the development of physics. During the decade that followed its discovery, steadily improving experimental tests continued to support his distribution law.[1] Simultaneously, attempts to prove that law without recourse to the odd relation $\varepsilon = h\nu$ regularly proved fruitless, as did attempts to derive the fixed energy element from classical principles. Beginning in 1905, new analyses and new applications of Planck's theory increasingly restricted the ways in which the relation $\varepsilon = h\nu$ could be interpreted, making its incompatibility with classical theory more and more apparent. With the advantage of hindsight, it is clear that there could have been no turning back. Assimilation of the papers examined in the preceding chapter would require fundamental reconstruction of well-established theory.

The bases for such an evaluation were not, however, available in 1901. Nor had Planck's theory then taken a form that brought it into explicit conflict with older views. To see how Planck's turn-of-the-century achievements came to mark a turning point, one must therefore ask, first, how Planck and others interpreted what he had done in his derivation papers, and second, what happened to change their initial interpretations. Since reactions from others were rare and, with one possible exception, of little significance before 1905, most of this chapter is restricted to an examination of Planck's own understanding of his results during the period to the first publication of his *Lectures on the Theory of Thermal Radiation* in 1906. Prerequisite to that examination is the suppression of numerous associations that the first view of Planck's constant h and of the energy element $h\nu$ inevitably now bring to mind.

The continuity of Planck's theory, 1894–1906

Note, to begin with, the relation between Planck's first derivation papers and the classical theory of black-body radiation he had developed to an apparently successful conclusion between 1894 and late 1899. When, in the latter year, Planck summarized for the *Annalen* the outcome of his earlier research, he acknowledged that his theory was in one respect essentially incomplete: it lacked a proof of the uniqueness of the function he had "defined" as resonator entropy. In March 1900, with the Wien law in doubt, he attempted to bridge that gap with the argument that eventuated in equation (IV-5), $\partial^2 S/\partial U^2 = -\alpha/U$, from which the Wien law followed. Presenting his alternate distribution law in October, he emphasized that that argument was the only portion of his previous work that need now be set aside, thus reintroducing the gap he had apparently bridged in March. The new combinatorial derivation papers he presented in December and January provided a new means of bridging it, for their product was a unique entropy function $S(U)$ to which his earlier theory could again be applied. From 1901 through 1906 that is what Planck and most readers took the role of his combinatorial argument to be. It provided a substitute for the inadequate uniqueness proof of March 1900. Though Planck's radiation theory raised problems that would require solution, those problems did not seem to threaten the integrity of his own earlier work, much less of classical physics.

That Planck himself took this view of the novelties introduced in his December and January derivation papers is strongly indicated by the last of the articles he prepared for the *Annalen* in 1901. Its title was "On Irreversible Radiation Processes," to which he added in parentheses the word "Addendum"[2]; its opening reference was to the summary paper he had published in the same journal under the same title at the beginning of 1900; after an opening statement of purpose, both its formulas and paragraphs were numbered to make the new paper a direct extension and continuation of the old. Planck's introductory remarks remind readers that he had originally defined as entropy a function that led to the Wien law. His supposition that that function was unique had, he acknowledged, proved to be mistaken, but he nevertheless reiterated his conviction that an examination of "the most general possible radiation process" would be compatible with only one form of entropy function. That way of finding a unique form did not, he continued, seem feasible "in the present state of our knowledge," but to demonstrate irreversibility one need only show that a given candidate for entropy function changes irreversibly with time. At that point, Planck referred to the derivation paper he had sent to the *Annalen* in January for an entropy function that "seems to be compatible with the facts determined to date by experiment."[3] In the form appropriate to a single resonator, the function had been

$$S = k\left\{\left(1 + \frac{U}{h\nu}\right) \log\left(1 + \frac{U}{h\nu}\right) - \frac{U}{h\nu} \log \frac{U}{h\nu}\right\},$$

and Planck at once proceeded to show (but now in the paragraphs numbered to follow his earlier paper, on the results of which he regularly called) that the corresponding total entropy S_t must satisfy $dS_t/dt \geq 0$. The entropy function he had derived from combinatorial techniques was thus fully assimilated to the radiation theory he had developed before the turn of the century.

Two aspects of Planck's subsequent publications on black-body theory make that assimilation especially impressive. First, there are none until 1906. Submitted in mid-October 1901, the paper just discussed is the last, or the last but one, that Planck prepared on this subject before the appearance of his *Lectures*. In a sense to be explored further below, Planck's insertion of his new entropy function into his older theory marked the successful conclusion of the research he had begun in 1894. He could and did then turn to other topics, though we shall see that the principal one he chose can be plausibly linked to what he saw as the puzzle his theory still posed. Second, when Planck did publish again on black-body theory, the volume he produced was essentially an expanded, self-contained, and much clarified version of material he had submitted to the *Annalen* between late 1899 and the end of 1901.

Of the five chapters in the *Lectures*, the first, "Fundamentals and Definitions," is an elementary account of black-body radiation and Kirchhoff's law. The second, "Consequences of Electrodynamics and Thermodynamics," opens with Maxwell's equations and from them, together with thermodynamics, derives radiation pressure together with the Stefan-Boltzmann and Wien displacement laws. With the latter written in the form $u = (\nu^3/c^3)F(T/\nu)$, where u is the density of radiant energy and F an unknown function, the chapter concludes by showing how entropy, temperature, and a few related quantities can be expressed in terms of the undetermined function F. So far,

Planck has presented background material, appropriate to a student audience but not to his original research reports. Chapter III, "Emission and Absorption of Electromagnetic Waves by a Linear Oscillator," presents results Planck had developed for himself, all from a period prior to the derivation of his own distribution law. The concept of a damped resonator in interaction with the field is introduced and used to derive appropriate equilibrium equations, including the fundamental relationship $u = (8\pi\nu^2/c^3)U$. The displacement law is rewritten in the special form $S = H(U/\nu)$, which is thereafter repeatedly deployed. Combinatorials enter for the first and only time in the next chapter, "Entropy and Probability," where Planck presents an extended and much clarified version of his second combinatorial derivation, thus at last fixing the form of the unknown function F. With that function determined, Planck returns, in a closing chapter, "Irreversible Radiation Processes," to the one element of his pre-1900 theory not previously introduced. There he develops non-equilibrium versions of the electromagnetic equations of his third chapter, applies the concept of natural radiation to them, and emerges with a proof, like that submitted to the *Annalen* in October 1901, of his electromagnetic H-theorem: entropy can only increase with time.

The structure of Planck's argument will gain in significance when the first edition of the *Lectures* is compared, late in this book, with the very different revised editions of 1913 and 1921, but its central lesson should already be clear. As of 1906, when Planck published the first full and mature account of his theory of thermal radiation, that theory still included all the main elements developed in the research program he had pursued from 1894 through 1901. They entered his text, furthermore, in very nearly the order, and to serve precisely the functions for which they had initially been developed. Except for the new importance they lent to the radiation constants, the events of late 1900 had not visibly changed Planck's view of the nature of the theory he had developed in the preceding years.

More than autobiography and pride of authorship account for the central position of Planck's pre-combinatorial achievements in the *Lectures*. They are an integral part of a sustained and coherent argument; in both obvious and subtle ways Planck needed them. The recourse to combinatorials provided information only about the equilibrium distribution of *resonator* energy with frequency. Planck's concern, however, had been and remained with *radiation*. His resonators were imaginary entities, not susceptible to experimental investigation.

Their introduction was simply a device for bringing radiation to equilibrium, and it was justified, not by knowledge of the physical processes involved, but by Kirchhoff's law, which made the equilibrium field independent of the equilibrium-producing material. Before his combinatorial arguments could be put to their intended use or to any other, Planck would have to convert resonator energy to field energy by means of the proportionality factor, $8\pi\nu^2/c^3$, which he had derived from Maxwell's equations. Though the use of those equations and that factor would shortly seem an inconsistency in Planck's theory, sometimes a reason for rejecting it, Planck and most early readers saw no such problem. Introduced at the start of his book, Maxwell's equations remained basic throughout. In their absence Planck could not have treated the interaction of field and resonators, attributed a specified entropy to the field, or produced an electromagnetic H-theorem, the last presented in 1906 as the crowning achievement of his book, though dropped soon thereafter. The point is not simply that Planck needed to use Maxwell's equations, but that he was apparently unaware of the slightest awkwardness in doing so.

Planck's need for the concepts of classical electrodynamics was not, in any case, restricted to the points where his argument required that he produce quantitative links between the behavior of resonators and of the field. Those concepts play an indispensable qualitative role within his combinatorial argument as well. In gas theory, whether approached through the H-theorem or combinatorials, only a collection of similar particles is characterized by entropy; attributing a value of that thermodynamic function to a single particle is nonsense. The radiation problem, on the other hand, requires only a single resonator at each frequency; if several happen to be present, each one must be in equilibrium with the field. Planck's early research had dealt exclusively with the one-resonator case. When, in March 1900, he first considered a problem involving n resonators at the same frequency, Wien had promptly challenged him to show that the result he obtained could be applied to a single resonator as well.[4] That challenge was still relevant at the end of the year, when Planck's use of combinatorials required that he again consider the many-resonator case. By itself, the combinatorial argument led only to equation (IV-12) for the entropy of a collection of N resonators. To put that result to use, Planck had first to produce a formula for the entropy of a single resonator, equation (IV-15). For that step there was no precedent in Boltzmann's work or elsewhere. What could the entropy of a single resonator mean?

Planck addressed himself to the problem directly in the second paragraph of his December 1900 lecture to the Physical Society:

> Entropy signifies disorder, and this disorder I believed I was required to recognize in the irregularity with which the vibrations of a resonator change their amplitude and phase even in a stationary radiation field.... The constant energy of a stationary vibrating resonator is therefore to be conceived simply as a time average, or, *what comes to the same thing*, as the momentary average of the energy of a large number of identical resonators which are sufficiently separated in that stationary field so that they cannot reciprocally influence each other.[5]

The same analysis is repeated early in Planck's second derivation, where it immediately precedes the appropriate but misleading clause, "the entropy, S_N, is a consequence of the disorder with which the total energy U_N is distributed over the individual resonators."[6] It recurs in his *Lectures*, where it takes on additional significance because it is explained in terms of the independent components in the Fourier expansion of the amplitude of a damped resonator:

> It is, therefore, these numerous independent partial vibrations which play the role with respect to elementary disorder that is played in a gas by the numerous molecules in constant interpenetrating motion. Just as one may not speak of the finite entropy of a gas if...the velocity of all its molecules are in some way ordered, so a resonator possesses no finite entropy if its vibrations are simply periodic or follow some determinate law which regulates all details [of its motion].... In short, for the thermal vibration of a resonator the disorder is temporal, while for the molecular motions of a gas it is spatial. For the computation of entropy, however, this difference proves less weighty than may at first appear; for it may be removed by a simple observation [the equivalence of spatial and temporal averages] which also marks an advance from the viewpoint of uniform treatment.[7]

Both those passages, but most explicitly the latter, indicate the extent to which the concepts underlying Planck's combinatorial theory remain, even in 1906, the ones with which he had begun his research more than a decade before. Interacting with an arbitrary field, the vibrations of a single damped resonator are still described by a Fourier series, which governs the continuous variation of its amplitude and phase with time. Under those circumstances, the time average of its energy can be calculated by known techniques, presumably those developed both in Planck's pre-1900 papers and again in the *Lectures*.[8†] Only because that average is the same as the average energy of a collection of N independent resonators at a single instant of time may the combinatorial definition of entropy be applied to the radiation problem at all. In the *Lectures*, as in the papers written six years before, Planck's conception of his theory remains classical.

Planck had recognized all these aspects of his theory when he presented his first combinatorial papers in 1900 and 1901. In the *Lectures*, however, he also noted a consequence he may not have seen at that time, and it explains what might otherwise seem oddities in his book. Immediately after the passage just quoted, Planck points out that, in the absence of equilibrium, the time average of the energy of a selected resonator need not, and generally will not, be the same as the space average of the energy of all resonators at the same frequency. Equilibrium is therefore a "necessary precondition" of his entire combinatorial approach. Combinatorial expressions representing non-equilibrium states (including those in his December 1900 Physical Society lecture) are therefore prohibited together with arguments that would demonstrate irreversibility by recourse to transitions from less to more probable states. Only after abandoning the initially classical basis of his theory would Planck reintroduce them. Simultaneously, he would drop the damping term in the resonator equation and eliminate his electromagnetic H-theorem from his text.

Natural radiation and equiprobable states

An additional illustration of the intimate interpenetration of the early electromagnetic and the subsequent combinatorial aspects of Planck's work is provided by the important use he continues to make of the concept of natural radiation. Originally introduced as a condition on the amplitudes and phases of permissible electromagnetic radiation, it is, from the start, basic also to his justification of his way of choosing and counting complexions. Early in his December lecture to the Physical Society, Planck remarked that to derive his distribution law "it is only necessary to give the hypothesis of 'natural radiation' which I introduced into electromagnetic theory, a somewhat extended interpretation."[9] What he had in mind begins to emerge together with other significant information towards the end of his paper where he discusses "the question of the necessity of the derivation [just] given." It rests, Planck says, upon a single proposition, which can be divided into two parts:

> 1. that the entropy of the system in a given state is proportional to the logarithm of the probability of that state, and 2. that the probability of any such state is proportional to the number of complexions which correspond to it, or in other words that any determinate complexion is

just as probable as any other. The 1st proposition, applied to radiation processes, probably amounts to a definition of the probability of a state since for radiation one possesses no means for defining probability other than a determination of entropy. Here lies one of the decisive differences from the corresponding circumstances of gas theory. The 2nd proposition provides the core of the theory just developed, and its proof can in the final analysis only be provided by experiment. It can also be regarded as a more precise version of my hypothesis of natural radiation which I previously embodied only in the statement that the radiant energy must be completely "irregularly distributed" over the individual partial vibrations which constitute it.[10]

Planck's second paper on his derivation includes similar, though far more concise, remarks, which add one element essential to an understanding of their author's intent. After repeating that experiment will have to determine the legitimacy of the hypothesis that attributes equal probability to each complexion, Planck continued:

Conversely, if experiment decides in favor of the hypothesis, it will be possible to draw further conclusions about the special nature of the resonator vibrations, that is, to use the words of J. v. Kries, about the character of "the indistinguishable elementary regions, comparable in their magnitudes" which enter into the [radiation] problem.[11]

To discover the viewpoint that underlies these passages, begin by noting that there is nothing new about using a definition to supply an hypothesis needed for the completion of statistical arguments. Boltzmann himself had introduced molecular disorder as the condition molecules must satisfy to permit a particular step in his derivation of the rate of collision between gas particles.[12] Planck had defined natural radiation in the same way, by means, that is, of a mathematical condition the Fourier components of the field must satisfy to permit the derivation of his H-theorem.[13] He cannot, under those circumstances, have been surprised by the discovery that his combinatorial argument demanded recourse to a similar device.

Nevertheless, as Planck also recognized, his new appeal to natural radiation did, in one respect, distinguish his argument from Boltzmann's. Though Boltzmann had required molecular disorder to derive the H-theorem for gases, no obviously similar hypothesis had been needed to complete his combinatorial derivation. Instead, Boltzmann had called, however intuitively and imperfectly, upon Liouville's theorem or on collision theory to justify the assertion that any molecule might, with equal probability, be found in equal volumes of phase or velocity space.[14] In his considerations of resonators in interaction with radiation, Planck could have recourse to no similar theorem. That is what he had in mind when, in the first of the quotations above, he wrote: "for radiation one possesses no means for defining probability other than a determination of entropy. Here lies one of the decisive differences from the corresponding circumstances of gas theory." In the absence of a substitute for Liouville's theorem, however, equiprobable configurations could not be specified a priori. The ultimate justification of any particular choice must inevitably be from experiment.

That point has a converse, which Planck makes explicit in the second of the passages quoted above. Experimental confirmation of the law he had deduced must supply information about "the special nature of the resonator vibrations." More precisely, experimental confirmation must provide information about the relative probability[15†] of the various possible sets of coefficients in the Fourier series that specifies the change of resonator configuration with time. This last characteristic is, of course, the one that renders Planck's "definition" of probability a refinement of his concept of natural radiation. An assertion about the relative probability of different sets of Fourier coefficients is an hypothesis about the relative frequency with which particular sorts of resonator motions occur in nature. It is thus an assertion of essentially the same sort as the one Planck had used to introduce natural radiation before.

Planck's original derivation papers, from which the preceding quotations are taken, were addressed, of course, to an audience familiar with one or more of the extended discussions of natural radiation he had presented since 1898. For an audience without that background, approaching the same papers now in search of quantum theory, his remarks on the subject are inevitably cryptic. But natural radiation also plays a major role in the chapter, "Entropy and Probability" in the *Lectures*, where Planck redeveloped his views at length from first principles. There, too, it emerges as a physical hypothesis about the distribution of microstates, and its role is to permit a definition of probability and thus a derivation of the second law. In the *Lectures*, furthermore, Planck extends to gases the argument he had developed five years before for radiation. No reference is there made to the possibility of deploying Liouville's theorem in the mechanical case nor to the corresponding "decisive difference" between gas and radiation theory. Instead, the two theories are developed in parallel; equiprobable distributions are in both cases specified by fiat; the justification

THE FOUNDATIONS OF PLANCK'S RADIATION THEORY 123

supplied by natural radiation for the black-body case is supplied by molecular disorder for gases.[16†]

Planck's fourth chapter, "Entropy and Probability," opens by presenting the same paradox with which, in 1897, his famous five-part series "On Irreversible Radiation Processes" had begun:

> Since the electromagnetic field equations together with initial and boundary conditions unequivocally determine the temporal course of an electromagnetic process, considerations which [like probability] lie outside of the field equations would seem unjustified in principle and in any case dispensable. Either, that is, they lead to the same result as the field equations—in which case they are superfluous; or they lead to different results—in which case they are wrong.[17]

To escape from "this apparently ineradicable dilemma," Planck recapitulates the argument he had developed with his electromagnetic H-theorem in 1899. A full electrodynamic treatment of any problem requires the specification, as initial conditions, of the amplitudes and phases of all the Fourier components of the field. Experimental evidence does not, however, permit any specification so full. On the contrary, one finds that almost all possible choices of amplitude and phase lead to the same values for the quantities that can be determined by experiment. Only a minuscule fraction of the possible boundary conditions lead to other results, for example to continuous absorption of incident energy without reradiation or even to negative absorption.[18] Gas theory, Planck emphasizes, presents the same paradox and leads to the same sorts of exceptional cases, cases that would violate thermodynamics.

> As a result, unless one is willing to renounce the attempt to grasp thermodynamics mechanically or electrodynamically, only one possibility remains: the introduction of a special hypothesis which restricts the initial and boundary conditions so that the equations of mechanics or electrodynamics lead to unique results which agree with experiment.[19]

The required special hypothesis (Planck says that it "will entirely fulfill its purpose if it says only that these extraordinary cases...do not occur"[20]) is, of course, molecular disorder in mechanics or natural radiation in electrodynamics. If the applicable one is not obeyed, the second law will be violated, and the concepts of entropy and temperature lose their meaning.

All of this is exceedingly familiar. Boltzmann had said very nearly the same things in Volume I of the *Gas Theory*; Planck had said them exactly in papers published in 1898 and 1899. Both men had, however, then been providing the bases for an H-theorem, whereas Planck now aims to provide a basis for combinatorial derivations. Though cryptic, incomplete, and very likely incompletable, the passage in which he makes the attempt leaves no doubt about his intention. Simultaneously, it shows both how far he has come since the late 1890s and how close he remains to the position he had taken at that time.

> Which mechanical or electrodynamic quantities shall now, however, represent the entropy of a state? Obviously [!] the magnitude in question relates somehow to the "probability" of the state. For since elementary disorder and the lack of any control over individual microstates [*der Mangel jeglicher Einzelkontrolle*] is of the essence of entropy, only combinatorial or probabilistic considerations offer the needed entry point for the computation of its magnitude. *Even the hypothesis of elementary disorder is essentially a probabilistic hypothesis*, since from an immense number of equally possible cases it singles out a determinate number and declares these to be nonexistent in nature.[21]

That thought becomes clearer in the following pages, where Planck introduces Boltzmann's relation between the entropy of a physical system and the probability of the corresponding "state." To specify further the latter notion, Planck continues:

> By the "state" of a physical system at a given time, we understand the totality of all those independent magnitudes which uniquely determine the temporal course of processes taking place in the system, insofar as these are subject to measurement.... In the case of a gas consisting of invariable molecules, for example, the state is determined by the law of space and velocity distribution, that is, by the specification of the number of molecules with coordinates and velocity components lying within individual small "intervals" or "regions."... On the other hand [since we are concerned only with quantities accessible to observation], the characterization of a state does not require our providing additional details about the molecules within individual elementary regions. The hypothesis of elementary disorder supplies what is missing and ensures the uniqueness of the temporal process despite the mechanical indeterminacy [of the "initial conditions" supplied by specifying only the number of molecules in each small region].[22]

Like Boltzmann in 1896, but now more clearly, Planck has recognized the difference between the molar and the molecular specification of states. Like Boltzmann, furthermore, he is preserving the second law by prohibiting the occurrence of just those special "ordered" molecular (or resonator-and-field) configurations that would lead to its violation. For Planck, however, prohibiting those configurations has somehow become a means of fixing the relative probability of the states that remain. Under those circumstances, the criterion of an appropriate choice of equiprobable states can only be that it yield experimentally observed regularities, first and foremost the second law of thermodynamics: "The decision about which hypothesis [concerning the exclusion of certain specified initial conditions] to prefer can only be made by testing the result to which the hypothesis leads against the experimental theorems of thermodynamics."[23] In short, an hypothesis governing the distribution of initial conditions within individual "intervals" or "regions" determines combinatorial probability and thus entropy. From the latter follows a unique energy distribution law, and experiments designed to check it therefore also test the hypothetical restriction on initial conditions. Developed initially for radiation, that analysis can be applied to equilibrium in gases as well.

Energy elements and energy discontinuity

As Planck's continuing emphasis on the close parallels between his theory and Boltzmann's suggests, his view of the radiation problem is still, in the *Lectures* of 1906, fully classical. Though his apparently more radical understanding of the energy element $h\nu$ remains to be discussed, what has already been said rules out any version of a long-standing historiographic tradition. Both in his original derivation papers and, far more clearly, in the *Lectures*, Planck's radiation theory is incompatible with the quantization of resonator energy. That theory does require fixing the size of the small intervals into which the energy continuum is subdivided for purposes of combinatorial computation, and the restriction to a fixed size does isolate the main respect in which Planck's theory diverges from Boltzmann's. But the divergence does not, as developed by Planck, make radiation theory less classical than gas theory, for it does not of itself demand that the values of resonator energy be limited to a discrete set. On the contrary, as this chapter has already shown, any such restriction would conflict both with the global structure and with multiple details of Planck's argument.

In Planck's theory, resonator emission and absorption are governed in full by Maxwell's equations. Variations of resonator energy with time are determined by the same sorts of differential equations and described by the same sorts of Fourier series that Planck had used for these purposes before 1900. Planck's H-theorem of 1899, presented in the closing chapter of the *Lectures* as its crowning achievement, demands those equations and series too, and it had to be abandoned when Planck gave up continuity after 1906. Thus, though the structure of the energy continuum is fixed by the energy element $h\nu$, the motion of Planck's resonators remains continuous, both within the elements constituting that continuum and from one to the next. With a single misleading exception, to be considered below, nothing in Planck's published papers, known manuscripts, or autobiographical fragments suggest that the idea of restricting resonator energies to a discrete set of values had even occurred to him as a possibility until others forced it upon him during 1906 and the years following.[24] My point is not that Planck doubted the reality of quantization or that he regarded it as a formality to be eliminated during the further development of his theory. Rather, I am claiming that the concept of restricted resonator energy played no role in his thought until after the *Lectures* was written. Could Planck responsibly have remained silent about it in that work if the idea had even crossed his mind as relevant to the theory he there presented?

To this historiographic heresy, Planck's treatment of natural radiation and molecular disorder lends essential support. Those special hypotheses are what, in Planck's theory, restrict the permissible microdistribution of resonators within the energy intervals $h\nu$ and of molecules within the phase-space cells $d\omega$. If Planck had wished to do so, he could have used them to prohibit a resonator or molecule from occupying any part of the interior of these small regions. Resonators, if thus restricted to the endpoints of the intervals into which Planck divides the energy continuum, could only have energies $nh\nu$. But, putting aside the improbability that Planck would have failed to mention a restriction quite so strange and quite so unlike the apparently parallel molecular case, the text of the *Lectures* prohibits any possibility of that sort. In the *Lectures*, after admitting the existence of special initial conditions that must be prohibited because they would lead to unobserved phenomena, like negative absorption, Planck continues:

> If, however, one examines more carefully the infinity of different cases, corresponding to the different possible values of C_n and θ_n compatible with a given [observed] radiation intensity, and if one compares the results of different choices, one finds that a huge majority of such choices lead on the average to corresponding [experimental] results, while those choices which result in noticeable deviations are, by comparison, negligibly small in number.[25]

Planck is again following Boltzmann. Some initial conditions must be prohibited to ensure the validity of the second law, but their number is small compared with that of the admissible initial conditions. Forbidden states are therefore not numerous enough to occupy the entire interior of a cell, restricting the admissible ones to its surface. Excluding rare singular cases, Planck's resonators, like Boltzmann's molecules, are to be found anywhere within the small cells or energy ranges required for combinatorial computations.

How then can anyone have found energy quantization in Planck's early discussions of his black-body law? Part of the answer is that the first edition of the *Lectures*, still the only unambiguous source of his position, was, from 1913, rapidly displaced by a series of better-known editions in which quantization does play a central role; for over sixty years Planck's original version has been read only by an occasional historian. His first derivation papers—the earliest source of the innovations from which energy quantization arose—have been far more widely read. But they are extremely brief and, in some respects obscure, so that readers aware of what happened next have been able to bridge the gap to later versions of black-body theory without realizing that it existed. More than obscurity however, is, responsible for the ease with which they have done so. Before the first appearance of the *Lectures*, Planck's first derivation papers were also misread, though not independently, by two of his contemporaries, men to whom none of the preceding explanations can apply. Those misreadings have another source, and its existence helps also to account for what has occurred since Planck's time. Two technical aspects of his presentation do suggest energy quantization. The first is easily disposed of and is not, in any case, likely to have affected his contemporaries.[26] The second is deep and will require careful explanation.

Though Planck does not, either in his derivation papers or the *Lectures*, ever equate the energy of a single resonator with an integral multiple of $h\nu$, he does repeatedly write expressions like $U_N = Ph\nu$, with P an integer. In such expressions, however, U_N is the total energy of N resonators. Restricting it to integral multiples of $h\nu$ does not impose any similar restriction on the energy of an individual resonator, which may vary continuously. Indeed, Planck's subdivision of *total* energy into an integral number of equal finite elements is entirely modeled on Boltzmann's. Though the size of the element employed by the latter was not uniquely fixed, it could not continuously approach zero, for it was required to be large enough to contain many molecules. If quantization is the subdivision of total energy into finite parts, then Boltzmann is its author.

A second aspect of Planck's presentation raises more basic difficulties. In both his early derivation papers Planck described the problem to be solved en route to his combinatorial form as "the distribution of the P energy elements over the N resonators." Each distribution was for him a "complexion," and he illustrated what that term meant with an example in which 7 elements were attributed to the first resonator, 38 to the second, and so on to a total of 10 resonators and 100 elements. Introducing his combinatorial form next, Planck described it as "the number of all possible complexions."[27] If this part of his presentation is taken literally, then his resonators can acquire only an integral number of energy elements $h\nu$, and they are therefore quantized. The passages in which these phrases and diagrams occur are the presumptive source of the traditional view, for which, in any case, they provide the only significant evidence.

Fortunately for the consistency of Planck's thought, these passages need not be read literally, and important sections of the *Lectures* show that they should not be. When Planck wrote them in late 1900 and early 1901, he was carefully following Boltzmann's 1877 paper. In that paper, described in Chapter II, Boltzmann had twice illustrated combinatorial derivations by distributing molecules over the subdivided energy continuum. In the first case, the energy of individual molecules was restricted to values $0, \varepsilon, 2\varepsilon, 3\varepsilon, \ldots$; in the second, molecules were described as lying in the range 0 to ε, ε to 2ε, 2ε to 3ε, and so on. Both cases led to the same combinatorial expression and, for large N and P, to the same distribution law. The two appear to be interchangeable, and Planck clearly thought that they were. As a result, he felt justified in simplifying his combinatorial derivation by describing a discrete energy spectrum when the physical situation he had in mind called for a continuum. In his *Lectures* the substitution is explicit.

Planck's presentation in the *Lectures* closely parallels the ones he had provided five years before. The meaning of "complexion" is again illustrated with a diagram that assigns an integral number of energy elements to each resonator. His combinatorial expression is described as providing "the number of individual orderings or complexions compatible with the distribution of the energy U_N over the N resonators." These and other echoes of his original papers strongly suggest that Planck had not in the interim changed his mind about the essentials of his theory. Nevertheless, either because he was taking special care or because he had discovered that his earlier way of putting his point could be misunderstood, he did pause long enough for an essential clarification. A few lines after the phrase just quoted, he spoke of "the number of resonators with energy of a given magnitude"

THE FOUNDATIONS OF PLANCK'S RADIATION THEORY

and then added at once, "(better: which lie within a given 'energy region')."[28] Only the omission of some equivalent parenthetical clause from his early papers makes it so difficult to discover what Planck had in mind.

A second passage in the *Lectures* reinforces the point, perhaps definitively. In it Planck shows how to compute "the number of complexions corresponding to a given state directly from the electromagnetic state of an individual resonator rather than from its energy, which is always a compounded quantity."[29] The result—an aside in this first presentation but later of great significance for the development of quantum theory—is a phase-space description of the equiprobable regions accessible to a resonator. To achieve it, Planck first rewrites equation (1-7) for resonator energy in terms of the resonator moment f and its conjugate momentum g:

$$U = \tfrac{1}{2}Kf^2 + \tfrac{1}{2}\frac{g^2}{L}.$$

Curves of constant energy are then readily shown to be ellipses of area U/ν, so that the equiprobable regions previously specified by equal energy increments $h\nu$ become elliptical rings of area h in the phase plane. Even before deriving that result, Planck had indicated the use he would make of it: "We conceive f and g as coordinates of a point in the phase plane [*Zustandsebene*] and inquire about the magnitude of the probability that the energy of a resonator lies *between* the values U and $U + \Delta U$."[30] When ΔU is later set equal to $h\nu$, resonators continue to lie within, not simply on the boundaries between, the elliptical rings thus formed.

These passages are, I suspect, decisive by themselves, but they do pose a puzzle. Why, if Planck conceived his resonators as lying within energy intervals $n h\nu \leq U < (n + 1)h\nu$, did he use a vocabulary that apparently restricted resonator energy to integral multiples of $h\nu$? That question, too, has an answer, one that depends upon an often overlooked difference between Planck's and Boltzmann's methods of determining equiprobable complexions. Both men began by dividing the energy continuum into P elements of size ε. Boltzmann's next step was to distribute molecules at random over the energy continuum, immediately labeling each one with the number specifying the energy on which, or the interval into which, it had fallen. Planck, on the other hand, separates the P individual elements of the divided continuum and distributes them at random over the N resonators. Only after the distribution process has been completed does he label each resonator with the number of elements it has received. Though the two methods are readily shown to be equivalent, they are at first glance clearly distinct.[31]

Planck could, of course, have used Boltzmann's method to derive his combinatorial form, a fact he notes in the *Lectures*. But, as he points out in the same place, the process would involve summing over all Boltzmann distributions compatible with the given energy and would be correspondingly cumbersome. His own method allows him "to go more quickly and conveniently...to the same goal"[32] and is therefore to be preferred. For the use of that shortcut, however, Planck paid an unnoticed price. Boltzmann's method of distribution could be used to place molecules either within energy intervals or at the boundaries between them. Planck's method, in the absence of explicit further specification like that given in the *Lectures*, could only leave each resonator with an integral number of the whole energy units, which had been distributed to it one at a time.

The quantum of action and its presumptive source

Despite its generally classical nature, Planck's statistical radiation theory did differ from Boltzmann's gas theory in one central respect. For Boltzmann, the subdivision of the energy continuum was a mathematical device,[33] and the size of the element employed to introduce it did not matter. For Planck, that subdivision was a physical necessity, and the size of the element was fixed by the relation $\varepsilon = h\nu$. Early in his December 1900 lecture to the Physical Society he had described that relation as "the essential point" of his theory.[34] Thereafter, until the appearance of the *Lectures*, he employed it without special comment, and even in 1906 his remarks on that relation were extremely brief. Immediately after introducing the relationship $\varepsilon = h\nu$, Planck continued:

> An immediately striking feature of this result is the entry of a new universal constant h of which the dimensions are a product of energy and time. It marks an essential difference from the expression for the entropy of a gas. In the latter, the magnitude of an elementary region which we call $d\omega$ disappears from the final result since its only effect is on the physically meaningless additive constant.... The thermodynamics of radiation will therefore not be brought to an entirely satisfactory conclusion until the full and universal significance of the constant h is understood. I should like to label it the "quantum of action" or the "element of action" because it has the same dimensions as the quantity to which the Principle of Least Action owes its name.[35]

Implicit in that passage is a subtle but extremely important change of emphasis. Though the difference between gas theory and radiation theory is the physical role played in the latter by a particular choice of cell size, what requires explanation is not the necessity for a fixed size but the "significance of the constant h," which determines its magnitude. If only h were understood, then fixed cell size might be seen to follow from it or even to be a misleading interpretation of some more fundamental aspect of radiation phenomena. For Planck, apparently, h has now become the "quantum of action," a phrase that, unlike its suggested alternate, "element of action," is henceforth standard usage in his writings on radiation theory. More important, it is standard also throughout his autobiographical writings, for it is this constant, and not a restriction on resonator energy or on continuous motion, that Planck regularly identifies as the novelty he introduced into physics. The quantum of action proved "cumbersome and refractory," he notes, when confronted by his efforts to assimilate it classically. Ultimately, it called forth "a break with classical physics far more radical than I had initially dreamt of."[36]

Doubtless the brevity of Planck's remarks in the *Lectures* was dictated by this interaction of circumstances with character. He had little to say about the quantum of action, and he was not one to speculate or waste words in scientific papers. But he did believe more needed to be said—"The thermodynamics of radiation will therefore not be brought to an entirely satisfactory state until the full and universal significance of the constant h is understood"—and he thought he knew the area in which work must be done to supply what his theory still lacked. The sentence replaced by ellipsis in the last long quotation reads: "Though contemporary theory offers no point of entry for its exploration, there can scarcely be a doubt that the constant h plays some role in the elementary oscillatory process at the center of emission." Attached to that sentence is a footnote calling attention to an earlier passage in which Planck had emphasized that his formulas for radiated energy apply only when radiation intensity is measured over periods T of sufficient length. "For smaller values of T," he had continued, "the simple linear equation [governing resonator vibration] may possibly need to be replaced by one better suited to natural phenomena."[37]

Those passages suggest that Planck expected, or at least hoped, that the puzzle posed by his theory would be solved by research on the microscopic detail of the emission process, thus by electron theory.

That suggestion is, in turn, fully confirmed by a letter Planck had written to the young physicist Paul Ehrenfest (1880–1933) during the summer before the *Lectures* appeared. Its importance and the use to be made of it in Chapter VI justify its reproduction in full:

Grunewald, 6. July 1905

Honored Herr Dr!

In response to your valuable letter of the first of this month, I will gladly give you my opinion about the question you have raised. First of all, I agree entirely with your principal point. Resonator theory (including the hypothesis of natural radiation) does not suffice to derive the law of energy distribution in the normal spectrum, and the introduction of the *finite* energy quantum $\varepsilon = h\nu$ is an additional hypothesis, foreign to resonator theory itself.

But perhaps it is not out of the question to make progress in the following way. If one assumes that resonator oscillations are produced by the motion of electrons, then a new element enters the theory in any case. Because the charge of the electron is proportional to div E, E cannot be increased by m^2 throughout the field unless the charge of the electron grows in the ratio $1:m^2$. Therefore, if the charges of electrons are constant, the process you describe [—] $E' = m^2 E$, $H' = m^2 H$, $f' = m^2 f$ [—] is impossible.

It seems to me not impossible that this assumption (the existence of an elementary quantum of electricity [the charge e]) offers a bridge to the existence of an elementary energetic quantum h, particularly since h has the same dimensions as e^2/c (e, elementary quantity of electricity in electrostatic units; c, velocity of light). But I am in no position to offer a definite opinion in this matter.

Your most humble
M. Planck[38]

Planck's localization of the problem of the constant h in electron theory, especially his view of its relation to the quantum of electricity e, was immensely plausible. Thiesen had emphasized in 1900 that any theory of black-body radiation would require two constants.[39] One of those used by Planck, the constant k, was related to the choice of a temperature unit and could be understood in terms of the mechanical theory of heat. To anticipate that the source of h would be radiation theory was more than reasonable. The only constant in Maxwell's equations is, however, the velocity of light, and it is dimensionally unsuited to the purpose. What remained was the charge of the recently discovered electron, at the time central to the most active and exciting area of physics. That the quantum of electricity might solve the puzzles of black-body theory was not, in any case, an idea that originated with Planck; indeed, he may well have borrowed it.

As early as 1900 the leading expert on electron theory, H. A. Lorentz, had argued that Kirchhoff's law could only be understood in terms of some property common to all matter. "In all probability," he added, "the likeness in question consists in the equality of the small charged particles or *electrons*, in whose motions modern theory seeks the origin of the vibrations in the aether."[40] The paper in which he announced that idea is also the source of the sorts of dimensional arguments to which the second paragraph of Planck's letter to Ehrenfest refers. (As will be seen in Chapter VI, that letter is part of a correspondence about issues that Ehrenfest was soon to explore in an important paper.) In 1903 Lorentz returned to electron theory to derive the form of Kirchhoff's universal function appropriate to long wavelengths, and in 1905 James Jeans used similar arguments to explain the universal character of the Wien displacement law, an explanation he required since he did not believe that black-body radiation was in an equilibrium state.[41] To these considerations Planck could add one other, the converse of the great achievement he had first announced at the end of 1900. At that time he had computed the value of e from experimental values of h and k. What, then, could be more natural than to explain the conceptual mysteries of h (k was in some sense understood) in terms of the electronic charge to which he had himself demonstrated its astonishing relation?

Electron theory, the field to which Planck was thus relating the problem of h, was characterized by numerous other unsolved problems, and great progress was being made with them. Since Planck himself had not as yet done research in this active area, he could have been forgiven if he had left the problem of explaining the quantum of action to electron theorists. In practice, however, he seems not to have done so. Though he published nothing explicitly concerned with black-body theory between 1901 and the appearance of his *Lectures* in 1906, the main topic of his research in the intervening years was the electromagnetic theory of optical dispersion, on which, from 1902, he produced a number of substantial papers. That topic did involve him for the first time with electron theory, and it seems probable that he took it up hoping that it would provide "a point of entry [to the]...elementary oscillatory process at the center of emission" and thus provide information about the constant h. The likelihood of that hypothesis is reinforced by Planck's later insistence that he had been deeply engaged with the problem of the quantum of action throughout the years when it seems otherwise to be virtually absent from both his published work and his correspondence.[42†] In a letter of 1910 to the physical chemist

Walther Nernst (1864–1941), Planck wrote: "I can say without exaggeration that for ten years, without interruption, nothing in physics has so stimulated, agitated, and excited me as these quanta of action."[43]

Needless to say, Planck's hopes for electron theory were frustrated, and in retrospect it is apparent that they could not in principle have been fulfilled within classical physics. There is, as Einstein would point out in a paper to be discussed in Chapter VII, a fundamental error in Planck's derivation. Though the energy continuum or phase space may be subdivided for purposes of computing combinatorial probabilities, the cells employed must be small enough so that varying the position of resonators or molecules within them produces no observable change in the physical state of the system being considered. For Planck's problem, that condition demands $h\nu \ll kT$, and it is not everywhere fulfilled in the frequency and temperature ranges explored by relevant black-body measurements. When it is not fulfilled, Planck's computation of probability takes account of only some of the states (those at or very near cell boundaries) available to his resonators. Noting this fact, Einstein would conclude that Planck's version of radiation theory demands, in effect, a redefinition of probability. Nothing of that sort could have been forthcoming from electron theory, though quantum theory has, in a sense, since supplied it.

Obvious in retrospect, Planck's mistake was, as we shall see in Part Two, everywhere overlooked for some time. Help in understanding its obscurity, particularly to Planck, may be provided by a last return to Boltzmann. The function of the condition $h\nu \ll kT$ is to ensure that the distribution function—Planck's $U(\nu, T)$ or Boltzmann's $f(u, v, w)$—does not vary significantly between neighboring cells. The necessity for such a condition is, however, a key point missed by Boltzmann when he discussed the transition from sums to integrals in his combinatorial paper of 1877.[44] For Boltzmann, as we have seen, that transition was a mathematical step, and its legitimacy did not depend on the physical condition of the gas. At a similar point in his own argument, Planck may once again have been following Boltzmann's lead.

Planck's early readers, 1900–1906[45]

Whatever Planck's own view of the extent to which his theory broke with the classical tradition, his work was not received as radical by most of his early readers. Partly for that reason, however, there were not many of them. Until after 1906, as before 1900, black-body theory remained an esoteric specialty. Nevertheless, Planck's papers on the subject were known. Reviewed during 1901 and 1902 in the standard British and German abstracting journals, they were there treated simply as further developing the line of research on which he had been reporting since 1895.[46] The only sign, probably without significance, of the recognition of something special about his most recent work is the extra but not unprecedented length of the German abstract of his 1901 article for the *Archives Néerlandaises*, in which values for the atomic constants were derived. Lorentz, too, cited Planck's new law and his first attempt to derive it in the paper, read at the end of 1900, in which he related electron theory to black-body laws. But he there simply coupled Planck's work with Wien's and discussed neither.

During the next five years, references to Planck's theory accumulated slowly in both the British and the German literature, but their character was somewhat different in the two countries. The German references were predominantly reportorial rather than analytic, and most of them appeared in books rather than articles. (There were also a number of articles dealing with the experimental adequacy of Planck's law.) The second volume of the famous *Handbook of Spectroscopy*, published in 1902 by Heinrich Kayser (1853–1904), includes many passages referring to Planck's work in the years 1897–1901. One of them sketches the derivation of the Wien law given in Planck's summary article for the *Annalen* in early 1900, adding that "in a more recent article Planck seeks a firmer foundation for his chosen expression for entropy."[47] Ten pages later, Kayser returns to the matter, reporting that: "For his derivation of the Wien law Planck chose an especially simple form of the entropy function. Since that law was not entirely [*durchweg*] confirmed, he tried a less simple expression and arrived at [his new] radiation formula."[48] That formula Kayser had previously characterized in a sentence that singled out Planck as the man who, "starting from electromagnetic theory, has apparently found the true radiation law."[49]

Two years later, in 1904, Planck's work was again mentioned in a widely circulated book: the two-volume treatise, *Thermodynamics*, by Woldemar Voigt (1850–1919). Most of its closing chapter deals with such standard topics as Kirchhoff's law and the displacement law; distribution laws occupy only a two-page section, which begins by mentioning Wien's attempt and its experimental inadequacy. Then Voigt continues: "By a most noteworthy combination of probability considerations with the theory of the emission of waves by electric resonators, M. Planck has arrived at a formula which satisfies experiment in the entire region that has been [experimentally] investigated."[50] Planck's law is then presented, and the experimental determination of the two constants is discussed. In a thirty-two–page chapter, "Thermodynamics of Radiation," nothing more is said about Planck's work.

During 1905, Planck's law was mentioned also by Albert Einstein (1879–1955) in a famous paper to be discussed in Chapter VII. But he appealed to it simply as the best experimental formula currently available, and he made use of it only at high frequencies where it becomes identical with the Wien law. Finally, a somewhat fuller sketch of Planck's new theory, drawn largely from the recently published *Lectures* was included in the second edition of the standard textbook of optics by Paul Drude (1863–1906). Drude does mention Planck's use of the formulas $S = k \log W$ and $\varepsilon = h\nu$. But it is the total energy U_N, not the energy of individual resonators, that he speaks of as "made up of a kind of atomistic energy elements."[51] Like Planck, too, Drude emphasizes both the significance of the computation of the electronic charge e from radiation measurements and the importance of discovering the still unknown "electro-dynamic significance of the elementary quantum of action h."[52]

The British references to Planck's theory during these early years were no more numerous than the German, but, coming from the country that was still the only center of significant interest in statistical mechanics, they were often more analytic and original. Burbury, in 1902, published a long, sympathetic, and ultimately important study of the techniques that had led Planck by 1899 to an electromagnetic H-theorem. Near the end he noted that Planck had recently, in the face of experimental counterinstances, introduced a new form of the entropy function "*without altering the general theory as developed in the former treatise.*"[53] In the same year Joseph Larmor (1857–1942) briefly indicated the general structure of Planck's new derivation in the article "Radiation," published in the supplementary volumes that transformed the ninth edition of the *Encyclopaedia Britannica* into the tenth.

For Larmor, unlike Burbury, the papers Planck had published from December 1900 did represent "a fresh start," but their novelty was only the use of Boltzmann's combinatorial definition of entropy. After sketching that departure, Larmor went on to say: "Whatever may be thought of the cogency of his [Planck's] argument, especially in view of the fact that his vibrators cannot change the types of the radiation, the result gains support from the fact that it involves determinations of the absolute physical constants of molecular theory that prove to be of the right order of magnitude."[54] A footnote to that sentence informed readers that "The argument has recently been recast by Larmor, so as to avoid the introduction of vibrators," a presumptive reference to a paper, published only in abstract, that he had read to the 1902 meeting of the British Association at Belfast.[55] Thereafter, Larmor, who appears to have been the first to take Planck's combinatorial derivations seriously, occasionally presented lectures on the subject, including a talk at Columbia University in 1905. But his first full publication was delayed until 1909, by which time his views could have little effect on the manner and rate at which the quantum theory developed.[56] Even then he said nothing to suggest that Planck's theory implied discontinuity, and in the next year he argued that it need not do so, since only the ratio U/ν had to be conserved.[57]

After these first references of 1902, there is no mention of Planck's theory in the British literature until 1905, when a remark by Lord Rayleigh (1842–1919) opened a controversy that marked the beginning of a continuous discussion in print. Some of the papers in the exchange will be considered at greater length in Part Two, but its beginning is relevant here. For some years James Jeans had been developing a theory of the transfer of energy between matter and the ether. His approach involved the assumption that the equipartition theorem could properly be applied to high frequency vibrations in the ether, and that assumption was questioned by Rayleigh in a letter to *Nature* during 1905. Assembling arguments against the general applicability of equipartition, Rayleigh noted that Planck's work appeared to be both empirically successful and also incompatible with the standard statistical basis of Jeans's approach. "A critical comparison of the two processes would be of interest," he continued, "but not having succeeded in following Planck's reasoning I am unable to undertake it.... My difficulty is to understand how another process also based on Boltzmann's ideas, can lead to a different result."[58]

Jeans responded at once in a famous critique of Planck's approach. Noting, among other shortcomings, Planck's failure to justify the choice of equal energy intervals as equally probable, Jeans especially emphasized Planck's having stopped short of setting $h = 0$, a relationship he mistakenly believed to be demanded by the principles of statistical mechanics. If only Planck had taken this required step, Jeans pointed out, his distribution law would be the same as the one

Jeans himself had recently derived from equipartition. Though he saw clearly the role of the energy element $h\nu$ in Planck's computation, Jeans regarded it as "a small quantity, a sort of indivisible atom of energy, introduced to simplify the calculations."[59] Not until 1910, when he somewhat ingenuously acted as though the discovery were his own, did Jeans suggest that Planck's theory required discontinuities in the classically continuous range of energies available to a physical body or to the radiation field.[60]

Turn, finally, to two anomalous readings of Planck. In 1903, as previously mentioned, Lorentz derived from electron theory the form of the distribution function appropriate to long wavelengths. Before doing so he mentioned Planck's distribution function, which he reregarded as "remarkable...[because it] represents very exactly the energy of the radiations for all values of the wavelength, whereas the following considerations are from the outset confined to long wavelengths." Planck's derivation was not described, but Lorentz devoted a paragraph to remarks about it. Among other things, he said:

> I shall not here discuss the way in which the notion of probability is introduced in Planck's theory and which is not the only one that may be chosen. It will suffice to mention an assumption that is made about the quantities of energy that may be gained or lost by the resonators. These quantities are supposed to be made up of a certain number of *finite* portions, whose amount is fixed for every resonator; according to Planck, the energy that is stored up in a resonator cannot increase or diminish by gradual changes, but only by whole "units of energy," as we may call the portions we have just spoken of.[61]

Two years later, Paul Ehrenfest (1880–1933) rephrased the relevant part of that description in a paper to be discussed at length in Chapter VI. His own concern was restricted entirely to Planck's electromagnetic H-theorem as it had been developed through 1899. Like Burbury, however, Ehrenfest closed by mentioning Planck's introduction of a special entropy function, adding that it was based upon a Boltzmann-like combinatorial analysis. Promising to discuss Planck's derivation in a later article, Ehrenfest meanwhile noted the two hypotheses on which, in his view, it depended. The first was, of course, the special choice of equiprobable states, the second "that the radiant energy of the various colors consists of minuscule energy particles of magnitude: $E_\nu = \nu \cdot 6.55 \times 10^{-27}$ erg·sec., where ν is the frequency of the color in question."[62]

Almost certainly these two non-standard readings of Planck's first quantum papers are mutually dependent, for Ehrenfest, whose

THE FOUNDATIONS OF PLANCK'S RADIATION THEORY 139

paper opens with a reference to Lorentz's black-body publications, had first learned of Planck's work from Lorentz's lectures at Leyden during 1903.[63] Perhaps one or both men recognized that, whatever Planck may have thought, his theory would not work if energy were absorbed and emitted continuously. But, in that case, they would probably have attributed the hypothesis of energy quanta not to Planck himself but to the demands of his theory. Given their phraseology and the universal difficulty in recognizing where Planck's derivation went astray, it seems far more likely that they were simply following Planck's misleading discussion of his way of populating states. Lorentz, in any case, soon recognized that Planck's own theory did not restrict resonator energy. Writing to Wien in 1908, he noted that "according to Planck's theory resonators receive or give up energy to the ether in an entirely continuous manner (without there being any talk of a finite energy quantum)."[64] And in 1913, during a discussion of remarks by Jeans at the Birmingham meeting of the British Association, he began by outlining his own approach to black-body theory and then continued:

> We might now suppose that the exchange of energy between a vibrator and the ether can only take place by finite jumps, no quantity less than a quantum being ever transferred to the medium or taken from it. Something may be said, however, in favor of the opposite hypothesis of a gradual action between the ether and the vibrator, governed by the ordinary law of electromagnetism. Indeed, it has been shown already, *in Planck's first treatment of the subject*, that by simply adhering to those laws, one is led to a relation between the energy of the vibrator and that of the black radiation, of whose validity we have no reason to doubt.[65]

For the elimination of Lorentz's initial misunderstanding, a likely cause is Planck's *Lectures* of 1906. It was, as previously noted, far clearer than his early papers, and, if the three known early reviews provide representative guidance, it was not misunderstood. The earliest of these reviews is the most interesting, for its author, Albert Einstein, had just published a paper demonstrating that Planck's combinatorial form can be derived only by assuming resonator energies restricted to integral multiples of $h\nu$. No hint of that idea however, is contained in his careful and generally laudatory summary of Planck's viewpoint.[66] Instead, he describes Planck's use of Maxwell's equations to develop relations between the energy of a resonator and that of the surrounding field, emphasizes the need to supplement those relations with Boltzmann's combinatorial definition of entropy, and identifies the difference between Boltzmann's and Planck's approaches as the use of an energy element of fixed size by the latter. The other two reviews contain even fewer hints that there has been a break with classical theory. Bryan, writing in *Nature*, wonders only whether Planck's choice of equiprobable energy ranges can be justified.[67] Clemens Schaefer (1878–1968), in the *Physikalische Zeitschrift*, views the key step in Planck's specification of entropy as the introduction of natural radiation.[68] Though attitudes towards the significance of Planck's work had begun to shift when these reviews were written, only two or three people were yet aware of any reason to suppose that a break with classical physics was implied. Planck himself did not publicly acknowledge the need for discontinuity until 1909, and there is no evidence that he had recognized it until the year before.

Part Two

THE EMERGENCE OF THE QUANTUM DISCONTINUITY, 1905–1912

VI

DISMANTLING PLANCK'S BLACK-BODY THEORY: EHRENFEST, RAYLEIGH, AND JEANS

If Planck's *Lectures on the Theory of Thermal Radiation* exemplifies classical radiation theory, then it is the culmination of that tradition. In 1905, the year before the book appeared, Jeans had argued and Einstein had remarked in passing that the only radiation law compatible with classical theory was very different from either Planck's or Wien's. That claim could be and was readily set aside, most explicitly by Rayleigh, from whose earlier work Jeans's argument derived. But the status of what has since usually been called the Rayleigh–Jeans law nevertheless began to change during 1906. Ehrenfest then argued that Planck's resonators could not fulfill the primary function for which they had been developed, a point Planck conceded in a strange paragraph added, probably in proof, to the end of the *Lectures*. Properly conducted, Ehrenfest continued, resonator theory should yield the same result as Jeans's theory of the behavior of radiation in an otherwise empty cavity. To achieve Planck's result, conservation of energy and of the number of vibrators or vibration modes would have to be supplemented by some additional constraint, foreign to classical theory. One such constraint, Ehrenfest pointed out in closing, would be to restrict the energy of each vibration mode to integral multiples of the energy element $h\nu$. Einstein, by a very different route, arrived simultaneously at a far stronger and more restrictive proof of the same result. Previously, Einstein declared, he had thought Planck's theory incompatible with the light-particle hypothesis he had himself introduced during the preceding year. Now he realized that, properly understood, Planck's theory required that hypothesis.

Though Einstein and Ehrenfest doubtless helped here and there to prepare the way for a new attitude towards the significance of Planck's work, only Max von Laue (1879–1960) appears to have found their analysis of Planck's theory convincing from the start. But conviction did follow rapidly in the aftermath of Lorentz's presentation of a new proof of Jeans's result in the early spring of 1908. By the end of the following year, Lorentz, Wien, and Planck himself had been persuaded that radiation theory required discontinuity. Arnold Sommerfeld (1868–1951) and Jeans, among others, were at least moving towards that position during 1910, the year in which Lorentz provided a particularly effective and widely read set of arguments for it. By the years 1911 and 1912, with which this volume closes, all or virtually all those physicists who had devoted significant attention to cavity radiation were persuaded that it demanded some Planck-like theory, which would, in turn, require the development of a discontinuous physics. Though no one claimed to know what the shape of the new physics would be, the men concerned all recognized that there could be no turning back.

This chapter traces the way in which Ehrenfest's recourse to the work of Rayleigh and Jeans prepared the way for recognition of the central role of energy discontinuity in Planck's theory. The next considers the breathtaking series of papers that led Einstein to insist in 1906 that discontinuity was the fundamental prerequisite for Planck's success. Chapter VIII then describes both the events that led Lorentz to embrace a discontinuous radiation theory during 1908 and also the impact of his newfound conviction on other physicists, especially Planck and Wien. The final chapter of Part Two then examines the status of the quantum discontinuity during 1911 and 1912, attempting in the process to set black-body theory in the context of the other applications that had meanwhile been proposed for the quantum. It suggests that by 1912 physicists had learned very nearly all they could from the examination of cavity radiation and that the leading edge of quantum research had suddenly shifted to the previously neglected problem of the specific heat of solids. Finally, a brief epilogue on Planck's so-called second theory concludes the volume.

The origin of the Rayleigh-Jeans law, 1900–1905

The claim that black-body radiation should conform to the distribution law that has since been variously attributed to Rayleigh and Jeans was not made until 1905. But the main conceptual foundations for that claim can be found in a two-page note published by Rayleigh in the June 1900 issue of the *Philosophical Magazine*.[1] Commenting on current attempts to discover "the law of complete radiation," Rayleigh called attention to a fundamentally implausible characteristic of the Wien distribution law, recently rederived by Planck and still generally supported by experiment. The Wien law, he said, makes the density of radiant energy at a particular wavelength proportional to $\lambda^{-5} e^{-a/\lambda T}$. If it were correct, then energy would no longer increase with temperature when λT was large compared with the known constant a. Furthermore, Rayleigh pointed out, the cutoff should occur within an accessible experimental range. If the Wien–Planck law held, the infrared intensity at 60 μ recently studied by Rubens should not increase further for temperatures above 1000° absolute. No such cutoff seemed likely, and Rayleigh therefore proposed that the law be modified. At the end of an argument partly theoretical and partly ad hoc, he suggested replacing the factor λ^{-5} in the Wien–Planck law with $\lambda^{-4}T$, yielding a distribution law with the form

$$u = b\lambda^{-4}T\, e^{-a/\lambda T}. \tag{1}$$

That proposal satisfied the displacement law and also permitted intensity to increase with temperature at all wavelengths.

Rayleigh's 1900 note is both cryptic and incomplete. For that reason and to avoid functionless repetition, the first part of the following sketch is based on the fuller argument he developed in 1905.[2] Completeness aside, the latter differs from its predecessor only in specifying the value of the proportionality constant in the distribution law. Though Rayleigh's expression for that constant was smaller by a factor of eight than the one derived below, the error was at once corrected by Jeans and acknowledged by its author.[3]

As an expert on the theory of sound, Rayleigh chose to represent the electromagnetic field in a cavity by the vibrations of an elastic medium, initially a vibrating string. If a string, fixed at both ends, has a length L, it can vibrate only in modes of wavelength $\lambda = 2L/k$, where $k = 1, 2, 3, \ldots$. The same condition applies to the vibration of an elastic fluid, for example air, in a rigid cubical box of side L, except that permissible modes are governed by three integers, k, l, m, with $\lambda = 2L/\sqrt{k^2 + l^2 + m^2}$. If each triple of integers k, l, m represents a point at distance $R\,(=\sqrt{k^2 + l^2 + m^2})$ from the origin, then the number of modes lying in the wavelength range λ to $\lambda + d\lambda$ is given by the number of points lying in the first octant of a spherical shell contained between radii R and $R + dR$. Since $\lambda = 2L/R$, the number of modes must be

$$\frac{4\pi}{8} R^2 \, dR = \frac{4\pi L^3}{\lambda^4} \, d\lambda. \tag{2}$$

As represented in equation (2), Rayleigh's count of modes yields the λ^{-4} in the formula toward which he was proceeding. To discover the dependence of energy density on temperature, he appealed next to a very different set of considerations, which lead to what he called "the Maxwell–Boltzmann doctrine of the partition of energy." Subsequently better known as the equipartition theorem, it specifies that in any mechanical system each degree of freedom will on the average possess the same kinetic energy. In terms of the constants Planck had introduced in January 1901, furthermore, that energy must be just $\frac{1}{2}kT$, i.e., one-third the total translatory energy of a molecule. If the theorem applies to the vibrations of an elastic solid, then each of the distinct modes enumerated by equation (2) corresponds to one of the infinite degrees of freedom of the vibrating medium. It should possess mean kinetic energy $\frac{1}{2}kT$ and mean total energy kT, since the average kinetic and potential energies of a linear vibrator are equal. Finally, for the electromagnetic case involving transverse vibrations, the number of modes indicated by equation (2) must be doubled to permit two independent states of polarization, and energy kT must then be attributed to each of the resulting modes. With the resulting energy for all modes divided by L^3 the energy per unit volume in the range λ to $\lambda + d\lambda$ is

$$u_\lambda \, d\lambda = \frac{8\pi kT}{\lambda^4} \, d\lambda \left(= \frac{8\pi \nu^2}{c^3} kT \, d\nu \right). \tag{3}$$

That equation, usually in its wavelength form, is what came after 1905 to be known as the Rayleigh-Jeans law, but it is not the law proposed in 1900 by Rayleigh. Long a critic of the equipartition theorem, which he thought valid only under restricted conditions, Rayleigh remarked early in his note that any general treatment of equilibrium energy relations "is hampered by the difficulties which attend the Maxwell–Boltzmann doctrine of the partition of energy." He also suggested, however, that, "although for some reason not yet explained the doctrine fails in general, it seems possible that it may apply to the graver modes." By "graver modes" he meant long wavelength vibrations, to which alone he thought a form like equation

(3) might apply. To obtain an equation that could apply also to the shorter wavelengths at which equipartition broke down, he suggested that the factor $\lambda^4 T$ from equation (3) be multiplied by the exponential term from the Wien distribution law. The result was equation (1), and Rayleigh's note concluded with the hope that the agreement of this quasi-empirical law with observation "may soon receive an answer at the hands of the distinguished experimenters who have been occupied with this subject."

The "distinguished experimenters" learned of Rayleigh's proposal with surprising speed, and some of them responded.[4] Lummer and his colleague E. Jahnke (1863–1921) cited the proposed new law in a paper submitted to the *Annalen der Physik* late in July 1900.[5] Rubens shortly found that for large values of λT the energy density of radiation did increase with temperature as Rayleigh's law required. Though it probably played no significant role in his work, Planck knew of that result before he presented his own newly invented law to the Physical Society in October. But he soon also knew the result of experiments for smaller values of λT. In a paper submitted to the Academy of Sciences late in the month, Rubens and Kurlbaum compared a number of proposed radiation formulas with their data and concluded that Rayleigh's was satisfactory only in the limit where it coincided with Planck's.[6] Since the law, as proposed, was almost totally ad hoc, there was no further reason to take it seriously. Less than six months after it had been suggested, it was set aside. Even the usually scrupulous Planck made no mention of Rayleigh's contribution until 1906, by which time a very different set of concerns had led to the rederivation, not of Rayleigh's law, but of equation (3), originally a step on the way to it.

What led physicists back to Rayleigh's mode count was not the black-body problem at all but anomalies in the specific heat of gases.[7] If equipartition were admitted, then kinetic theory predicted that γ, the ratio of the specific heat at constant pressure to that at constant volume, should be given by $(n + 2)/n$, with n the number of degrees of freedom of a gas molecule. To bring that result into agreement with experiment it was necessary to suppose that molecules were composed of atomic mass-points rigidly connected, and even the agreement achieved in that way was far from satisfactory. In addition, it was difficult to conceive of atoms as dimensionless points, especially because the existence of spectra suggested that they must have a complex internal structure. Since there were many observed spectral lines, each presumably corresponding to a separate degree of freedom, n ought to be very large and γ close to unity, which it was not for any known gas.

The problem was an old one, and Boltzmann himself had commented on it in the well-known letter he had sent to *Nature* in 1895:

> But how can the molecules of a gas behave as rigid bodies? Are they not composed of smaller atoms [the vibrations of which cause radiation through the ether]? Probably they are; but the *vis viva* of their internal vibration is transformed into progressive and rotatory motion so slowly, that when a gas is brought to a lower temperature, the molecules may retain for days, or even for years, the higher *vis viva* of their internal vibrations corresponding to the original temperature. This transference of energy, in fact, takes place so slowly that it cannot be perceived amid the fluctuations of temperature of the surrounding bodies. The possibility of the transference of energy being so gradual cannot be denied, if we also attribute to the ether so little friction that the earth is not sensibly retarded by moving through it for many hundreds of years.[8]

Despite its plausibility, no one attempted to explore Boltzmann's idea until 1901. Then it was taken up or redeveloped by James Jeans, who thereafter worked on the subject for a number of years. In his first paper Jeans referred to the possibility of interaction between molecules and the ether, and he then continued: "That such an interaction must exist is shown by the fact that a gas is capable of radiating energy." On that assumption he undertook to demonstrate that

> a slight deviation from perfect conservation of energy [among the molecules of a gas] may result in a complete redistribution of the total [molecular] energy, and it will appear that this new distribution of energy will lead to values for the ratios of the two specific heats which are not open to the objections mentioned above.[9]

Jeans's technical argument, which has no relevance here, was expanded and refined at considerable length in Chapters 8 through 10 of the first edition of his well-known *Dynamical Theory of Gases*, published in 1904. Those chapters constituted the most original, if also the least enduring portion of his book, and Jeans continued to develop the theory they presented after the volume appeared, preparing two major papers on the interaction between matter and ether for publication in 1905.[10] In the first of these he redeveloped Rayleigh's method of mode counting in order to find the number of ether vibrations with periods comparable to or less than the time occupied by a collision between molecules. Only those modes would, he thought, take up energy rapidly; higher modes would respond extremely slowly; and the total energy in the ether could be shown to be very small.

Even before those papers appeared, Rayleigh, in a letter to *Nature*, opened a continuing discussion with Jeans by questioning the methods the latter had used in the *Dynamical Theory of Gases* to reconcile equipartition with observed specific heats.[11] A slow radiation of molecular energy into the ether might, he remarked, have the result Jeans desired if the molecules radiated into empty space. But if ether and molecules were confined together in a perfectly reflecting enclosure, there would be no net dissipation; all the infinitely many modes of ether vibration would receive their share of energy, so that the total energy would be dissipated over the infinity of higher modes. Since black-body measurements showed that nothing of the sort occurred, equipartition must break down for short wavelengths. "A full comprehension here would," Rayleigh concluded, "probably carry with it a resolution of the specific heat difficulty." Equipartition, not interaction with the ether, was, in his view, the source of the difficulty examined by Jeans.

The discussion that followed is neither uninteresting nor entirely one-sided, but only a few of its details are relevant here. To show that equipartition must break down, Rayleigh rederived his previous formula in a second letter, now with the proportionality constant and without the exponential term.[12] (This is the point at which he produced a constant eight times too large, and it is the resulting discrepancy between his formula and the long wavelength limit of Planck's that led him to wonder "how another process also based on Boltzmann's ideas, can lead to a different result."[13]) If the formula he had derived were accurate, Rayleigh pointed out, then a continuous ether, characterized by an infinite number of vibration modes, would absorb infinite energy. An atomic ether, he conceded, might reach equilibrium with finite total energy, but it would nevertheless absorb essentially all the initial translational energy of the molecules of the gas. Jeans responded at once, refusing to acknowledge difficulties with equipartition and insisting that the only possible equilibrium distribution for the energy of the ether was the one he had introduced, equation (3) above.[14] Though he agreed that such a distribution could never be physically realized, Jeans maintained that no paradox need result. Millions of years might, he thought, be required to transmit energy from the lower to the higher modes of ether vibration; a genuine equilibrium might never be achieved. For him, the physical situations studied in black-body experiments were therefore not cases of equilibrium at all. That is the position he redeveloped systematically in a paper read to the Royal Society on 16 November 1905, the first of his technical works in which he transferred the skills developed through research on specific heats to the problem of radiation alone.[15] His concerns and those that motivated Rayleigh's 1900 note had at last begun to intersect.

This debate, conducted in the columns of *Nature* during 1905, is the source of the Rayleigh–Jeans law together with the claim that no other equilibrium distribution of radiant energy can be compatible with classical theory. But Jeans was the only physicist except Einstein to make or accept such a claim at the time, and the strength of his convictions should probably be understood in terms of the subject's relation to his own earlier work on the specific heat of gases. His contributions to the latter subject had been since 1901 a primary source of his still developing professional identity, and the results he had presented during 1904 in his *Dynamical Theory of Gases* were impressive and in most respects plausible. His radiation theory, on the other hand, though apparently required to preserve his work on gases, had a very different status when evaluated separately. Its theoretical basis was in several respects doubtful, and much experimental evidence spoke against it as well.

By 1905 Planck's law was known to be in excellent quantitative agreement with experiment, and nothing in Jeans's argument suggested how that could possibly be the case. Indeed, when Lorentz briefly defended Jeans's position in 1908, experimentalists insisted at once that long-available experimental data permitted its rejection out-of-hand, an episode to be considered in Chapter VIII. Nor was it only experiments on the distribution of radiant energy that became, in principle, inexplicable from Jeans's viewpoint. By denying that these or related experiments dealt with equilibrium situations, Jeans also denied the relevance of thermodynamic arguments to them. New derivations would have to be found for Kirchhoff's law, the Stefan–Boltzmann law, and the Wien displacement law. Like the Rayleigh–Jeans law itself, the old derivations were valid only under unrealizable conditions; they could not apply to any of the situations actually encountered in laboratory research. With respect to the Stefan–Boltzmann law, Jeans acknowledged the difficulty explicitly, and he quickly began to seek a new, non-equilibrium derivation of the displacement law, from which the Stefan–Boltzmann law followed.[16] Though his arguments were of interest and developed with considerable skill, their acceptance would have exacted a high price for very little positive achievement.

Given such grounds for skepticism, Jeans's arguments for his radiation law could readily be set aside. His derivation presupposed the equipartition theorem, and that theorem was, as Rayleigh's position indicates, a source of much question and debate.[17] The standard routes to equipartition depended upon the explicit consideration of the process of collision between complex molecules. The resulting theorem was therefore applicable only to gases, and it, in any case, depended on special hypotheses and approximations which were themselves often doubted. At the beginning of the century those doubts were of special urgency because, as previously noted, both spectra and specific heats appeared to provide direct evidence against the applicability of equipartition even to gases. To consider still more general mechanical systems, including the elastic ether postulated by Rayleigh and Jeans, required some form of ergodic hypothesis: the phase point of the system in question must in time densely fill all the space compatible with the system's constraints. That hypothesis had generally been considered, even by its authors, to be extremely dubious.[18]

Besides, even if equipartition could be shown to apply to general mechanical systems, one might legitimately doubt its applicability to the electromagnetic displacements postulated by Maxwell's theory. No attempt to provide a mechanical model of the ether had succeeded, and physicists had increasingly stopped trying to provide one. The elastic fluid considered by Rayleigh and Jeans was not, in any case, a suitable model for electromagnetic phenomena. Indeed, their model did not even provide a coupling mechanism for the transmission of energy from one mode to the next. It could therefore represent radiation only in a perfectly conducting enclosure, one which was known to preserve the frequency distribution of the radiation initially introduced. How could it provide any basis at all for a theory of black-body radiation?

Planck's model, in contrast, seemed free of such difficulties, at least until the middle of 1906. It deployed Maxwell's equations rather than an elastic fluid ether; it had no apparent recourse to any doctrine like equipartition;[19] and it provided, or seemed to, the coupling mechanism absent in the theory developed by Rayleigh and Jeans. There were important questions about the way Planck's law was to be derived from his model, but they did not at all detract from the adequacy of the model itself. It was apparently founded on secure physical principles, as the Rayleigh–Jeans model clearly was not. Not surprisingly, therefore, when Planck briefly described the Rayleigh–Jeans law in his *Lectures* he treated it as simply "another interesting corroboration of the black-body radiation law for *long wavelengths* as well as of the relation between the radiation constant k and the absolute mass of ponderable molecules."[20] Planck mentioned Jeans's theses that the law represented the only possible equilibrium distribution for all frequencies and that experimental black-body radiation could therefore not be in an equilibrium state, but he quickly dismissed them for many of the reasons outlined above. When that eminently reasonable evaluation was prepared, no one but Jeans himself, and probably also Einstein, would have been inclined to quarrel with it. In any case, the experimentalists who, in 1900, had responded at once to Rayleigh's ad hoc proposal remained for some time entirely silent about Jeans's. They could have refuted it at once, but apparently did not think it worth the effort. The Rayleigh–Jeans law and what came to be called the "ultraviolet catastrophe"[21] did not yet pose problems for more than two or three physicists.

Ehrenfest's theory of quasi-entropies

That the theses first enunciated by Jeans finally did become central to physics is due to their repeated rederivation by a variety of different techniques, often applied to more realistic radiation models, including Planck's. Among such rederivations, the earliest were supplied by Einstein and Ehrenfest. Here we consider Ehrenfest's first, since it resulted from a critical study of Planck's radiation theory, and it made use as well of the earlier work of Rayleigh and Jeans. Einstein's more incisive results were, by contrast, the end product of his independent development, from 1902, of a generalized statistical thermodynamics. Their consideration will occupy the whole of the next chapter.

Paul Ehrenfest first learned of the black-body problem and of Planck's theory from Lorentz during a student visit to Leyden in the spring of 1903.[22] As a student of Boltzmann's he found the subject attractive, a fact attested by a series of notebooks entries made in late June after his return to Vienna.[23] But his notebooks also suggest that more urgent concerns—he completed his dissertation and married during 1904—postponed its persistent pursuit until the spring of 1905. Scattered references to the subject appear from early April of that year and are virtually continuous from mid-June. By the latter date, furthermore, Ehrenfest's attention had focused upon an especially puzzling feature of Planck's work, the one he would discuss in a paper presented to the Vienna Academy in early November. In addition, he had established most of the main elements of the approach to cavity radiation, that would lead him to the more consequential conclusions he submitted to the *Physikalische Zeitschrift* late in June 1906.²⁴

Ehrenfest's principal approach to the study of radiation theory, and especially of Planck's puzzling conclusions, is disguised in his papers but emerges explicitly in his notebooks around the beginning of May 1905. Though he occasionally considered Maxwell's equations or electron theory,²⁵ he preferred to investigate kinetic or mechanical analogues of the radiation problem, evaluating their adequacy by their success in producing forms equivalent to Kirchhoff's law, the Stefan–Boltzmann law, and the Wien displacement law.²⁶ One early model, considered in mid-May, involved a weightless leaf spring, loaded with a mass M at one end and forced to vibrate when that mass was struck by molecules of mass m.²⁷ Other models of the same general type recur until the end of the year, sometimes involving non-uniform or composite elastic strings and sometimes a network of particles connected by coil springs.²⁸ Though Ehrenfest regularly introduced some mechanism, usually gas molecules, to redistribute energy over the various vibration modes of his model, his problem otherwise closely resembled the mechanical one considered by Rayleigh and Jeans. Apparently, furthermore, he was aware of the resemblance. A first cryptic reference to Jeans appears in his notebooks in mid-July though it was not for another eight months that Jeans's theory became central to his thought.²⁹

By mid-June, also, Ehrenfest had discovered an intriguing puzzle in Planck's theory. The earliest hint of its recognition is the phrase, "Noteworthy that [there is] only *one* maximum," which constitutes the initial entry under a heading, "About black radiation."³⁰ What Ehrenfest had in mind is partially indicated by a series of entries that quickly follow, the first of them immediately after an item dated 18 June: "Planck found many functions that always increase—and accordingly [produce] different stationary states—Clarify!!... Boltzmann's H corresponds to entropy in those cases where the latter is defined—but what is the case with Planck? // Planck's stationariness-formula is sufficient to make $dS/dt = 0$ but [is] not necessary!"³¹ With only one brief item intervening, Ehrenfest then turned for the first time to the role of "complexion theory" in Planck's work, noting that, even in the absence of any possible recourse to Liouville's theorem, Boltzmann's combinatorial definition permits the introduction of "Abstract Entropy Theory," a phrase he placed in quotation marks and underlined.³² At about the time he wrote it, he must also have entered into the correspondence with Planck of which the only surviving portion is the letter of 6 July quoted in Chapter V. (Reasons to suppose the exchange was extended will emerge below.) A notebook entry just before 22 July records Ehrenfest's reaction to what Planck there said: "Can one seriously believe that the magnitude of the electronic charge can by itself provide that a pre-established quantum [of] total energy will be pulled about [zerzaust] in a certain manner[?]—Try to prove the contrary."[33]

Ehrenfest's notebook entries suggest that, when the last remark was written, his attention had temporarily shifted to other research topics. Taking up the black-body problem again in late October, he would rapidly proceed towards a deeper resolution of the puzzles he had posed between mid-June and mid-July. But that earlier month of concentrated work had already given him a command of both the problem and the concepts basic to the paper he presented to the Vienna Academy in early November.[34] There, under the title "On the Physical Presuppositions of Planck's Theory of Irreversible Radiation Processes," he argued that Planck's proof of the electromagnetic H-theorem lacked an essential step present in Boltzmann's equivalent theorem for gases. Planck has produced, Ehrenfest said, an entropy function Σ that can only increase or remain constant for all possible "natural" radiation states. But he has failed to prove that, "for given total energy, Σ will remain constant only if the stationary state reached by the radiation conforms to a distribution uniquely determined by the total energy." In fact, Ehrenfest continued, no such proof can be given in the absence of additional assumptions. Under the conditions required by Planck's theory, "it is possible to specify infinitely many radiation fields which: 1. all possess the same total energy; 2. [are] 'natural'; 3. [are] stationary, so that, in particular, the value of Σ does not increase when they are present; but which nevertheless, 4. correspond to infinitely many different distribution functions."[35] At the very end of his paper Ehrenfest specified, though without discussion, the additional conditions that appeared to ensure that Planck's theory would yield a unique distribution. They were, of course, complexion theory and the relationship $\varepsilon = h\nu$.[36] Ehrenfest stated he would devote a later paper to them.

In the event, Ehrenfest's argument proved more important than his conclusion, and it took, in outline, the following form. During 1900 Planck had made use of two different expressions for entropy, $\Sigma_1(U)$ and $\Sigma_2(U)$, equations (III-20) and (IV-15). Each specifies entropy as a function of the energy (and wavelength) alone. But the two are different: maximizing the first yields the Wien distribution, $\phi_1(\lambda)$; maximizing the second leads to Planck's distribution, $\phi_2(\lambda)$. Now imagine two different states, Z_1 and Z_2, of the radiation and resonators, each specified by a particular set of electromagnetic field functions, $\mathbf{E}(x, y, z, t)$ and $\mathbf{H}(x, y, z, t)$, together with the set of functions $f_i(t)$ for the various resonator moments. Both sets of functions are assumed to satisfy the conditions of natural radiation as well as Maxwell's field equations and those governing Planck's resonators. What differentiates the two is that the first, specifying Z_1, corresponds to the Wien distribution $\phi_1(\lambda)$, a total density of radiant energy $\rho_1 (= \int \phi_1 \, d\lambda)$, and a total energy E_1; whereas the second corresponds to the Planck distribution $\phi_2(\lambda)$, a total density of radiant energy ρ_2, and a total energy E_2. Using the function $\Sigma_1(U)$, Planck has shown that, if the state Z_1 ever occurs, then Σ_1, and consequently U, must thereafter remain constant. But if U is constant, then Σ_2 must be constant as well, for both Planck's entropy functions increase with energy. Precisely the same argument applies to the state Z_2. If it ever occurs, then not only Σ_2 but Σ_1 must thereafter be stationary. It follows that, if radiation in an initial state Z evolves to Z_1 in accordance with the equations governing the field and resonators, then, while Σ_1 rises to an absolute maximum, Σ_2 will rise to a stationary value, and conversely for an initial state that evolves to Z_2. Since Planck's criterion for an entropy function has been simply that it increase steadily to a stationary state, he has no basis on which to choose between them.[37†]

What made this situation especially serious, in Ehrenfest's view, is that the two states Z_1 and Z_2 may readily be adjusted so that they correspond to the same density of radiant energy and thus, by the Stefan–Boltzmann law, to the same temperature. The relevant arguments are the dimensional ones traceable to Lorentz and mentioned by Planck in his July letter. Multiplying all the field and resonator functions specifying Z_1 by m^2 yields a new set of functions specifying a new state Z_1'. It satisfies all the preceding conditions on Z_1 except that the total energy, density of radiant energy, and distribution function are given by $E_1' = m^4 E_1$, $\rho_1' = m^4 \rho_1$, and $\phi_1'(\lambda) = m^4 \phi_1(\lambda)$. If m^2 is now chosen so that $\rho_1' = \rho_2$, two simultaneously stationary radiation states, Z_1' and Z_2, have been determined, which correspond to the same temperature but to different distributions of radiant energy in the field. Unlike Boltzmann's arguments, Ehrenfest concludes, Planck's cannot determine a unique distribution function without recourse to some special assumptions.

A preliminary version of this argument and of the conclusions drawn from it are presumably what Ehrenfest communicated to Planck at the end of June. Planck's answer suggests that he was unimpressed, and one can easily see why. Excepting their generality, which the article does not altogether justify,[38†] Ehrenfest's conclusions add nothing to what Planck had known in 1900. By March of that year he had persuaded himself that any function $S(U)$ would have the characteristic properties of entropy if it satisfied the equation, $\partial^2 S/\partial U^2 \propto -f(U)$, with f any positive function that approaches zero as U goes to infinity. To find a unique distribution law would therefore require some additional argument to determine f. Planck's use of combinatorials was, in fact, his second such additional argument within the year. But Planck had very likely failed to recognize that the radiation state that maximized one entropy function was also a stationary state for all the others. That circumstance posed a further puzzle which, between late October 1905 and mid-February 1906, led Ehrenfest to a set of conclusions far more important than those described in his Academy paper. What aspects of the physical situation could account for the existence and strange interrelationships of the various functions that could serve as entropy? By the time his first paper was presented, Ehrenfest, in his notebooks, had begun to call for a "general theory of [such] quasi-entropies," one that would explain both their "always increasing" and the "uniqueness of the final state" to which each led. That theory was also to relate somehow to "complexion theory."[39]

Roughly two weeks before he first used the term "quasi-entropies," Ehrenfest recorded in his notebooks an important clue to the puzzling behavior of the functions for which he coined the name, and a similar clue appeared as an aside in his November paper. In the latter he wrote that, when several possible entropy functions exist with maxima corresponding to different radiation states, then the state actually reached "is not determined [simply] by the total energy but depends in some scarcely specifiable manner on the other initial conditions of the motion."[40] The equivalent notebook entry is quickly followed by a sketch of a model that would, by stages, lead him to understand how initial conditions might function. In that model, resonators are represented by numerous small masses, and these are interconnected by a network of springs that represent the ether. In addition Ehrenfest introduced a multitude of gas molecules able to excite the resonators by collision and thus to transfer energy between the various vibration modes of the ether-resonator net.[41]

What processes within the model, Ehrenfest quickly asked, correspond to increases of entropy? The first part of his notebook entry is a list of processes that, like diffusion, produce no such increases. Then he continued:

> H increases only if the mol[ecules in the phase-space cell] $do\ d\omega$ experience different events—
>   a) Collisions with each other
>   b)   ″   with a very rough wall
>   ??!! c) Enorm[ously] long time—[42]

In that alphabetized list, the first two items (the third will be considered in the next section) are independent mechanisms that can produce entropy increase. Either can cause H to change in the absence of the other. The effect of each can therefore be examined separately, a fact Ehrenfest immediately exploited by considering molecules that do *not* ordinarily collide and, in the process, significantly changing his model. Five items after the one just quoted, Ehrenfest wrote:

> Heuristic for a Thermod[ynamic]-Kinetic Theory of Cavity Radiation:
>
> | Partial oscillations [i.e., ether vibrations at individual frequencies] pass each other without [interaction] | Molecules of different sort travel without any collisions |
> |---|---|
> | – o – | – o – |
> | Resonators | Catalytic substances[43] |

Individual *non-colliding* molecules here take on a new role as analogues to individual modes of free ether vibration. The function of transmitting energy from mode to mode is transferred to resonators, on the one hand, and to a catalytic mechanism that promotes interaction between otherwise mutually transparent molecules, on the other.

First recorded in early November, the representation of radiation in a reflecting enclosure by the molecules of a "collision-free" gas played an important role in Ehrenfest's thought from the end of the following month. Then, in two notes written just before 1 January 1906, he explicitly employed the model to explore the behavior of entropy when a radiation-filled cavity expands without doing work.[44] The same model reappeared in February, at which time the functions previously attributed to rough walls were performed by spheres fixed in the cavity. Together, these spheres and the collision-free gas constituted, for Ehrenfest, a "Simplified Model for [a] Quasi H-Theorem," and that is the role in which the model was introduced in the paper Ehrenfest submitted to the *Physikalische Zeitschrift* at the end of June.[45]

What he then pointed out he must have recognized at least five, and very possibly eight, months before.[46†]

If the molecules of a collision-free gas interact only with rough walls or fixed elastic spheres, but not with each other, then only their directions of motion, not their individual speeds, will change over time. In the absence of a mechanism for redistributing energy among them, whatever velocity distribution they possess initially will be preserved. Boltzmann's H and other suitably selected functions may nevertheless increase with time to a stationary value as molecules are redistributed in position and in the directions of their motions. Such functions may, that is, behave as quasi-entropies. But only for some specially chosen initial velocity distribution—itself varying with the choice of entropy function—will a stationary value thus reached be an absolute maximum. Boltzmann's H, in particular, will reach such a maximum only if the molecules conform to the Maxwell–Boltzmann distribution from the start.

That was precisely the behavior that had so puzzled Ehrenfest when he discovered it in Planck's model of radiation interacting with resonators. Placing it in the more familiar gas-theoretic context provided clues to its source. A collision-free gas could proceed to any of an infinite number of different final states (and thus possess no unique equivalent for Boltzmann's H) because no mechanism was available to change the value of one of the parameters characterizing the state of a molecule, in this case its energy or speed. At least for a collision-free gas, the existence of numerous quasi-entropy functions results from the incompleteness of the available mechanisms of equilibration. If only Planck's model were similarly incomplete, its strange behavior could be at once explained. By the time he submitted his June paper, Ehrenfest had identified the required incompleteness. Once it was recognized, he went on to claim, Planck's model could be seen to be equivalent to the one employed by Rayleigh and Jeans. Properly conducted, even recourse to complexion theory should lead to the Rayleigh–Jeans law.

The impotence of resonators

Burbury's scrupulous 1902 analysis of Planck's pre-1900 radiation theory, last mentioned in Chapter V, may well have helped Ehrenfest to identify the element that that theory lacked. In an addendum attached to his Vienna Academy paper in proof, Ehrenfest called special attention to Burbury's three-year-old article, which he had not previously

DISMANTLING PLANCK'S BLACK-BODY THEORY 159

seen. Close to its end Burbury noted in passing what he took to be an inadequately developed aspect of Planck's proof:

> Planck has given no account of interchanges of energy between systems of different vibration periods. His method is in fact based on the assumption or proof (art. 6) that waves of different period from that of a resonator pass the resonator unaffected, so that no interchange of energy takes place. This, however, is not quite rigorous. If the difference of periods, though not zero, be very small, some very small interchange of energy between the wave and the resonator will consistently with the equations of page 433 take place....
> Now Planck does not investigate the law of these slow interchanges. He assumes that an entropy function exists for them, and that it is precisely the same function (but with variable ν) which has been defined above for systems having the same period. That may be true, but it cannot, I think, be accepted as an axiom. It seems to me that this branch of the subject requires further elucidation.[47]

Though the second sentence of the passage is not quite fair to Planck, Burbury's central point is sound. Planck had recognized that a damped resonator responds to incident frequencies near but not coincident with its resonant frequency, and he had implicitly relied on such interactions to redistribute energy among the various vibration modes of the field, thus opening a path to equilibrium. But Planck had not investigated the field-mediated interaction between resonators at slightly different frequencies, and his use of an analyzing resonator tuned precisely to the frequency of the field resonator under investigation precluded his doing so. Though resonators at nearby frequencies might, in fact, exchange energy, Planck's theory did not take such exchanges into account. Under those circumstances, if Planck had been able to exhibit a function that could only increase to a stationary value with time, the increase must be due to field alterations taking place at a single frequency, for example to changes in phase, direction, or polarization of individual partial vibrations.

With or without Burbury's intervention, Ehrenfest's June paper opened with a related but significantly stronger point:

> 1. The frequency distribution of the radiation introduced into the model [described by Planck] will not be influenced by the presence of arbitrarily many Planck resonators, but will be permanently preserved.
> 2. A stationary radiation state will [nevertheless] result from emission and absorption by the oscillators in that the intensity and polarization of all rays of each color will be simultaneously equilibrated in magnitude and direction.
> In short: radiation enclosed in Planck's model may in the course of time become arbitrarily disordered, but it certainly does not become

blacker.—For the discussion to come the following formulation is especially suitable: Resonators within the reflecting cavity produce the same effect as an empty reflecting cavity with a single diffusely reflecting spot on its wall.[48]

Ehrenfest here announces the incompleteness of the equilibrating mechanism that permits Planck's radiation model to support numerous quasi-entropies. Fixed linear resonators cannot alter the frequency distribution of the energy in the radiation field.

Ehrenfest's notebooks are frustratingly elusive with respect to the date at which he recognized this surprising and initially counter-intuitive theorem. Its first explicit enunciation, to be considered near the end of this section, does not appear until late May 1906, but it is then attached to a particular model of a problematic process with which Ehrenfest had been concerned for at least the preceding six months. He had surely been aware of potential difficulties about frequency redistribution by November 1905, at the latest, and he may have recognized the general theorem at that time. Reasons for wishing that his recognition of the impotence of resonators could be dated more precisely will shortly appear.

As early as June 1905, Ehrenfest had recorded the general solution of the differential equation for a damped resonator of frequency ν_0 driven by a sinusoidal field of frequency ν.[49] The resulting resonator moment consisted, he noted, of two superimposed vibrations, one of fixed amplitude at the exciting frequency ν, the other a damped oscillation at a field-independent frequency almost coincident with ν_0. Presumably, at this point, he took the latter to represent the new frequency to which interaction with the resonator had shifted some field energy previously at the frequency ν.[50†] Doubts about that manner of understanding energy redistribution however, began to appear almost at once.

One likely source of these doubts, to which Ehrenfest alluded in his paper of June 1906, derived from two of the models for resonators he had mentioned in his notebooks twelve months before. Two nearby entries point out that entities defined simply by Planck's resonator equation can be physically represented either by tiny tuned perfect conductors or by suitably chosen bits of dielectric.[51] In print, that remark was accompanied by highly general theoretical reasons why no change in frequency distribution could result from the introduction of resonators of this sort. But even in the absence of such reasons, existing research on dispersion and reflection would have suggested that neither way of modeling resonators could produce significant alterations of the distribution of energy with frequency except conceivably over very long periods of time. At best the effect would be vastly more gradual than the rapid changes of phase, direction, and polarization due to interaction between a resonator and the incident field. Possibly Ehrenfest had such slow changes in mind when he closed the list of items that could cause changes in H with "??!! c) Enorm[ously] long time."[52]

Ehrenfest could have seen these difficulties as early as the summer of 1905, and by November he had apparently recognized others. Almost immediately after his first list of entropy-increasing factors, he wrote in his notebook: "[How] to specify resonators which transform each incident wave into a small spectrum (dep[endent] on temp–) (Diff[erential] equ[ation] non-linear)," and four items later, "Black body: changes any given energy quant[um] into a precisely det[erminate] spectrum (dependent on T)."[53] These entries read like proposals for the introduction of resonators with properties, like non-linearity, not possessed by Planck's. Very possibly, when he wrote them, Ehrenfest had realized that the apparently "new" frequency near ν_0, introduced when a linear resonator is first struck by an incoming "monochromatic" wave of frequency ν, is actually present through all past time in the Fourier representation of that wave and is thus not really a new frequency at all. Late in the following month, December, his notebook entries display his awareness of another problem of only slightly less force. The energy spectrum of a given field is not obtained simply by squaring the amplitudes of the field's individual Fourier components. One must instead, as Planck had shown in 1898, average the square of the new field over a time that is long compared with the period of its significant components. The corresponding measure of field intensity demands analyzing resonators with bandwidths too large to discriminate between a resonator's natural frequency ν_0 and those field frequencies ν sufficiently near ν_0 to cause significant resonator excitation.[54] In his published paper Ehrenfest would emphasize that the plausibility of Planck's frequency-changing mechanism was due, in part, to confusing the properties of the distribution of field strength, with those of the distribution of energy.[55]

Early in 1906 Ehrenfest had thus developed a far deeper understanding of the "Physical Presuppositions of Planck's Theory of Irreversible Radiation Processes" than was presented in the paper with that title he had read to the Vienna Academy at the beginning of the preceding November. By February, at the latest, he knew that the existence of quasi-entropy functions could be due to an incomplete equilibrating mechanism, and he had earlier had reason at least to suspect the nature of the particular incompleteness that characterized Planck's model. Those discoveries need not, for him, have suggested that no classical model of black-body radiation could succeed. A footnote in his published paper points out that molecular collisions could redistribute energy between fixed resonators and that a non-linear resonator equation would have the same effect.[56] But the discoveries did raise fundamental questions about the adequacy of Planck's theory, and Ehrenfest is likely to have responded to them by renewing the correspondence he had initiated with its author during the preceding June or July. That he, in fact, did so is strongly suggested by a surprising "Conclusion" added, presumably in proof, to the first edition of Planck's *Lectures*. Planck specially directed his readers' attention to it at the end of his brief preface, dated "Easter 1906." The first numbered paragraph of Ehrenfest's June article refers to it as well.[57†]

> § 190. **Conclusion:** The theory of irreversible radiation processes developed here explains why, in an irradiated cavity filled with oscillators of all possible frequencies, the radiation, regardless of its initial conditions, reaches a stationary state: the intensities and polarizations of all its components are simultaneously equilibrated in magnitude and direction. But the theory still is characterized by an essential gap. It treats only the interaction between radiation and oscillator vibrations at the same frequency. At a given frequency the continuous increase of entropy to a maximum value, required by the second law of thermodynamics, is therefore proven on purely electro-dynamic grounds. But for all frequencies taken together the maximum reached in this way is not the absolute maximum of the system's entropy, and the corresponding state of the radiation is not in general the [state of] absolutely stable equilibrium (cf., paragraph 27). The theory does not at all elucidate the manner in which the radiation intensities corresponding to different frequencies are simultaneously equilibrated, that is, the way in which the initial arbitrary distribution settles in time to the normal distribution characteristic of black radiation. The oscillators which provide the basis for the present treatment influence only the intensities of the radiation corresponding to their own natural frequencies. They are not able, however, to change its frequency if their effects are restricted to the emission and absorption of radiant energy.[58]

When Planck wrote that passage, the article in which Ehrenfest announced the identical discovery had not yet been composed, and Planck may therefore conceivably have been led to it independently by his work on dispersion theory. But, if so, the discovery must have come very late, after Planck's research had shifted from dispersion theory to relativity and after his manuscript of the lectures delivered in the Winter Semester, 1905–1906, was largely completed. Had Planck made the discovery earlier, he would surely have presented it within the body of his book, the structure of which would have required significant revision. In its published form, the *Lectures* opens with two chapters on the behavior of radiation in a cavity that contains no resonators. Paragraph 27, to which Planck calls attention in the passage just quoted, emphasizes that, though entropy will increase in such a cavity, it will not reach an absolute maximum "since the given total energy can be arbitrarily distributed over the various colors of radiation."[59] Resonators are then introduced in the third chapter to provide a mechanism for redistributing energy over frequency, and they retain that role until the last two paragraphs of the book.

Those paragraphs, which constitute Planck's new "Conclusion," must have been devastating to write, for they invalidate, not the details of his presentation, most of which remain worthwhile, but the overall structure of the argument in which those details had been embedded through the preceding two hundred and twenty pages. Much of the awkwardness that inevitably results could have been avoided if Planck had recognized the impotence of resonators in time. Under these circumstances, it seems probable that Ehrenfest's intervention was required to bring Planck to his "Conclusion," and that likelihood is reinforced by a footnote attached to the close of the preceding quotation. It cites Ehrenfest's 1905 paper to the Vienna Academy, an article in which, as previously emphasized, nothing whatsoever is said about the mechanisms that lead to entropy change. Except as an inadvertently disguised acknowledgment of Ehrenfest's role, that citation is extraordinarily difficult to understand.[60†]

If, however, Planck accepted Ehrenfest's still unpublished theorem, he said nothing about how it was to be proven or why his initial intuition of the effect of damped resonators was wrong. Ehrenfest's remarks on that subject in his June paper are too cryptic to be helpful, a prelude to the quite different subject to which we shall shortly find his article primarily devoted. Soon after it appeared, with the increasing recognition of discontinuity and the consequent elimination of the damping term from the resonator equation, the problem vanished from physics, only to reappear in a quite different context, scattering theory. As a result, the claim that damped resonators can alter the frequency distribution of radiant energy is still widely believed, even among physicists. A brief discussion of the considerations that invalidate it may therefore eliminate confusion and facilitate understanding of the concepts that connect the concerns of Ehrenfest's notebooks with his brief published remarks.

A radiation-damped resonator will oscillate naturally at its resonant frequency ν_0, but it can also absorb energy from a wave at a nearby frequency ν. That such a resonator will alter the frequency distribution of radiant energy is plausible, for, if it absorbs energy from an incident wave of frequency ν, it will, one supposes, reradiate that energy at its natural frequency ν_0 after the initial wave has passed. Two sets of considerations, both adumbrated in Ehrenfest's notebooks, are relevant to showing that nothing of quite that sort occurs. A damped resonator will reradiate only at frequencies already contained in the exciting wave. At each of the frequencies to which it does respond, furthermore, it will reradiate only as much energy as it absorbed at that frequency.

Planck's equation for the moment f of a damped resonator, equation (I-8b), may for convenience be rewritten in the form,

$$\ddot{f} + 2\alpha \dot{f} + \omega_0^2 f = \beta Z(t), \tag{4}$$

with $Z(t)$ the electric field parallel to the resonator axis, ω_0 the resonator's natural angular velocity, α its decay rate (equal to $\sigma\omega_0/2\pi$), and β a constant equal to $3c^3\sigma/2\pi\omega_0$. The right-hand side of that equation, $\beta Z(t)$, may generally be represented by a Fourier integral, assumed convergent,

$$\beta Z(t) = \frac{1}{\sqrt{2\pi}} \int_{-\infty}^{+\infty} E(\omega)\, e^{i\omega t}\, d\omega, \tag{5}$$

with $E(\omega) = E^*(-\omega)$ to ensure that $Z(t)$ is real. The general solution of equation (4) is then given by

$$f(t) = A\, e^{-\alpha t} \cos(\omega_0 t + \theta) + \frac{1}{\sqrt{2\pi}} \int_{-\infty}^{+\infty} \frac{E(\omega)\, e^{i\omega t}\, d\omega}{\omega_0^2 - \omega^2 + 2i\alpha\omega}, \tag{6}$$

where A and θ are arbitrary constants.

In equation (5), the function $E(\omega)$ may be chosen so that there is no net field $Z(t)$ during all time prior to some selected instant t_0. By complex integration clockwise about a contour consisting of the real axis and an infinite semicircle in the lower half plane, it is then easily shown that, for $\alpha > 0$, the integral on the right of equation (6) must vanish for all $t \leq t_0$. Therefore, unless the resonator has been excited by some source other than the field, equation (6) provides a physically permissible solution of the resonator equation only if the arbitrary constant A is set equal to zero. The resonator's natural angular velocity ω_0 does not then appear in the spectrum of $f(t)$ unless it is already represented in the spectrum of the driving field. Among those frequencies that are present in the incoming wave, the resonator does, of course, respond most strongly to the ones closest to ω_0. But that is only to say that it *both* absorbs *and* reradiates more strongly near its resonant frequency than it does elsewhere. Since the techniques required to show the conservation of the energy of wave plus resonator can be applied separately to each frequency, no net redistribution of energy can occur.

Some cryptic entries in Ehrenfest's notebooks suggest that he reached this conclusion by a somewhat different route, one less likely to be familiar to most readers. Characteristically, he again called on a mechanical model the motion of which was to be represented in what are usually called normal coordinates.[61] Such coordinates are always available for motions consisting exclusively of displacements from equilibrium provided that the restoring forces are linear functions of the displacements. For present purposes, the special feature of normal coordinates is that the variation of each of them occurs at a single frequency and can be represented by a form like $C_i \sin(2\pi\nu_i t + \theta_i)$. Here, ν_i is some linear combination of the original displacement frequencies, and both C_i and θ_i are constants determined by the masses of the system and their equilibrium positions. The total energy is then given, in normal coordinates, by one-half the sum of the squares of the individual coordinates and of their first derivatives, so that a certain unchanging energy, $\frac{1}{2}C_i^2$, is attributed to each of the coordinates over all time. In the original displacement coordinates, on the other hand, the energy attributed to each coordinate oscillates slowly at a frequency given by some linear combination of the various displacement frequencies.

Reduced to standing wave solutions, the Rayleigh–Jeans problem of radiation in an empty reflecting cavity is clearly equivalent to a mechanical vibration problem treated in normal coordinates. Ehrenfest appears to have recognized their relation by late March 1906 when he wrote in his notebooks, "Energy distribution over normal vibrations.... An H-theorem on this basis."[62] If a linear resonator weakly coupled to the normal modes of the empty cavity is introduced, the new system thus formed can again be reduced to normal coordinates, and the problem treated as before. In either case, the energy in each mode remains constant; no energy is transferred from one vibration frequency to another. Some such theorem is what Ehrenfest must have anticipated when he wrote the following entry in his notebook just before 30 May: "By returning to the normal vibrations of the system [consisting of] ether plus resonators, show that no change in the 'color distribution' can ever be brought about in this way."⁶³ That is his first fully explicit expression of his conviction about the impotence of resonators, though it presumably does not represent the basis for his earlier doubts on the subject of energy redistribution.

It is against this background that Ehrenfest's very brief published remarks about the problem of redistributing energy must be read. Early in his June paper he refers readers to the closing paragraphs of Planck's *Lectures* (which can scarcely have appeared) and proceeds to his own, previously quoted, statement about the equivalence of Planck's model to a perfectly conducting cavity containing a diffusely reflecting spot. Next, he mentions the treatment of the Rayleigh–Jeans cavity in the *Lectures*, and he quotes from it the statement that in such an empty cavity "there can be no talk...of a tendency towards the equilibration of the energy allotted to individual partial vibrations."⁶⁴ Then he continues:

> This conclusion applies immediately to the Planck model. So long as the oscillators are defined only by the linear homogeneous differential equation,* which Mr. Planck establishes for them, they are essentially identical with small specks of complete conductors or suitable dielectrics. In that case every state of motion of the Planck model is again [like the empty cavity] a superposition of the normal vibrations of this more complicated system. There can, therefore, in this case also, be no talk of a tendency towards the equilibration of the energy allotted to individual partial vibrations.⁶⁵

The asterisk in that passage leads to a footnote that admits the possibility of energy redistribution in the presence of moving molecules or with a non-linear oscillator equation; Planck noted an equivalent escape route in the last paragraph of his "Conclusion." To Ehrenfest, however, these possibilities were not of current interest; his immediate object was to analyze Planck's theory rather than produce one of his own. Having pointed out, at the start, that Planck's success could not be due to the use of resonators, he therefore devoted the body of his paper to exploring the potential of another apparently special aspect of Planck's approach. It was the use of combinatorials or what Ehrenfest had previously labeled "Abstract Entropy Theory."

Complexion theory and the Rayleigh–Jeans law

In his papers on radiation and again in his *Lectures*, Planck had evaluated entropy by applying complexion theory to resonators. But complexion theory, thus applied, was simply probability theory, and its utility should therefore be independent, Ehrenfest pointed out, of recourse to resonators. Developing a representation "that corresponds more nearly to the methods of Rayleigh and Jeans" than to those used by Planck,[66] Ehrenfest therefore proceeded in the third part of his paper to apply Boltzmann's probabilistic definition of entropy directly to the field.

For radiation in any empty cavity, each of the independent modes of oscillation may be conceived as an undamped oscillation with energy

$$\varepsilon_\nu = \tfrac{1}{2}(\alpha_\nu f^2 + \beta_\nu \dot{f}^2) = \tfrac{1}{2}\left(\alpha_\nu f^2 + \frac{1}{\beta_\nu} g^2\right). \tag{7}$$

In these equations, f is the moment of a field oscillator, g its conjugate momentum $(= \partial \varepsilon_\nu / \partial \dot{f})$, and ν its oscillation frequency $(= \sqrt{\alpha_\nu/4\pi^2 \beta_\nu})$. Now let $F(\nu, f, g)$ be the still unknown distribution function that specifies the fraction of field oscillators with frequency ν and coordinates in the range f to $f + df$ and g to $g + dg$. In addition let $N(\nu) d\nu$ be the number of field oscillators or of vibration modes in the range ν to $\nu + d\nu$. The standard generalization to phase space of Boltzmann's probabilistic definition of entropy[67] then yields for the entropy of the field,

$$S = -k \int_0^\infty N(\nu) \, d\nu \int\!\!\int_{-\infty}^{\infty} F(\nu, f, g) \log F(\nu, f, g) \, df \, dg. \tag{8}$$

In that expression $N(\nu) d\nu$ is simply the Rayleigh–Jeans mode count, equation (3), rewritten in terms of frequency and multiplied by two for application to perpendicularly polarized transverse waves. Continuing to follow Boltzmann's technique, Ehrenfest finds the equilibrium state by maximizing equation (8) subject to the constraint on total energy E_t and an appropriate normalization condition on F:

$$\int\!\!\int_{-\infty}^{\infty} F(\nu, f, g) \, df \, dg = 1$$

$$E_t = \int_0^\infty N(\nu) \, d\nu \int\!\!\int_{-\infty}^{\infty} \varepsilon_\nu F(\nu, f, g) \, df \, dg. \tag{9}$$

Finally, with F thus determined, the total energy of the radiation with frequency between ν and $\nu + d\nu$ can be written

$$E(\nu)\, d\nu = N(\nu)\, d\nu \int\!\!\!\int_{-\infty}^{\infty} \varepsilon_\nu F(\nu, f, g)\, df\, dg.$$

Given the examples provided by both Boltzmann and Planck, the required manipulations are straightforward. Their result, however, is a demonstration that F can be maximized only by attributing equal mean energies to each of the vibration modes. If, furthermore, those modes interact with the molecules of a gas, that constant mean energy is just kT. Complexion theory applied directly to the field thus yields the same impossible radiation law that Jeans had derived directly from the equipartition theorem in 1905.[68]

If, however, Ehrenfest's adaptation of complexion theory to the field led only to a known result, his reformulation suggested how other results might be achieved. Planck's introduction of resonators had only confused the fundamental physical issue. Equations (7) and (8) together with the associated maximization techniques are, Ehrenfest emphasized, common to Planck and Boltzmann. If they nevertheless can lead to different distribution functions, he continued, the source of the difference must lie in the choice of constraints, and indeed choices different from equations (9) may be justified.

> Suppose [for example] that radiation in nature is produced only through the intervention of electrons and that these electrons everywhere possess the same definite structure. This structure—perhaps merely the eternally fixed electron cross-section—may in principle be capable of preventing the excitation by natural means of some of the conceivable normal vibration modes of our cavity.[69]

During the year before that passage was written, Ehrenfest's view of Planck's hopes for electron theory had clearly changed.

Ehrenfest ends his paper with a brief general discussion of constraints, showing that they are not ordinarily uniquely determined by the distribution function that their application yields. Other constraints besides Planck's might thus have led to his distribution law. But one can at least demonstrate, Ehrenfest concludes, that Planck's law can follow from complexion theory applied to the field alone. One suitable device is an additional constraint that restricts the value of ε_ν to integral multiples of the energy quantum $h\nu$, "just as if, for each frequency, the vibration energy consisted of 'energy atoms' of numerical magnitude, $\varepsilon_\nu^0 = 6.548 \cdot 10^{-27} \cdot \nu$ erg."[70] That condition, he adds, can be restated in a form more nearly standard in statistical mechanics. In the two-dimensional phase space f, g,

> the phase point of a proper vibration of frequency ν cannot occupy any position on the surface: instead it may lie only on a family of curves, namely the family of ellipses
>
> $$\tfrac{1}{2}\left(\alpha_\nu f^2 + \frac{1}{\beta_\nu} g^2\right) = mh\nu,$$
>
> where m runs through the series of integers to a value such that $mh\nu$ would exceed the previously specified total energy if m increased further.[71]

Though Ehrenfest notes in passing that this formulation is not quite the same as the one provided in Planck's *Lectures*, nothing in his paper suggests that he thought that more than a slip of Planck's pen might be involved. He was still reading Planck through the spectacles provided by Lorentz and had not, in any case, had the book long enough to assimilate Planck's developed approach.

Ehrenfest's paper includes no proof that his quantization of field oscillators results in Planck's law, but one can easily be supplied. In eq. (7), ε_ν must be equated to $nh\nu$; in the equations that follow, F becomes a function of n rather than g, and the corresponding integrals over g become sums over n. Then $F(\nu, f, n)$ becomes the fraction of those modes with frequency ν that possess energy $nh\nu$ and lie between f and $f + df$, and $N(\nu)$, the number of modes at any fixed frequency, is everywhere equal to two. With these changes, Planck's law results straightforwardly from the manipulations Ehrenfest has sketched. The elaborate relationships between damped resonators and field—the core of Planck's approach from late 1894 through 1906—is thus shown to be irrelevant to their author's principal achievement. His law can be derived without recourse to resonators.

VII

A NEW ROUTE TO BLACK-BODY THEORY: EINSTEIN, 1902–1909

Though the implications of the paper Ehrenfest had submitted in late June 1906 were startling, they were not by that date still altogether new. Very differently expressed, they had also been educed in an article "On the Theory of the Emission and Absorption of Light" received by the editor of the *Annalen der Physik* more than three months before. Its author was Albert Einstein, another little-known young physicist who, unable to obtain an academic position, wrote from the Swiss patent office in Berne. Analyzed in classical terms, Einstein said, Planck's black-body model could lead only to the Rayleigh–Jeans law. Planck's radiation law could be derived instead, but only by decisively altering the concepts its author had employed for that purpose. Midway through his paper Einstein wrote:

> We must, therefore, recognize the following position as fundamental to the Planck theory of radiation: The energy of an elementary resonator can take only values which are integral multiples of $(R/N)\beta\nu$ [where R is the gas constant, N Avogadro's number, and β a constant]. During absorption and emission the energy of a resonator changes discontinuously by an integral multiple of $(R/N)\beta\nu$.[1]

That passage is the first public statement that Planck's derivation demands a restriction on the classical continuum of resonator states. In a sense, it announces the birth of the quantum theory.

Though their conclusions largely overlapped, Einstein's paper was in several respects quite different from the one Ehrenfest was to submit a few months later. Its argument was both more general and more compelling. Unlike Ehrenfest, furthermore, Einstein did not suppose that he was simply restating Planck's own premise, and the structure of his argument highlighted the impossible difficulties to be encountered in developing the proposal that fixed size h be attributed to phase-space cells. Still more important, whereas Ehrenfest's paper had been a study of Planck, Einstein's was primarily a study of nature. What brought Einstein to the black-body problem in 1904 and to Planck in 1906 was the coherent development of a research program begun in 1902, a program so nearly independent of Planck's that it would almost certainly have led to the black-body law even if Planck had never lived. To understand Einstein's involvement with black-body theory one must begin by briefly retracing his steps.

Einstein on statistical thermodynamics, 1902–1903

As Klein has emphasized, Einstein was deeply impressed from the start of his career by the simplicity and scope of classical thermodynamics. In that respect he was like Planck, but for Einstein thermodynamics included the statistical approach he had learned from Boltzmann's *Gas Theory*.[2] His first two papers, published in 1901 and 1902, were attempts to investigate intermolecular forces by applying phenomenological thermodynamics to such phenomena as capillarity and the potential difference between metals and solutions of their salts.[3] Finding his results inconclusive, Einstein quickly abandoned that approach and began instead to develop a statistical thermodynamics applicable not only to gases, the main concern of earlier workers, but to other states of aggregation as well. Presumably, he felt that the statistical approach provided a firmer basis than the phenomenological for conclusions at the molecular level. Whatever his motive, however, the result of Einstein's effort was a series of three brilliant papers, published successively during each of the years 1902, 1903, and 1904. These papers are now little remembered because their principal results were simultaneously established by Gibbs in his *Statistical Mechanics* of 1902.[4] Nevertheless, they provided the starting point for much of Einstein's future work, especially for that on Brownian motion and on the quantum, both traceable from 1905.

The first of Einstein's statistical papers developed a theory of statistical thermodynamics for mechanical systems governed by Lagrangian equations of motion with an explicit potential function. Those mechanical equations were required, however, only to justify recourse to Liouville's theorem and to energy conservation, a fact that suggested, Einstein said, that his theory might be redeveloped for systems of a far more general sort.[5] That generalization was supplied in the second paper of Einstein's series. Since in other respects the two papers are closely parallel, attention is here restricted to the second.[6]

Einstein opens that paper by directing attention to a system of which the state can be specified by n independent variables p_i. (For the mechanical systems he had previously treated, these variables were the generalized coordinates and velocities of a Lagrangian description.) If such a system is isolated, its state at one instant must determine the values of the state variables at the next. The behavior of the system over time, that is, must be governed by a set of n equations,

$$\frac{dp_i}{dt} = \phi_i(p_1, p_2, \ldots, p_n), \tag{1}$$

which play the role taken by Lagrange's equations in the mechanical case. Einstein supposes, in addition, that the system of equations (1) possesses one *and only one* independent integral,

$$E(p_1, \ldots, p_n) = \text{Constant}, \tag{2}$$

which is a powerful condition equivalent to an ergodic hypothesis.

For systems of this very general sort, Einstein next develops a series of theorems for which gas theory had had no special need but which are required for the transition to a more widely applicable statistical mechanics or statistical thermodynamics. In particular, Einstein produces, in terms of the functions ϕ_i and E, expressions for those quantities that must correspond to the temperature, entropy, and probability of a state. (In gas theory, temperature could be defined by recourse to the ideal gas thermometer, and an expression for entropy then followed.) These are, of course, precisely the conceptual elements most obviously missing from Planck's discussion of the black-body problem. He had been forced to "define" the probability of a state, and he had noted that his definition, though plausible, could be further justified only by experiment. That uncertainty about probability was transmitted to entropy, the logarithm of probability, and thence to temperature, the derivative of energy with respect to entropy. Einstein's paper, written before he showed any signs of concern with black-body theory, bridged these gaps.

Einstein begins by asking what conditions the trajectories specified by equations (1) and (2) must satisfy in order that the corresponding systems be physical, i.e., possess observable properties. "Experience teaches us," he says, "that an isolated physical system settles after a time into a state such that none of its observable magnitudes changes further; we call such a state stationary."[7] Observables, he goes on to say, are represented by time averages of functions of the microscopic coordinates p_i (consider the pressure of a gas), and his discussion implies that the time interval over which the average is taken must be long enough so that the coordinates take on all possible combinations of values. Two measurements made at different times will then yield the same value if the trajectories specified by equations (1) return to their starting points with some constant frequency, for the system will then, during each cycle, spend the same proportion of time in the vicinity of any given point specified by the p_i's. Indeed, a slightly weaker condition will produce the same result. Imagine a region Γ in the n-dimensional space of the p_i's; observe the system over some time interval T; and determine the portion τ of that interval during which the system lies in Γ. Then, if for each selected Γ the fraction τ/T approaches a limit with increasing T, the system will possess fixed, observable properties. It will be a physical system.

Though Einstein's argument, to this point, is in several respects defective, it is readily salvaged if observables are taken to be averages not over time but over the members of a suitably arranged set or ensemble of identical systems.[8†] That is the step to which Einstein at once proceeds, and the properties of the set had probably guided his thoughts as he sought the conditions that make an individual system physical. The collection of systems he envisaged is very close to what, following Gibbs, has since been called a microcanonical ensemble.[9] It consists, that is, of a large number N of identical systems, all governed by the equations of motion (1), all independent, and all with the energy constant of equation (2) in the narrow range E^* to $E^* + \delta E^*$. If the systems of this ensemble are distributed through p_i-space in such a way that the number of systems m in a region Γ remains constant during their motion, the ensemble is said to be stationary, and it then possesses the two following properties. The value of m/N is at all times the same as the previously specified limit of τ/T for individual members; all members of the stationary ensemble are therefore physical. In addition, time averages of functions of the p_i's over long intervals T may be replaced by averages over the N members of the ensemble at any instant. Though the statement of these properties is obscure in Einstein's paper, he was clearly aware of both, and he used them.

Examining the properties of his ensemble, Einstein first points out that, if g is an infinitesimal volume in p_i-space, then the number of systems in g at a given instant is

$$dN = \varepsilon(p_1, \ldots, p_n) \int_g dp_1\, dp_2 \cdots dp_n,$$

where $\varepsilon(p_1, \ldots, p_n)$ is the density of system-points in the space of the p_i's. If the ensemble is to be stationary, then this density must obey standard hydrodynamic continuity conditions, and Einstein introduces them to show that ε can depend on the p_i's only through the energy, which he had previously restricted to an infinitesimal range. This part of his analysis he can therefore complete by rewriting the preceding equation in a form fundamental to all that is to come:

$$dN = \text{Constant} \int_g dp_1 \, dp_2 \cdots dp_n. \tag{3}$$

Though it occurs near the beginning of his paper,[10] Einstein's equation (3) displays the aspect of his thought which made it impossible for him to accept Planck's version of black-body theory. For him, as for Planck, the state of a system is specified in terms of the small cell g in which the system's coordinates lie. But, through the physicality condition, Einstein's concept of state carries with it a concept of probability which differentiates it from Planck's. The probability W_g of finding a given system in the state g must be the fraction (τ/T) of time which the system spends in g or, equivalently, the fraction (dN/N) of the members of the ensemble that are to be found in g at a given time. Thus, W_g is simply the right-hand side of equation (3), appropriately normalized, and it is necessarily proportional to the volume of g. Planck's hope that some physical mechanism yet to be discovered would account for the need to keep cell size constant is therefore unrealizable in principle. To fix the size of cells while retaining continuous trajectories does violence to the concept of probability.

Even before explicitly applying the notion of an ensemble to considerations of probability, Einstein had employed it to define temperature and entropy.[11] Each of the N systems of the ensemble may be conceptually subdivided into interacting sub-systems, a large one Σ and a small one σ, the latter shortly to be called a "thermometer." The large system is specified by the subset $\Pi_1, \ldots, \Pi_\lambda$ of the original n variables, and it has energy H; the small one is specified by the remaining variables $\pi_1, \ldots, \pi_l$ and has energy η. Since the two subsystems interact, only $E(=H + \eta)$ is a constant, but Einstein assumes that their construction requires $H \gg \eta$, so that H is very nearly constant.

Next Einstein asks how many of the N systems of the ensemble will have the thermometer variables π_i in a specified region, the remaining variables being unspecified. The answer, he shows, is given by the equation

$$dN_2 = \text{Constant } e^{-2h\eta} \, d\pi_1 \cdots d\pi_l, \quad (4)$$

where h is a parameter that depends only on the total energy E^* and on the structure of the large system Σ. With that structure specified, a new function $\omega(E^*)$ may be defined by

$$\omega(E^*) = \int_{E^*}^{E^* + \delta E^*} d\Pi_1 \cdots d\Pi_\lambda. \quad (5)$$

When that function is known, the parameter h required by equation (4) is given by

$$h(E^*) = \frac{1}{2} \frac{\omega'(E^*)}{\omega(E^*)}. \quad (6)$$

The quantity $h(E^*)$ thus defined has, Einstein next shows, the following properties: With the energy and structure of the large system known, h fully specifies, through equation (4), all effects of Σ on the observable properties of the smaller system σ, the thermometer. If two large systems Σ_1 and Σ_2 have the same effect on a thermometer σ, they will also have the same effect on any other thermometer σ'. Finally, two large systems Σ_1 and Σ_2 can have the same effect on a thermometer σ only if that effect is identical as well with the effect on σ of the composite system $\Sigma_1 + \Sigma_2$. These properties are precisely those of the observed quantity temperature, which must therefore be some function of the quantity h. Einstein suggests that an appropriate definition of T (now a symbol for temperature rather than for a time interval) is given by $T = 1/(4h\chi)$ with χ "a universal constant." Shortly afterwards he shows that a molecule of a perfect gas must, according to his theory, have mean energy $3/4h$ so that χ must be equal to $R/2N$ (or one-half of Planck's k).[12]

With temperature thus defined, Einstein proceeds finally to a representation of entropy. In order that work may be done on the previously isolated system Σ, he supposes that the function E of equation (2) depends not only on the p_i's but also on a set of slowly varying parameters λ_i. The change in E corresponding to any small change in the state of the system is then given by

$$dE = \sum \frac{\partial E}{\partial \lambda_i} d\lambda_i + \sum \frac{\partial E}{\partial p_i} dp_i,$$

and Einstein readily shows that the first summation is the work done on the system during the variation of the λ_i's, the second the added heat. With entropy S defined as $\int dQ/T$, further manipulation yields the formula

$$S = \frac{E^*}{T} + 2\chi \log \int e^{-2hE(p_1,\ldots,p_n)} dp_1 \cdots dp_n. \qquad (7)$$

In deriving that formula, Einstein supposes that, both before and after variation of the λ_i's, the system under examination interacts with a system far larger than itself, at the same temperature, thus making it susceptible to the treatment he had previously provided for thermometers. As a result, the p_i's in the integral of equation (7) range over the entire p_i-space, and h ceases to be implicitly dependent on δE^*, as it had been before. Finally, Einstein concludes by proving what for present purposes may be taken for granted: S must increase as a system moves from less to more probable states. That is his version of the second law of thermodynamics.

For physical systems of an extraordinarily general sort, Einstein had thus, by the summer of 1903, produced both a generalized measure of the probability of states and also corresponding measures for temperature and entropy. In an eighteen-page article he had shown how to transform a field that had seldom transcended gas theory into a fully general statistical thermodynamics. Only Gibbs's book, published the year before, offers a significant precedent, and the systems considered by Gibbs are less general than Einstein's. That Einstein nevertheless felt the need to go farther is an example of his extraordinary ability to discover and explore problematic interrelationships between what others took to be merely factual generalizations about natural phenomena.

Fluctuation phenomena and black-body theory, 1904-1905

By the time he finished his 1903 paper, Einstein had recognized that his "universal constant" χ could be evaluated in terms of the values of the gas constant and of Avogadro's number. The theory that had led him to the constant was, however, applicable to systems far more general than gases, and it should therefore have a correspondingly general physical basis. That basis, Einstein is likely to have thought, should reflect the statistical nature of the approach that had led him to the constant, thus explaining not only its role as scale-factor for thermodynamic temperature but also its position as a multiplier in the probabilistic definition of entropy.[13†] Establishing the physical significance of χ was, in any case, the central problem attacked in Einstein's third statistical paper, submitted to the *Annalen* in the spring of 1904. Its solution lay in the phenomenon of energy fluctuation, which had usually been either ignored or dismissed as unobservable in the earlier literature.[14†] Once again Gibbs provides the only precedent for Einstein's treatment, but the precedent is in this case partial. Einstein makes fluctuations physically central and at once suggests their quantitative application. That last step is what brought him, for the first time in print, to the black-body problem.

Einstein's 1904 paper opens with a minor but consequential redefinition of the function $\omega(E^*)$. For its original definition, equation (5), he substitutes

$$\omega(E^*)\,\delta E^* = \int_{E^*}^{E^* + \delta E^*} \mathrm{d}p_1 \cdots \mathrm{d}p_n.$$

That substitution eliminates the implicit dependence of $\omega(E^*)$ on the size of the interval δE^* and thus on the particular structure of the imagined ensemble. It is therefore available not only for ensembles but also for individual systems.[15†] Einstein at once applies it in a new definition of a system's entropy and then, in the course of a new derivation of the second law, to the behavior of a system in interaction with a large heat bath at temperature T. The bath, he points out, determines only the average, not the instantaneous energy of the system. The latter will fluctuate about the average in a manner governed by the thermometer equation (4). With the aid of the newly defined ω, that equation may be rewritten to specify the probability $\mathrm{d}W$ that the system will at a given instant have energy between E^* and $E^* + \mathrm{d}E^*$:

$$\mathrm{d}W = C\,\mathrm{e}^{-E^*/2\chi T}\omega(E^*)\,\mathrm{d}E^*,$$

where the normalizing constant is readily determined by equating the integral of $\mathrm{d}W$ over all possible energies to unity.

From the last equation it follows that the average energy of the system at temperature T is just

$$\bar{E} = \int_0^\infty CE\,\mathrm{e}^{-E/2\chi T}\omega(E)\,\mathrm{d}E,$$

so that

$$\overline{(\bar{E} - E)} = 0 = C \int_0^\infty (\bar{E} - E) e^{-E/2\chi T} \, dE.$$

Differentiating the last equation by T and equating to zero the average value of the multiplier of the exponential term within the resulting integral yields

$$\overline{\varepsilon^2} \equiv \overline{(\bar{E} - E)^2} = \overline{E^2} - \overline{E}^2 = 2\chi T^2 \frac{d\bar{E}}{dT}. \tag{8}$$

That result is especially remarkable, Einstein points out, because it no longer contains "any magnitudes reminiscent of the hypotheses basic to the theory." The mean square fluctuation of the energy of any system in contact with an infinite heat bath has been expressed in terms of the measurable quantities T and $d\bar{E}/dT$ together with the absolute constant χ. The physical significance of that constant is therefore at last determined. In Einstein's words: "The magnitude $\overline{\varepsilon^2}$ is a measure of the thermal stability of the system; the larger $\overline{\varepsilon^2}$, the smaller this stability. The absolute constant χ thus determines the stability of the system."[16]

Einstein's recognition of the physical role of the constant χ is what directed his attention to the black-body problem. The words with which he introduces the transition are worth examining, for they appear to imply more than they actually say. Immediately after having related χ to thermal stability, Einstein continues:

> The equation just found would permit an exact determination of the universal constant χ if it were possible to determine the energy fluctuation of a system. In the present state of our knowledge, however, that is not the case. Indeed, for only one sort of physical system can we presume from experience that an energy fluctuation occurs. That system is empty space filled with thermal radiation.[17]

What does Einstein have in mind when he states that "for only one sort of physical system can we presume from experience that an energy fluctuation occurs"? Part of the answer he supplies at once by relating χ quantitatively to the displacement-law constant $\lambda_m T$, where λ_m is the wavelength of the radiation of maximum intensity at temperature T. It seems likely, however, that Einstein had in mind a point of greater generality. The equations governing a phenomenon significantly affected by fluctuations should contain an additional constant not derivable from the macroscopic laws applicable to that phenomenon, e.g., the laws of mechanics or the electromagnetic field. The need for two natural constants in black-body radiation laws had, however, been a recognized puzzle since at least 1900, one to which Lorentz in particular had repeatedly referred.[18] Einstein may well be hinting in the passage just quoted that the entry of a second constant is due, not to the universal electronic charge, but rather to the existence of fluctuations. It is, in any case, by introducing fluctuations in the analysis of a previously unexplained black-body regularity that Einstein illustrates his point.

In a black-body cavity, Einstein points out, the fluctuation of total field energy will be very small if the cavity's dimensions are large compared with the dominant wavelength. (Total energy is proportional to cavity volume; fluctuations, due to the interference of different partial waves at some point, are independent of cavity size and may be in opposite senses at different locations in a large cavity.) But if the dimensions of the cavity equal the wavelength corresponding to maximum intensity, then the mean total energy should be of the same order of magnitude as the mean fluctuation, $\bar{E}^2 \doteq \overline{\epsilon^2}$. The total energy in such a cavity is given by the Stefan–Boltzmann law as $E = aVT^4$, where V is the cavity volume ($= \lambda_m^3$) and a is an experimentally determined constant. Applying equation (8) at once yields

$$\lambda_m = \frac{2}{T} \sqrt[3]{\frac{\chi}{a}} = \frac{0.42}{T}, \tag{9}$$

where the numerical constant on the right is computed from existing measurements of a and R together with established estimates of N. That expression Einstein compares with experimental results that have shown the wavelength of maximum intensity of black-body radiation to be $\lambda_m = 0.293/T$. In view of the order-of-magnitude methods employed to obtain equation (9), the agreement is extraordinary.

By the time he achieved that result Einstein was reading Planck, for he mentions the latter's definition of entropy in the introduction to his paper. About Planck's radiation law, however, he still says nothing at all. As previously noted, Einstein had reasons of his own to doubt that law's derivation, but he was as yet unable to replace it or to understand why it should yield so successful a result. Two more steps in the development of his own research program were needed before he would reach that understanding, and Einstein took the first of them in a famous paper published in the next year, 1905.[19] Both its structure and its content strongly suggest that Einstein had, following his discovery of equation (9), begun to seek a black-body law of his own, that he had quickly encountered paradox, and that he had then dropped the search for a law in favor of an exploration of the paradox itself.

Einstein's new black-body paper was submitted for publication in March 1905, a month before the beginning of the correspondence in *Nature*, through which Rayleigh and Jeans produced the law since known by their names. That law might, therefore, with equal justice be attributed to Einstein, since a few sentences at the start of his paper anticipate their result. Resonators fixed in a black-body cavity that also contains gas molecules should, Einstein points out, acquire mean energy $U = (R/N)T$ when repeatedly struck by those molecules. Planck has shown from electromagnetic theory, Einstein continues, that the equilibrium density u_ν of field energy is related to the resonator energy U_ν by the proportionality factor $8\pi\nu^2/c^3$. The Rayleigh–Jeans law for u_ν results, and Einstein pauses over it only long enough to note its impossible consequence: infinite energy in the radiation field.

Having reached this point, Einstein temporarily abandons the search for a black-body theory. Instead, he introduces Planck's law as the one that "satisfies all experiments to date,"[20] and he then proceeds to explore its high and low frequency limits. For low frequencies, where classical theory and experiment agree, he derives the relationship between black-body and atomic constants, thus rendering Planck's most striking result "to a certain extent independent of his theory of 'black radiation.'"[21] For high frequencies, where experiment and theory diverge to infinity, he develops an argument designed to give physical structure to paradox. The Wien law, he suggests, though clearly not exact, has been well confirmed for large values of ν/T. In the region where it applies, he goes on to show, the entropy of radiation behaves, not like that of waves, but like that of particles. No basis for such behavior can be found in Maxwell's equations. Presumably it is their failure at high frequencies that accounts for the impossible difficulties of the Rayleigh–Jeans law.

To develop these points Einstein first supposes that radiation in a cavity of volume V and temperature T obeys the Wien distribution law, $u_\nu = \alpha\nu^3 \exp(-\beta\nu/T)$. By conventional arguments, familiar from Chapters III and IV,[22] he then shows that, if E is the total radiant energy at frequency ν, its entropy can be determined from the distribution law and must be given by

$$S = -\frac{E}{\beta\nu}\left(\log\frac{E}{\alpha\nu^3 V} - 1\right), \tag{10}$$

a form equivalent to the one Planck had introduced by definition in 1899. Examining that formula, Einstein directs attention to the manner in which entropy varies for fixed energy and changing volume. If S_0 is the entropy corresponding to volume V_0, then equation (10) may be rewritten

$$S - S_0 = \frac{E}{\beta\nu}\log\left(\frac{V}{V_0}\right). \tag{11}$$

That relationship, Einstein at once points out, is precisely the one that governs "the variation with volume of the entropy of an ideal gas or of a dilute solution."[23]

To show what he has in mind, Einstein next introduces Boltzmann's probabilistic definition of entropy in the form

$$S - S_0 = \frac{R}{N}\log W, \tag{12}$$

where W is the *relative* probability of the state with entropy S compared to that of the state with entropy S_0. If a single molecule is known to be contained in a volume V_0 ($W_0 = 1$), then the probability that it is, in fact, located in a smaller volume V of the same container is just V/V_0. Correspondingly, if a gas has n molecules somewhere in V_0, then the probability that they are all in the smaller volume V is given by $(V/V_0)^n$, so that eq. (12) becomes

$$S - S_0 = n\left(\frac{R}{N}\right)\log\left(\frac{V}{V_0}\right).$$

That equation is identical in form with equation (11). High frequency radiation with energy E therefore behaves like a collection of n particles each with energy $\beta\nu R/N$. Furthermore, though Einstein does not say so for another year, the constant $\beta R/N$ has the same value as Planck's h, so that the energy of Einstein's light-particles is identical with the size of Planck's energy elements. A link between Planck's and Einstein's formulations of statistical radiation theory has at last appeared.

So, of course, has a great deal else. Einstein's light-particles, introduced as an "heuristic viewpoint" useful for the analysis of high frequency radiation, were ultimately to become the photons of modern physical theory.[24] But, for the entire period between their introduction in 1905 and the discovery of the Compton effect in 1922, very few theoretical physicists besides Einstein himself believed that light-particles provided a basis for serious research. The distinguished scientists who joined forces in 1914 to urge the creation of a special Berlin chair for Einstein even thought it necessary to explain his persistence in defending light-particles as the inevitable price to be paid for creative genius.[25] The evolution of photon theory therefore contributes only indirectly to the early development of the quantum theory, and nothing more will be said about it until the penultimate chapter of this book. For present purposes, light-particles play only two primary roles, neither of which depends upon their ultimate vindication. Their conception was prerequisite to Einstein's reinterpretation of Planck's theory, to be considered in the remainder of this chapter. In addition, as later chapters will show, they greatly influenced the reception of that reinterpretation and of others related to it.

Einstein on Planck, 1906–1909

Though Einstein, in his light-particle paper of 1905, discussed Planck's radiation law, he had nothing whatsoever to say about Planck's theory. That omission he explained in a paper submitted during the spring of the following year. After a one-paragraph summary of his earlier "heuristic viewpoint," Einstein continued:

> At the time [when I published that viewpoint] Planck's theory of radiation seemed to me in a certain respect the antithesis of my own. New considerations, which are presented in §1 of this paper, demonstrated to me, however, that the theoretical bases on which Planck's radiation theory rests are different from those of Maxwell's theory and of electron theory. The difference, furthermore, is precisely that Planck's theory implicitly makes use of the light-quantum hypothesis sketched above.[26]

Einstein's demonstration begins with the general formula for entropy, equation (7), he had developed in the second of his papers on statistical thermodynamics. That formula, he points out, can be applied to the black-body problem only by assuming that many resonators are present at each frequency. But if they are, then the only significant contribution to the value of the integral term comes from a small range

δE^* of energies near the average energy E^*. For the case of n resonators, equation (7) then reduces to[27]

$$S = k \log \int_{E^*}^{E^* + \delta E^*} dp_1 \cdots dp_{2n}, \tag{13}$$

providing that δE^*, though small, may be chosen large enough so that $k \log \delta E^*$ is also small. If such a δE^* exists (an assumption which Einstein does not trouble to prove but which follows directly from an application of equation (8) to his results), the value of S in equation (13) is independent of the choice of δE^*.

To evaluate equation (13), Einstein continues, the coordinates of the ith resonator may be taken to be x_i and ξ_i, where x_i is the resonator's displacement from equilibrium, and $\xi_i = dx_i/dt$. If the energy E_i is some quadratic function of x_i and ξ_i (an essential but very general condition), then the integral may be transformed so that

$$\int_{E_i^*}^{\overline{E_i^*} + \delta E_i^*} dx_i \, d\xi_i = \text{Constant } \delta E_i^*.$$

Equation (13) for the entropy of n resonators at frequency ν can therefore be put in the form $S = k \log W$, with

$$W = \int_{E^*}^{E^* + \delta E^*} \cdots \int dE_1 \cdots dE_n.$$

If resonator energy varies continuously, Einstein now asserts, these equations necessarily lead to the Rayleigh–Jeans distribution law. "But one can gain the Planck distribution," he continues, by supposing that the energy E_i of a resonator cannot take any arbitrary value, but only values which are integral multiples of ε, with $\varepsilon = (R/N)\beta\nu$."[28]

To discover the force and generality of Einstein's assertion, look briefly at the mathematical argument he omits. Define a function $\Psi(E^*)$ by the equation

$$\Psi(E^*) = \int_0^{E^*} dE_1 \cdots dE_n. \tag{14}$$

The equation for the entropy of n resonators may then be written

$$S = k(\log \Psi'' + \log \delta E^*), \tag{15}$$

where the second term may be neglected by virtue of the previous hypothesis about the size of δE^*. If the E_i vary continuously, then equation (14) readily yields $\Psi(E^*) = (E^*)^n/n!$, from which it follows that $\Psi'(E^*) = (E^*)^{n-1}/(n-1)!$. Inserting the latter value in equation (15) and differentiating yields

$$\frac{1}{T} = \frac{\partial S}{\partial E^*} = \frac{k(n-1)}{E^*},$$

so that for large n the average energy of each resonator can only be kT. That value multiplied by $8\pi\nu^2/c^3$ is just the Rayleigh–Jeans law for the distribution of radiant energy. Einstein has provided by far the most general of the early proofs that that law alone is compatible with classical theory.

On the other hand, if the energy E_i can take only values that are integral multiples of ε, and if the total energy E^* contains exactly p of these elements, then equation (14) becomes

$$\Psi(E^*) = \varepsilon^n \sum_{j_n=0}^{p} j_n \sum_{j_{n-1}=0}^{p-j_n} j_{n-1} \cdots \sum_{j_1=0}^{p-j_n-\cdots j_2} j_1.$$

That multiple sum is, however, by inspection, just the number of ways of distributing p identical elements into n boxes, and it must therefore be the function of n and p that Planck calls the number of complexions. From it his distribution law follows by the arguments he had himself employed. From a single set of equations, Einstein has shown how to produce at will either the Rayleigh–Jeans or the Planck distribution. The former follows if the energy spectrum is continuous, the latter if it is discrete.

More accurately, Planck's law follows for a discrete energy spectrum, $\varepsilon = nh\nu$, provided one makes an extraordinary additional assumption to which Einstein at once turns. The preceding equations yield distribution laws only for resonator energy. To provide the related laws for the field requires the introduction of the familiar factor $8\pi\nu^2/c^3$, derived by assuming that resonators emit and absorb energy continuously. No such assumption is compatible with Einstein's version of Planck's theory, and he therefore suggests adopting the following substitute: "Although Maxwell's theory is not applicable to elementary resonators, the *average* energy of such a resonator in a radiation field is the same as that which one would compute from Maxwell's theory."[29] That statement marks the emergence of the basic paradox of the old quantum theory. The theory has recourse to both Maxwell's equations and those of classical mechanics, but its further formulation is incompatible with one or both of those classical theories. Other physicists were to exploit the resulting inconsistency as an argument against any form of quantum discontinuity, and Einstein himself was deeply disturbed by it. For some years, he sought a revised, non-linear set of field equations that would reduce to Maxwell's equations at low frequencies and would represent discontinuities as singularities of the field.[30] But neither he nor anyone else was successful in finding so nearly classical a resolution of the quantum paradox. When, two decades later, Bohr and others found a way to resolve it, Einstein was unable to accept their fundamentally non-classical interpretation.

If, however, Einstein's derivation of Planck's law demanded a paradoxical assumption, his derivation of the Rayleigh–Jeans law from Planck's premises did not. Nor was the latter derivation subject to the charge of physical artificiality, which could so effectively be directed against the fluid-ether versions of Rayleigh and Jeans. After 1906 their work had to be taken far more seriously than it had before. Einstein himself made the point forcefully in a paper of 1909 where he again showed how to derive the Rayleigh–Jeans law from Planck's model:

> Against Jeans's conception, one can maintain that it is perhaps not permissible to apply the general contents of statistical mechanics [directly] to a cavity filled with radiation. One can, however, reach the law inferred by Jeans in the following way....
>
> That our current theoretical views lead necessarily to the law advocated by Jeans cannot, in my view, be doubted. But we must also acknowledge, as proven with almost equal certainty, that formula (III) [the Rayleigh–Jeans law] is irreconcilable with the facts.[31]

A break with tradition, Einstein thought, had become inevitable.

That analysis is especially significant in the context of this volume because Einstein immediately insists that Planck himself had not noted the need for such a break. Probably he would have done so, Einstein suggests, if he had recognized the necessity of justifying his choice of equiprobable elements by statistical consideration of his theoretical representation of resonators and field. (Note that such justification can, in principle, be provided only if resonator trajectories are continuous.) Then Einstein continues:

> Delighted as every physicist must be that Planck in so fortunate a manner disregarded the need [for such justification], it would be out of place to forget that Planck's radiation law is incompatible with the theoretical foundations which provide his point of departure.

It is easy to see in what ways the foundations of Planck's theory must be modified in order that the Planck radiation law become really a consequence of those theoretical foundations.... [The factor $8\pi\nu^2/c^3$ must be preserved by supposing that electro-magnetic theory yields correct time-average values of field quantities. In addition, the statistical theory of heat must be altered in the following manner.] A structure which may vibrate with the frequency ν and which, because it possesses an electric charge, can convert radiation energy into energy of matter and vice versa, may not occupy vibration-states with any arbitrary energy, but only those states with energies an integral multiple of $h\nu$.[32]

These are the points which Planck, in Einstein's view, had missed.

Speaking at Salzburg later in 1909, Einstein capped his critique of Planck by isolating with precision the point at which Planck's derivation had gone astray:

One may regard the number of complexions... as an expression for the multiplicity of the possible ways of distributing the total energy over the N resonators only if every imaginable mode of distribution of the energy is counted in computing W, at least to a suitable approximation. It is thus necessary that, for all ν which yield an appreciable energy density u, the energy quantum ε be small compared with the average resonator energy U. One finds, however, by simple calculation that for the wavelength $0.5\,\mu$ and absolute temperature $T = 1700$, the quantity ε/U is actually very large, not small, compared to 1.... It is clear that this procedure makes use only of a vanishingly small portion of the distributions which we must see as possible according to the foundations of the theory. The number of these complexions is thus no measure of the probability in Boltzmann's sense. To adopt Planck's theory is, in my opinion to reject the foundations of our radiation theory.[33]

More than their intrinsic cogency gave authority to Einstein's closing words. In 1909 he was no longer the unknown he had been in 1905 and 1906. On the contrary, one of the two main themes of the papers and discussions at the Salzburg meeting of the Naturforscherversammlung where he spoke, was the theory of relativity (the other being radioactivity). Doubtless that is why Einstein was invited to deliver there the special address from which the preceding quotation is taken. Published under the rubric "The Development of Our Views on the Nature and Constitution of Radiation," it dealt with relativity and the quantum together.

In considering these topics, Einstein, of course, argued for more than the necessity of introducing discontinuities into black-body theory. He asked, in particular, whether those discontinuities could be restricted to the interaction of radiation and matter—thus preserving the validity of Maxwell's equations for propagation in empty space—and he concluded that they could not. One must, he insisted, accept the particle-like behavior of high frequency radiation, a position he defended by a considerable extension and generalization of the fluctuation argument he had developed in 1905.³⁴† But, as previously noted, even his extended argument persuaded almost no one. Unknown before 1902, thermodynamic fluctuations were an unfamiliar and a not yet quite physical subject matter. No one, including Einstein himself, saw how particulate properties could be reconciled with the vast range of interference effects the explanation of which had marked the steady advance of the wave theory of light for one hundred years. As a result, the primary impact of these portions of Einstein's talk was to license uncertainties about what the content of Planck's black-body theory might be. The concepts of particles of light and of resonators restricted to energy $nh\nu$ had entered physics together in Einstein's papers of 1905 and 1906, and they remained for him parts of a single, if entirely unfinished, theory. Since the first was abhorrent even to those theorists persuaded of the second's necessity, disentangling the two or finding a substitute for both was, by 1909, a central task for those concerned to develop or apply the quantum.

VIII

CONVERTS TO DISCONTINUITY, 1906–1910

Return now to mid-1906, the period in which were published, first, Einstein's "Theory of the Emission and Absorption of Light" and, then, Ehrenfest's "On Planck's Radiation Theory." Both papers coupled two convictions that were individually new and in combination unique to their authors. Like Jeans (and unlike Rayleigh), Einstein and Ehrenfest insisted that the Rayleigh–Jeans law provided the only distribution function compatible with classical theory. Unlike Jeans, however, both men felt that Planck's law must represent at least approximately the equilibrium distribution of radiant energy and that no such law could be derived without a fundamental break with classical theory. About the nature of the break required, Ehrenfest at least was less certain, but he joined Einstein in demonstrating that Planck's law could be derived if resonator energies were restricted to integral multiples of $h\nu$, and he had no concrete alternative to offer. Both men thus associated Planck's theory with discontinuity and by doing so placed themselves in an isolated position. Planck himself was not to accept discontinuity for perhaps another two years, and even Larmor, the only other physicist known to have taken Planck's combinatorial derivation seriously, was not persuaded as late as 1910 that it demanded a restriction on resonator energies.[1]

Einstein's 1906 paper appears to have brought a third member into the discontinuity camp. He was Max von Laue, a young student of radiation thermodynamics and from 1905 to 1909 Planck's Assistant in Berlin.[2] Impressed by the first report to the Berlin colloquium concerning special relativity, Laue took an early opportunity to call upon Einstein in Berne, a visit from which dates a long continuing correspondence and exchange of manuscripts and printer's proofs. In his first extant letter to Einstein, Laue thanks his correspondent "for the proofs of your article [on emission and absorption] which has since appeared in the *Annalen*. I have," he continues, "read it with much interest and, as I shall shortly show, entirely agree." After a brief paragraph concerning his own activity, Laue continues:

> If, at the beginning of your last reply, you state your heuristic viewpoint [i.e., the light-particle hypothesis] in the form, radiant energy can only be emitted and absorbed in certain finite quanta, then I know nothing to which to object; also all your applications correspond to this mode [of conceiving your theory]. Only, this is not a characteristic of electromagnetic processes in a vacuum, but of the absorbing or emitting material. Radiation does not consist of light-quanta, as [you say] in § 6 of the first [i.e., the light-particle] paper but only behaves during energy exchange with matter as though it did.[3]

When those lines were written, Laue had not yet discussed Einstein's radiation theory with Planck.[4] Possibly he did so later, but Planck remained unconvinced for at least another year. Nor is there evidence that Laue or Einstein or Ehrenfest convinced anyone else at this time. All three were too young and little known for their opinions to carry much weight on so potentially controversial a point.[5] Even when Einstein's reputation grew, as it quickly did, his views on the necessity of the quantum discontinuity remained suspect because they were repeatedly coupled with the generally rejected light-quantum hypothesis. If the physics profession was to recognize the challenge of Planck's law, better established figures would need to be persuaded that it demanded a break with classical physics.

In the event, several of them quickly were. During 1908 Lorentz produced a new and especially convincing derivation of the Rayleigh–Jeans law. Shortly thereafter he was persuaded that his results required his embracing Planck's theory, including discontinuity or some equivalent departure from tradition. Wien and Planck quickly adopted similar positions, the former probably and the latter surely under Lorentz's influence. By 1910 even Jeans's position on the subject had been shaken, and he publicly prepared the way for retreat. These are the central events through which the energy quantum and discontinuity came to challenge the physics profession.

Lorentz's Rome lecture and its aftermath

Lorentz's overt concern with the black-body spectrum dates from 1900 when he published a paper asking how the existence of a temperature-dependent but matter-independent wavelength of maximum intensity might be explained.[6] Electromagnetic theory alone could not provide an answer, he insisted, for the only fixed quantity it involved was c, the velocity of light. Probably the required explanation would be found in some characteristic common to all matter, and the most likely candidate was the recently discovered electron.

A second Lorentz paper on the black-body problem appeared in 1901 and a third during 1903.[7] In the latter, which was frequently cited in the contemporary literature, Lorentz derived a black-body distribution law at long wavelengths from the first principles of electron theory. An advantage of his approach, he emphasized at the start, was that it permitted a treatment of the black-body problem "by means of the heat-motion of its [a metal's] free electrons, without recurring to the hypothesis of 'vibrators' of some kind, producing waves of definite periods."[8] The result he obtained for long wavelengths was, of course, what can now be recognized as the Rayleigh–Jeans law. Lorentz, however, made no reference to Rayleigh's note of 1900, a further reminder that, except as an amplitude factor in an ad hoc formula, that law did not yet exist. But he did point out that "A comparison of my formula with that of Planck is also interesting.... There appears...to be a full agreement between [the results of] the two theories in the case of long waves, certainly a remarkable conclusion, as the fundamental assumptions are widely different."[9]

For long waves, at least, Lorentz had thus successfully recast the black-body problem as a problem in his own special field, electron theory. What remained to be done was to remove the restriction on wavelengths, thus deriving a general Planck-like formula, which would explain the wavelengths of maximum intensity in terms of the characteristic properties of the electron. "Over this problem," Lorentz reported to Wien in 1908, "I have ceaselessly racked my brains,"[10] and some clues to the nature of his efforts are contained in papers he published during 1905. While they continued, Lorentz remained skeptical of Planck's theory, though he cited the radiation law repeatedly and with growing respect. Until at least 1905, however, he continued to regard the long wavelength agreement of his result with Planck's as "a happy coincidence,"[11] and in April 1908, when he read a long report on the radiation problem to the Fourth International Congress of Mathematicians meeting in Rome, he still found it only "exceedingly curious."[12] His view of its status was, however, to change at once. Lorentz's widely reprinted Rome lecture, "The Division of Energy between Ponderable Matter and the Ether," turned out to be the last as well as the culmination of his series of attempts to provide a classical account of cavity radiation.

Lorentz's lecture opened with a brief summary of the relevant thermodynamic theorems derived by Kirchhoff, Stefan–Boltzmann, and Wien (the displacement law). "We shall see in what follows," Lorentz continued, "that certain considerations could lead us to believe after all that these remarkable laws do not conform to reality. Nevertheless, their deduction is certainly among the most beautiful achievements of theoretical physics."[13] Following that now incongruous remark, Lorentz applied Gibbs's statistical mechanics to an ensemble of identical systems consisting of free and bound electrons, atoms, and radiation contained in a perfectly reflecting enclosure. By restricting the field to wavelengths greater than some arbitrary cutoff λ_0, he kept the system's number of degrees of freedom finite. Rigorous and straightforward application of the laws of mechanics and electromagnetic theory then enabled him to show that the Rayleigh–Jeans law must describe the distribution of energy in the field for all $\lambda > \lambda_0$, where λ_0 could be chosen arbitrarily close to zero.

"It is this result," Lorentz immediately pointed out," that I had in mind when I said that perhaps the [radiation] laws of Boltzmann and Wien could not be maintained." Those laws presuppose equilibrium between matter and radiation; the result just reached is equivalent to the statement that no such equilibrium can exist. If the Jeans law is correct, then any energy initially possessed by matter will in time be entirely taken up by the field, where it will increasingly be concentrated in the shortest wavelength modes. Under those circumstances, no purely thermodynamic derivations could apply, and Lorentz therefore continued:

> All this initially seems very strange, and I admit that, when Jeans published his theory, I hoped that by examining it more closely one would be able to demonstrate the inapplicability to the ether of the theorem of "equipartion of energy" on which it is based; thus one would find a true maximum for the [distribution] function $F(\lambda, T)$. The preceding considerations seem to me to prove that that is not the case and that one cannot escape Jeans's conclusion, at least not without profoundly modifying the fundamental hypotheses of the theory.[14]

About the need for such profound modifications, Lorentz professed himself uncertain. He emphasized that the transmission of energy to short wavelength modes might occur extremely slowly and that experimental cavities might not be black for the wavelengths thus achieved. Jeans's explanation of the observed maximum in radiation intensity—as an experimental artifact due to failure to achieve equilibrium—he described as "indeed the only one which can be given," and he clearly thought that it might also be correct. Recognizing that he had provided no solution to the problem his paper had posed, he concluded his presentation to the assembled mathematicians with the following words:

> If one compares the theories of Planck and Jeans one finds that they both have their merits and their drawbacks. Planck's is the only one that supplies a formula in agreement with the results of experiment, but we can adopt it only by altering profoundly our fundamental conceptions of electromagnetic phenomena.... Jeans's theory, on the other hand, obliges us to attribute to chance the presently inexplicable agreement between observation and the laws of Boltzmann and Wien. Fortunately one may hope that new experimental determinations of the radiation [distribution] law will permit us to decide between the two theories.[15]

For reasons shortly to be examined, Lorentz's proof of the Rayleigh–Jeans law made a far deeper impression on his contemporaries than any that had appeared before. But the conclusion he drew from it was another matter, and among experimentalists his concluding sentence was greeted as absurd. If the choice between Jeans and Planck could be dictated by experiment, then no new experiments were required. Wien, a gifted experimenter as well as a theoretician, put the point strongly in a letter to Arnold Sommerfeld (1868–1951) written not quite six weeks after Lorentz's talk:

> The lecture which Lorentz gave in Rome has disappointed me greatly. That he developed nothing except the old Jeans theory without adding even a new point of view seems to be rather shabby. Besides, the question of whether or not Jeans' theory should be regarded as discussable is to be decided on experimental grounds. In my opinion it is not discussable, because observations show immense deviations from the Jeans formula in a [wavelength] range where one can easily control the deviation of the radiation source from a black body. What is the purpose of presenting this question to the mathematicians, none of whom is equipped to make a judgment on this sort of point. In addition, it seems to me somewhat comical to locate the advantage of the Jeans formula—in spite of its corresponding [experimentally] with nothing—in its permitting one [theoretically] to retain the whole unbounded multiplicity of electron vibrations. And the spectral lines? This time Lorentz has not shown himself to be a leader of physics.[16]

Those feelings Wien transmitted also to Lummer and Pringsheim as well as, doubtless in different terms, to Lorentz. The former pair quickly responded with a short article, "On the Jeans–Lorentz Radiation Formula," which appeared in the 15 July issue of the *Physikalische Zeitschrift*. If politer about Lorentz than Wien had been in his letter to Sommerfeld, they nonetheless showed that they shared the strong feeling his remarks had raised among experimentalists.

If one examines the Jeans–Lorentz formula, one sees immediately that it leads to impossible consequences which are in crass conflict not only with the results of all observations of radiation, but also with everyday experience. It would therefore be superfluous to consider this formula further if it had not been defended by two theoretical physicists of such distinguished reputation and authority.[17]

Among Lummer and Pringsheim's several arguments, the most effective was that from everyday observation. Lorentz was to develop the same point himself in his acknowledgment. Set aside entirely the question of the distribution law and ask simply about the temperature dependence of the radiation emitted at a given frequency. Molten steel at a temperature of 1600–1700° absolute emits a blinding light, and the emission from a black body would inevitably be greater. Since the Jeans law makes radiation intensity proportional to absolute temperature, the emission from a black body at room temperature (300° absolute) must be at least one-sixth that from molten steel. A black body should therefore be clearly visible in the dark, as should steel itself, and also the large number of experimental substances that, whatever their behavior at very high frequencies, are clearly black in the visible range. The failure of these materials to glow in the dark shows, Lummer and Pringsheim concluded, that the Jeans law is experimentally impossible. The evidence called for by Lorentz was already at hand.

Lorentz responded at once in a note mailed to the *Physikalische Zeitschrift* four days after the publication of Lummer and Pringsheim's paper. A letter from Wien had already, he wrote, persuaded him of the points made by his experimental critics; he had revised his article accordingly and would also add a note to the version to appear in the Congress' *Proceedings*.

> Now that the considerable difficulties which one encounters in this way [of defending the Jeans law] have become clear to me, I can only conclude that a derivation of the radiation law from electron theory is scarcely possible without deep-seated changes in its foundation. I must therefore take Planck's theory as the only tenable one. We shall have to acknowledge that the exchange of energy between matter and ether takes place by means of the resonators assumed by Planck or of similar particles which somehow escape the application of Gibbs's theorem.[18]

That public concession does not yet embrace discontinuity. But the letter in which Lorentz had gratefully acknowledged Wien's criticism shows that he was ready for that step and simultaneously explains his hesitancy in taking it. After agreeing that Jeans's theory must be abandoned, and some bold new assumption adopted instead, Lorentz continued:

> The elementary quanta of energy provide precisely such a new assumption. In and of itself, I have nothing against it; I concede at once that much speaks in its favor and that it is precisely with such novel views that one makes progress. I would, therefore, be prepared to adopt the hypothesis without reservation if I had not encountered a difficulty. It is that, according to Planck's formula, those resonators with λ significantly smaller than λ_m never obtain a single [whole] energy element. In other words, some of these resonators (under appropriate circumstances the majority) must possess no energy, and yet they are exposed to the continual excitation of the electromagnetic ether waves just as the others are. It is to be noted that, according to Planck's theory, resonators receive or give up energy to the ether in an entirely continuous manner (without there being any talk of a finite energy quantum). At this time I'd rather not [attempt to] go more deeply into the matter; I hope soon to determine Professor Planck's own position about it.[19]

Though Lorentz's enquiry has been lost, the letter in which Planck responded survives, and it includes his first known concession of the need to restrict resonator energy. After receiving it, Lorentz unequivocally took up energy quanta himself. Lecturing at Utrecht in April 1909, he first described Planck's novel hypothesis and then demonstrated its fruits in application. According to Planck's theory, he said, the exchange of energy between matter and the ether is mediated by "certain particles," which Planck calls resonators. "To such a resonator he attributes the property that it cannot receive or give up energy in infinitely small quantities but only in suitable finite quantities proportional to its frequency."[20] After adopting that position, Lorentz quickly became a leader in developing and propagating the quantum theory.

Indeed, in a more restricted circle, Lorentz's influence appears to have been much felt even before he began to use the quantum in his own work. Twenty years later, looking back at Lorentz's career, Planck singled out for special mention his demonstrations "that classical theory, developed in an internally consistent manner, leads to a unique energy distribution, namely the Rayleigh radiation law." In addition, Planck credited Lorentz with the argument that, if that law held, "a silver plate at 0 °C should be plainly visible in the dark, in sharp contrast to reality."[21] Together with material to be treated in the next two sections, those attributions suggest that Lorentz's proof of the Rayleigh–Jeans law was important for those physicists who had been moved little or not at all by the earlier demonstrations of Jeans,

Einstein, and Ehrenfest. Only after Lorentz's Rome lecture does the physics profession at large seem to have been confronted by what shortly came to be called the ultraviolet catastrophe and thus by the need to choose between Jeans's theory and a non-classical version of Planck's.

Part of the reason for Lorentz's special effectiveness was doubtless his great personal authority. He had founded, and was the world's leading expert on, electron theory. In addition, unlike most German physicists in the first decade of the century, he had long been an expert on the techniques of statistical mechanics, with which he had done important work since 1887.[22] Perhaps equally important, Lorentz was widely admired as a wise, penetrating, and humane judge in matters scientific. Planck particularly stressed that Lorentz's "rich knowledge and experience in all fields of physics together with his skill in handling men and situations equipped him far more than others for the role of mediator during discussions at scientific meetings."[23] Einstein, midway through a debate with Lorentz about the light-particle hypothesis, reported to a friend: "I admire that man more than anyone else; I might even say I love him."[24] Only in intensity and expression were Einstein's feelings different from those of other members of the profession.

More than the status its author had earned by past performance is, however, likely to have been responsible for the special impact of Lorentz's Rome lecture. His proof of the Rayleigh–Jeans law was, in its own right, the fullest, most general, and most compelling yet presented. Jeans's demonstration had been based on an acoustic analogy to the electromagnetic field. Even when recast, as by Planck in his *Lectures*, to deal with normal modes of electromagnetic oscillation in a specified cavity, the demonstration provided no mechanism for interaction between modes and thus for energy redistribution. Einstein's argument depended on currently unfamiliar abstractions and was, in any case, directed primarily to the distribution of energy among fixed resonators, the factor $8\pi\nu^2/c^3$ being introduced only at the close. Ehrenfest's treatment shared the disadvantages of Jeans's and demanded besides recourse to complexion theory. None of these difficulties were present in Lorentz's demonstration. His radiation field satisfied Maxwell's equations; his cavity contained not fixed resonators but moving atoms and electrons, both free and bound; except for his use of the Gibbs's ensemble, his analytic techniques were standard if advanced. That the Rayleigh–Jeans law was nevertheless forthcoming gave his audience good reason to be impressed. In particular, one could no longer believe, as both Planck and Ehrenfest had, that recourse to moving resonators or to molecular collisions might, by providing a mechanism for frequency redistribution, open a direct route from electron theory to Planck's radiation law.[25]

Planck on discontinuity, 1908–1910

How did Planck respond to Lorentz's Rome lecture, and how had he responded to the earlier black-body papers of Einstein and Ehrenfest? Unfortunately, little information relevant to those questions is available for the crucial thirty months between April 1906, when he signed the preface to his *Lectures*, and October 1908, when cavity radiation first entered as more than a passing reference into his ongoing correspondence with Lorentz. Planck's publications deal with other topics, and only one is even indirectly revealing.[26] If, as is likely, he wrote Einstein or Ehrenfest to acknowledge their contributions, those letters are no longer to be found. Nevertheless, what evidence is available suggests that until 1908 Planck's position remained close to what it had been in the *Lectures*. Not until the following year, furthermore, does he seem clearly to have acknowledged the fundamental inconsistencies of the classical black-body theory he had presented there.

An odd first sentence in one of Planck's early articles on relativity theory suggests that, as late as June 1907, he still saw no need to revise classical mechanics or electromagnetic theory. In that paper he attempted to find an alternative for the distinction, untenable in relativity theory, between the kinetic and the internal potential energy of a moving body. To make the problem concrete, he examined the effect of motion on radiation within a cavity, and he justified his choice of so unusual a special case by the introductory assertion that "recent experimental and theoretical research on thermal radiation has shown that a system consisting only of electromagnetic radiation, stripped of ponderable matter, conforms completely not only to the principles of mechanics but also to the two laws of thermodynamics."[27] His own theory of the entropy of radiation was presumably among those he took to be both successful and also compatible with classical theory.[28] If so, he can scarcely yet have accepted the central claims of the papers that Einstein and Ehrenfest had published the year before. Further reason to suppose that he had not done so is provided by the total absence of the black-body problem in his ongoing correspondence with Wien and Lorentz. After 1908, when he did take the new challenges

CONVERTS TO DISCONTINUITY, 1906-1910 197

seriously, cavity radiation is a persistent topic in his exchanges with both.[29]

The topic first enters in a letter Planck sent to Lorentz in early April, shortly before the latter's departure for Rome. Acknowledging a communication just received, Planck wrote:

> Naturally it will interest me immensely to learn from your lecture in Rome what you think about the great question of the distribution of energy between ether and matter. That electron theory, in the absence of additional hypotheses, necessarily leads to Jeans' conclusion is to me very plausible, and I think it can only be useful if this point is expressed with all [possible] force.[30]

That attitude towards the status of the Rayleigh–Jeans law is far more positive than the one Planck had taken in his *Lectures* two years before.[31] In this respect Planck probably had been influenced by the papers of Einstein and Ehrenfest. But the "additional hypotheses" that Planck thought were required to evade that law need not have gone beyond the ones developed in the *Lectures*, first and foremost the fixed size of the energy elements used in combinatorial formulations.

Planck's next letter to Lorentz, written in early October, was a belated reply to one he had received two months before. In the latter, Lorentz had apparently amplified his new-found recognition of "Planck's theory as the only tenable one," asked how an electromagnetic field could fail to excite resonators exposed to it, and indicated his conviction that the difficulties must be rooted in properties of the ether. Planck objected vehemently to that localization of the difficulties, pointing out that it would necessarily "lead to the abrogation of Maxwell's field equations.... I still see," he continued, "no compelling reason to abandon the supposition of the absolute continuity of the free ether and of all events in it. The action element h is therefore a characteristic of the resonators."[32] Doubtless that attitude had also conditioned his reaction to Einstein's and Ehrenfest's analyses of his theory. The latter had derived Planck's radiation law by quantizing the field's vibration modes; the former had attributed to Planck the implicit "use of the light-quantum hypothesis."

Planck's letter continues by quoting part of the passage in which Lorentz had asked about the resonators that failed to respond to stimulation. His own view, Planck goes on to say, is as follows:

> If a resonator is placed in such a stationary radiation field [i.e., one in which the instantaneous value of the field fluctuates rapidly due to interference of partial waves], it will be set into vibration by the field. But—

and now comes the essential point—this excitation does not correspond to the simple known law of the pendulum; rather there exists a certain threshold: the resonator does not respond at all to very small excitations; if it responds to larger ones, it does so only in such a way that its energy is an integral multiple of the energy element $h\nu$, so that the instantaneous value of the energy is always represented by such an integral multiple.

In sum, I might therefore say, I make two assumptions:
1) the energy of the resonator at a given instant is $gh\nu$ (g a whole number or 0);
2) the energy emitted and absorbed by a resonator during an interval containing many billion oscillations (and thus also the average energy of a resonator) is the same as it would be if the usual pendulum equation applied.

These two assumptions do not seem to me to be incompatible. I have indicated these thoughts in a note to §109 of my lectures on thermal radiation.[33]

Because it includes Planck's first known acknowledgment that his theory demands a restriction on resonator energy, that passage is striking. But it is, in two respects, correspondingly subject to overinterpretation. First, Planck need not have learned the need for the restriction by interaction with Lorentz. Though the opening of his 1907 relativity paper would then become difficult to understand, Einstein's and Ehrenfest's papers of 1906 remain a possible source. More important, though Planck's admission that resonator energy must be a multiple of $h\nu$ is a clear departure from his *Lectures*, his closing reference to the book suggests that he had not yet seen the magnitude of the break his theory required. The passage of text to which the closing sentence in the last quotation leads, contains the derivation of the differential equation (I-8a), governing a radiation-damped resonator. It emphasizes that that equation applies only to average behavior over a long interval T and that, "In order to have an exact solution of the radiation problem the nature of the oscillator would have to be known in full detail, both at its surface and in its interior." The footnote mentioned by Planck then points out that, "For smaller values of T another vibration law, better suited to natural processes, might replace the simple linear [resonator] equation."[34] Two chapters later, in the passage attributing the puzzles surrounding h to the as yet unknown processes occurring at emission centers, Planck calls special attention to that footnote, and he refers readers to this passage in the very last sentence of his book.[35] In all these places he is suggesting that the existence of h will be explained by the use of electron theory supplemented by some *additional* hypothesis about microprocesses at emission centers. Even after conceding discontinuity, Planck apparently retained the hope that electron theory itself could be preserved.

Interaction with Lorentz soon persuaded him that that hope was vain. After conceding that the source of h could not be the ether, Lorentz continued to insist that free electrons, i.e., those not restricted to oscillation at a specified frequency, could be treated in full by the canonical techniques he had himself established.[36†] Planck's resonators must therefore be some special sort of particle to which electron theory did not apply. The notion that some special additional particle would be required to account for Kirchhoff's and other radiation laws was, however, anathema to Planck. In mid-1909 he first sketched for Lorentz a viewpoint he would be forced to urge upon him repeatedly thereafter:

> Now a free electron exerts an influence on the ether only when it alters its velocity in magnitude or direction. This occurs mainly by collision, either with ponderable molecules or with other electrons. About the laws of such collisions we know virtually nothing at all, and in my opinion it is an unfounded hypothesis that they are governed by Hamilton's equation. Rather, in this case, only an hypothesis which leads to consequences in conformity with experience is permissible. Such an hypothesis I take to be the following. *The energy exchange between electrons and free ether always occurs only in whole numbers of quanta hv.* This holds both for free electrons and for those which, as in my resonators, vibrate about an equilibrium position.[37]

Forced to choose between abandoning electron theory and positing a special sort of particle, Planck without apparent hesitation elected the former course.

That was the position that Planck at last made public in late 1909 and early 1910. During the discussion of Einstein's lecture at Salzburg in October of the former year, he firmly rejected the light-particle hypothesis, but acknowledged a large gap in contemporary electron theory:

> Perhaps one may suppose that an oscillating resonator does not have a continuously variable energy, but that its energy is a simple multiple of an elementary quantum. I believe that the introduction of this proposition can lead to a satisfactory radiation theory. But the question remains: how is anything of the sort to be done? We require, that is, a mechanical or electrodynamic model of such a resonator. But mechanics and contemporary electrodynamics possess no quantum of action, and we therefore cannot produce a mechanical or electrodynamic model. Mechanically, the task seems impossible, and *one will just have to get used to it.*[38]

Three months after speaking those words Planck extended them in article submitted to the *Annalen der Physik*. Within a brief review of contemporary opinions, "On the Theory of Thermal Radiation," he urged the search for a modification of existing theory which would, unlike Jeans's proposal, "do justice to the new facts," but which would not, like Einstein's, "sacrifice their most valuable parts."[39] Such a modification, he insisted, would demand recognition that "certain elementary radiation processes which in Jeans's theory are assumed to be continuous, in reality occur discontinuously.... In my opinion," he continued, "one will not for this purpose have to give up the Principle of Least Action, which has so strongly attested its universal significance, but [one will need to abandon the hypothesis] of the universal validity of the Hamiltonian differential equations."[40] By the beginning of 1910 Planck was at last firmly and publicly committed to the entry of discontinuity and the abandonment of some part of classical theory.

That that commitment was for Planck both new and consequential is indicated by its immediate effect on his research as well as by a shift in his vocabulary. Except for lectures and a brief note on new experiments, he had published nothing on the quantum or black-body theory during the eight years after 1901. Beginning in 1910, however, he returned to quantum problems, and his research dealt with virtually nothing else until 1926. The first product of that research was his so-called second theory, and the first signs of its development appear in the same correspondence and publications where he began to acknowledge discontinuity and the need to restrict energy levels. Only the emergence of those concepts can plausibly explain his development of the second theory, for, as the last chapter of this book will show, discontinuity is the only element of the theory not available to Planck a decade before. The intervention of others, most notably Einstein and Lorentz, had given Planck's problem a new form after 1908, opening new areas of investigation which he quickly entered.

Some aspects of the new view Planck took of his problem are indicated also by two significant changes in his choice of words. From 1897 through 1901 he had regularly referred to the entities that promoted disorder in radiation as "resonators." In his *Lectures* the more general term "oscillator" was sometimes used as well, but "resonator" remained an acceptable choice until the end of 1909.[41] After that, however, it was banished. Sending a preliminary draft of his 1910 paper to Lorentz, Planck commented in his covering letter, "Of course, you are entirely right to say that such a resonator no longer deserves its name, and that has moved me to strip it of its title of honor and call it by the more general name 'oscillator.'"[42] Probably Planck had an argument like the following in mind: an oscillator is simply an object that vibrates, whatever the cause; but a resonator (the term is from acoustics) vibrates sympathetically in a tuned response to an external vibrating stimulus. Once Planck had begun to speak of thresholds for excitation and of discontinuity, the connotations of "resonator" were excessively specific, while the broader term "oscillator" continued to apply. Planck restricted himself to the latter in his publications from the beginning of 1910. Whatever his reasons for doing so, the change was clearly a conscious response to his new understanding of his theory.

A second, more gradual, but also more consequential shift of vocabulary began to characterize Planck's writings in the same year. Before 1909 he had referred repeatedly to the quantum of electricity (the charge e), to the quantum of matter (the atom), and, since 1906, to the quantum of action (the constant h).[43] But the quantity $h\nu$ he regularly called an "element" rather than a "quantum" of energy. The two known exceptions are in letters to men who themselves did use the latter term for reasons to be described below.[44†] Again, a plausible reason for the terminological distinction is easily found. As the preceding examples indicate, a quantum is a fixed amount that can exist by itself in isolation; it cannot, that is, be simply an imagined part of an entity conceivable only as whole (e.g., a line or surface). While $h\nu$ was merely the size of a subdivision of the energy continuum, it was not a quantum.

Not surprisingly, therefore, Planck's regular use of the term "quantum" for $h\nu$ begins in the italicized sentence quoted from his June 1909 letter to Lorentz: "*The energy exchange between electrons and free ether always occurs only in whole numbers of quanta $h\nu$.*" He next uses similar terminology the following October in his comment on Einstein's Salzburg lecture ("an elementary quantum [of energy]") and then in the 1910 version of lectures delivered at Columbia University in the spring of the preceding year.[45] These terms enter again at the start of his first paper on the second theory, to be discussed below, and they are used regularly thereafter. What is novel about such usage is simply that it comes from Planck. Ehrenfest had spoken of "energy particles" in 1905 and of "the energy quantum" and "the energy atom" in 1906. Einstein used terms like "energy quanta" repeatedly from 1905; Laue adopted that term in the following year; and Wien used it in a paper published during 1907.[46] Their references are, however, invariably to separable units of energy that matter can emit or absorb one at a time. Planck's adoption of their terminology during 1909 is thus one additional bit of evidence about the nature and extent of his change of view.

The consolidation of expert opinion: Wien and Jeans

Lorentz and Planck were not the only experts on black-body theory to have had second thoughts during the years 1908 to 1910. Nor, during that period, was cavity radiation the only area to which the quantum was being applied. But, as the next chapter will show, most of those who first introduced Planck's constant h in other areas were initially unaware of the extent of the break their recourse to it implied. Until after 1910 almost no one who had not wrestled strenuously with the black-body problem was persuaded of the need for a new, discontinuous physics. Einstein, Ehrenfest, Laue, Lorentz, and Planck had all studied radiation thermodynamics in depth. That is true also of Willy Wien and James Jeans, the next two men to face the prospect of discontinuity publicly.

Wien's case is especially important because of his position in the profession. His name had been associated with the black-body problem since his announcement of the displacement law in 1893, and he was shortly to receive the Nobel prize for that and subsequent contributions. In addition, from 1907, he was editor-in-chief of the principal journal of German physics, the *Annalen der Physik*. When Planck in 1900 had derived the radiation law that Wien himself had introduced, Wien had quickly identified the missing elements in the argument,[47] and he is not likely to have been more fully satisfied with Planck's subsequent derivation of his own distribution law. But Wien is likely to have felt that Planck had revealed some fundamental aspect of nature, and he was in any case soon impressed by what he described as Einstein's application of "Planck's theory to photo-electric phenomena."[48] For him, the content of Planck's theory was that matter and radiation exchange energy only in whole quanta, and in two papers published in 1907, he showed how that idea could be extended to other phenomena. In one of those articles, Wien suggested that the still uncertain wavelength of x-rays might be computed by assuming that the entire kinetic energy of an exciting electron was converted into a single quantum of x-radiation.[49] In the other, he suggested, on the basis of experiments done in his own laboratory, that free and bound electrons were in dynamic equilibrium and that the emission and absorption of spectral lines might occur when bound electrons became free and vice versa.[50]

That a man who could describe his proposed computation of x-ray wavelength as "an easy extension of Planck's theory of radiation"[51] should have found Lorentz's Rome lecture "rather shabby" is not surprising. What Wien was saying in the letter to Sommerfeld quoted above is that Lorentz was behind the times. But Wien's next letter to the same correspondent, written immediately after Lorentz's renunciation of the Rayleigh–Jeans law, indicates that Wien, too, had failed to assimilate recent discoveries in radiation theory.

> Lorentz has recognized his error over radiation theory and that the Jeans law is untenable. Now, however, the situation is not so simple, for it in fact appears that Maxwell's theory must be abandoned for [application to] atoms. I therefore have another problem to pose for you: namely, to examine the soundness of Lorentz's statistical-mechanical proof that a system which obeys Maxwell's equations (including electron theory) must also conform to the theorem of "equipartition of energy" from which the Jeans law would follow.... If that is really so, one will not need to continue breaking one's head in order to treat the energy element and spectral theories on an electromagnetic basis. Instead, one must try to supplement Maxwell's equations for [application to] the interior of atoms. I am not myself sufficiently fluent with the whole of statistical mechanics to be able to form a secure judgment about the extent of its [i.e., Lorentz's theory's] reliability.[52]

Only through the intervention of Lorentz did Wien recognize what accepting Planck's theory implied.

Wien's remaining doubts were soon resolved, whether by Sommerfeld or someone else, and his new convictions were publicly acknowledged during 1909 in a major review article entitled "Radiation Theory." Planck's theory is there treated carefully and at length; Jeans's theory very briefly. Lorentz's Rome lecture and its aftermath are sketched, and the penultimate paragraph of Wien's paper isolates the importance of these events: "Lorentz's proof that Jeans's theory must be universally valid if one presupposes the equations of electron theory and statistical mechanics shows that alterations in the foundations of the theory must be undertaken."[53] Wien was also at pains to rehearse the internal problems of Planck's theory: the lack of a mechanism for altering the frequency distribution of energy; the incompatibility of the hypothesis of energy elements with Maxwell's equations; and the difficulties of reconciling Planck's approach with classical dispersion theory.[54] But these were for him problems to be worked out, and he implicitly accepted Planck's approach as the best currently available guide. In the closing paragraph of his paper, furthermore, he tied these problems to another one, which his contemporaries would begin to take up in the following year and which would, after 1913, transform the development of the quantum theory:

> That Maxwell's theory is inadequate to represent processes within the atom follows also from the impossibility of deriving from it a satisfying account of spectral series. If, therefore, the theory of spectral lines seems possible by a modification of Maxwell's equations, then the energy curve for black bodies will probably only be derivable by the statistical treatment of a large number of molecules which simultaneously emit and influence each other.[55]

By 1910 the considerations that had led Lorentz, Planck, and Wien to admit discontinuity had apparently also decisively influenced James Jeans. In 1908 he, too, had written an answer to Lummer and Pringsheim, but unlike Lorentz he had simply denied the force of their experimental evidence, insisting that it would be relevant only if his theory dealt with the equilibrium case.[56] In 1910, however, he submitted to the *Philosophical Magazine* a paper, "On Non-Newtonian Mechanical Systems, and Planck's Theory of Radiation," which opens as follows:

> Planck's treatment of the radiation problem, introducing as it does the conception of an indivisible atom of energy, and consequent discontinuity of motion, has led to the consideration of types of physical processes which were until recently unthought of, and are to many still unthinkable. The theory put forward by Planck would probably become acceptable to many if it could be stated physically in terms of continuous motion, or mathematically in terms of differential equations....
> The question discussed in the present paper...is in brief as follows:—Can any system of physical laws expressible in terms of continuous motion (or of mathematical laws expressible in terms of differential equations) be constructed such that a system of matter and aether tends to a final state in which Planck's law is obeyed? It will be found that the answer is in the negative.[57]

Jeans's route from question to answer need not be followed here, for it is virtually identical, including some notational detail, with the one followed in Einstein's papers of 1904 and 1906. At the end of the paper, Jeans even notes, still without reference to Einstein, that the Planck distribution can be used to explain the known anomalies of the specific heat of gases, the problem which his own theory had originally been designed to solve, and which had been the primary reason for its plausibility. Though Jeans presented his own theory once more in a talk on "The Kinetic Theory of Specific Heats" at the first Solvay Congress in 1911, his attitude towards it had altered considerably. In the discussion, he emphasized that his paper had not been an attempt "to defend that theory in particular or any other, but simply an effort to discover whether one could not base a consistent theory on the ideas of Maxwell and Boltzmann."[58] Though Jeans did not follow Lorentz and announce his acceptance of Planck's energy elements, he had clearly prepared himself to do so.

By 1910, however, more than an acknowledgment of past error would have been required to bring Jeans abreast of Lorentz. Since 1908 the latter had been actively engaged in ordering his own thoughts about the quantum. The result, significant more for its clarity and authority than for its essential novelty, was the concluding three lectures in a series of six delivered at Göttingen in October 1910 under the auspices of the Wolfskehl Foundation. Arranged for publication by Max Born (1882–1970), who himself first began to publish on quantum topics two years later, the lectures soon appeared in the *Physikalische Zeitschrift* under the title "Old and New Questions of Physics."[59]

Of the three lectures on the radiation problem, the first deals entirely with the results to be gained from classical theory, and the second announces the impossibility of finding an adequate classical distribution formula. Lorentz argues that "an entirely new hypothesis must be introduced: Hamilton's principle may not be applicable to radiation. The [required] hypothesis is the introduction of the *energy elements* invented by Planck."[60] That step is, however, in Lorentz's view the only one which physicists are currently required to take. Einstein's light-particle hypothesis, though of considerable heuristic value, must be rejected for its incompatibility with many well-known optical phenomena, including interference and reflection. Other attempts to extend Planck's theory show some promise, Lorentz feels, but raise great difficulties as well. Finally, in his concluding lecture, Lorentz presents his own derivation of the radiation formula. Unlike Planck in December 1900, he restricts attention to N resonators at frequency v rather than considering $N + N' + N'' + \cdots$ resonators at frequencies $v, v', v'', \ldots,$ respectively. Correspondingly, when maximizing probability, Lorentz varies the energy attributed to individual resonators rather than the energy of the entire set of N. What results is the since standard combinatorial derivation of Planck's law, sketched more fully in Chapter IV.[61] During the five years 1906–1910, black-body theory was rapidly taking the form Planck was once thought to have given it in 1900. As it did so, furthermore, most of the experts on the theory of cavity radiation were persuaded to take it seriously.

# IX

BLACK-BODY THEORY AND THE STATE OF THE QUANTUM, 1911–1912

In the preceding pages the early history of the quantum has been detailed as though it were coextensive with the evolution of black-body theory. Inevitably, such a view is partial, but little distortion results before the end of 1910. Until that date most of the scientific articles that invoked the quantum at all were devoted to cavity radiation. The few quantum papers on other topics were uniformly speculative, either by their nature or for lack of experimental evidence, and the areas to which their authors applied the quantum were almost as numerous as those authors themselves. Though some of these papers later proved important, black-body theory appeared through 1910 to be the only area in which solid science demanded recourse even to Planck's constant, much less to a still unknown discontinuous physics. With the exception of Einstein, and to a lesser extent, of Wien, the physicists who worked on radiation theory either rejected or ignored the scattered quantum literature on other topics.

During 1911, however, that situation changed quickly. Articles that applied the quantum to other topics then outnumbered those on black-body radiation for the first time, and some were backed by impressive experimental evidence. In part because of that evidence, physicists like Planck and Lorentz, who had previously taken the constant h to be characteristic only of the radiation problem, began to consider additional areas in which others had earlier staked quantum claims. Specific heats and thermodynamics especially attracted their attention, but their papers on these topics also included references to such subjects as x-rays, the photoelectric effect, and atomic models. In the event, the future development of the quantum theory was associated primarily with these newer topics and others related to them. After 1911 the study of cavity radiation plays only an occasional role in the evolution of quantum concepts.[1]

As a result, this attempt to trace the black-body problem's role in the development of quantum concepts cannot usefully be extended further. The present chapter brings it to a close with a preliminary survey of the state and status of the quantum during 1911 and 1912. To what extent was the quantum by then an established subject of professional concern? To what problems in physics was it then thought to apply, and what position did the black-body problem occupy among them? Though necessarily tentative, answers to those questions conclude discussion of the themes that have been central to Part Two of this book.

Part One, however, had another main concern, the evolution of Planck's thoughts about the black-body problem. Ironically, after a long period of stability, they began to develop rapidly again just when other physicists were increasingly turning to other topics. Planck's so-called second theory was the result, and its relevance to the concerns of this book requires that it be analyzed more closely than some other equally important development of the years 1911 and 1912. A separate chapter, constituting Part Three, is therefore devoted to the second theory. Because that epilogue extends the themes of Part One, simultaneously relating them to the concerns of Part Two, it provides an appropriate termination to the volume.

The decline of black-body theory

Though central to earlier chapters, the detailed analysis of individual scientific papers can contribute little to an understanding of the status of the black-body problem or of the quantum. For that purpose, an overview of quantum publications is required, one that permits consideration of their changing numbers, topics, and audience. Inevitably, information on such questions is scattered and to some extent unreliable. Nevertheless, the results of even a preliminary survey are striking. In particular, as this section and the next will show, they highlight the declining importance of the black-body problem from 1911 and its rapid replacement during that year by the problem of specific heats.

To establish a basis for comparison, examine the upper, broken curves in Figures 1a and 1b. They represent—the first on a linear, the second on a logarithmic scale—the total number of *authors* publishing on quantum topics during the years from 1905 to 1914, inclusive.[2†] Drawn from the standard abstracting journal, the *Fortschritte der Physik*, the count of authors is doubtless incomplete.[3†] The *Fortschritte's* coverage is not quite exhaustive, especially outside of Germany; besides, by 1911

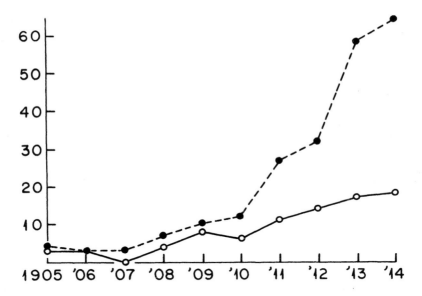

Figures 1a and 1b: Total and Black-Body Authors, 1905–14. The solid circles indicate how many authors published on quantum topics each year. The open circles show how many of them dealt with black-body theory. Figure 1a, above, is plotted on a linear scale; Figure 1b on a semilog scale to display the rate of exponential growth.

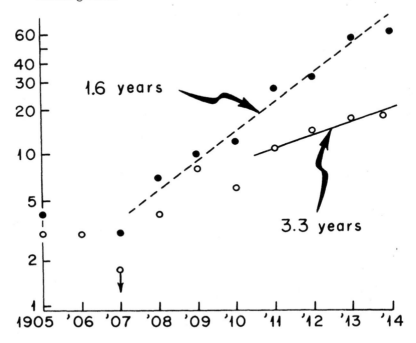

the quantum had become a sufficiently standard element in physics papers so that brief abstracts do not always refer explicitly to its appearance in a given article. As a result, the rates of growth recorded below may be very slightly too low for the years after 1909, but nothing of relevance here is likely to be affected by that error.

Before 1905 the only quantum author is Planck himself, and he appears in that role only during 1900 and 1901. Except for passing references, usually by experimentalists reporting on his law, regular publication on the quantum begins only in 1905 with the intervention, mostly critical, of Einstein, Ehrenfest, Lorentz, and Jeans. Publication is thereafter continuous and at a steadily increasing rate. Furthermore, from as early as 1907, the data for the total number of quantum authors conform remarkably closely to the canonical form, exponential growth, with a doubling period, in this case, of approximately 1.6 years.[4] That result is of particular interest when the count of total quantum authors is compared with that of authors publishing on the black-body problem (the solid curve in Figures 1a and 1b). Until 1909 or 1910 most quantum authors published on the black-body problem, and their number grows as rapidly as the total literature. Beginning in 1911, however, blackbody authors become a minority of total authors, and their rate of increase slows to a doubling rate of not less than 3.3 years.

A survey of the sources and nature of the black-body literature in and after 1911 suggests both the reasons for its quantitative decline and also that the decline was just beginning in the period before the outbreak of the first World War. Through 1910, the rapidly increasing black-body literature had, excepting for occasional British contributions by Jeans and Larmor, come from Germany, Holland, and Switzerland. After that date contributions from these countries increased very little; fully one-half of the black-body papers published from 1911 through 1914 were produced in Britain, France, Russia, and the United States, countries previously un- or under-represented. Their physicists had yet to convince themselves of what earlier black-body theorists were already realizing: while the nature of Planck's oscillators and of the corresponding emission process remained a mystery, the black-body problem could provide no further clues to physics. Though research on cavity radiation did continue in Germany as it increased rapidly elsewhere, after 1910 it everywhere proceeded along previously well-established lines.

Three early and especially well-known examples will illustrate what this later black-body literature was like at its very best. Peter Debye

210 THE EMERGENCE OF THE QUANTUM DISCONTINUITY

(1884–1966), following leads provided by the early papers of Rayleigh and Jeans, showed in 1910 how to derive Planck's law, including the factor $8\pi\nu^2/c^3$, by quantizing the vibration modes of the electromagnetic field without recourse to oscillators.[5] Years later he recalled that Paul Langevin (1872–1946) had particularly liked the clarity of the derivation, a reaction others are likely to have shared.[6] In the same year, Ehrenfest produced a fuller proof than any previously given that no radiation law which agreed with Planck's in the high and low frequency limits could be derived without recourse to discontinuity.[7] Henri Poincaré (1854–1912), a skeptic through much of 1911, independently proved a somewhat more general theorem just before the close of the year. Because of their author's international reputation, both inside and outside the profession, Poincaré's papers were especially important in spreading belief in the quantum beyond Germany's borders.[8] But, even at their increasingly occasional best, papers like these could only refine, reinforce, and spread convictions that men like Planck, Lorentz, Einstein, and Wien had already acquired. They produced, in short, new converts rather than new physics.

The emergence of specific heats

Turn therefore to a second category of quantum literature, one which, beginning early in 1911, very suddenly displaced the black-body problem as the central concern of quantum physicists. Like so much else in the history of the quantum theory through 1925, the new topic emerges first in a paper of Einstein's, this one published early in 1907 and concerned with the specific heats of solids.[9] Its argument is now familiar, but not the approach, which starts from black-body radiation. Not only the vibrations of electrons but also those of charged ions must, Einstein points out, contribute to the black-body spectrum. If these ions are bound in a solid lattice, their vibrations are, following Drude's widely accepted theory, the presumptive source of the well-known infrared resonance frequencies determined by dispersion measurements. Since the corresponding infrared radiation is a component of the black-body spectrum, Planck's law demands that the energy of ionic vibrators, like that of Planck's resonators, be restricted to integral multiples of $h\nu$.

Each ion, Einstein continues, possesses three vibrational degrees of freedom, so that its mean energy is given by

$$\bar{E} = \frac{3h\nu}{e^{h\nu/kT} - 1}.$$

Its heat capacity must therefore be

$$c = \frac{d\bar{E}}{dT} = \frac{3k(h\nu/kT)^2 e^{h\nu/kT}}{(e^{h\nu/kT} - 1)^2}.$$

Multiplying by Avogadro's number and summing over the various types of ion in the solid yields for the molecular heat capacity, in calories per degree,

$$c = 5.94 \sum_i \frac{(h\nu_i/kT)^2 e^{h\nu/kT}}{(e^{h\nu_i/kT} - 1)^2}. \tag{1}$$

Except that he still writes β for h/k, equation (1) is Einstein's formula for the specific heat of a solid. It predicts that, at sufficiently high temperatures, the molecular heat capacity of all elements in their solid state will be the same, a result in accord both with classical kinetic theory and with the long-standing empirical regularity known as the Dulong-Petit law. For very low temperatures, however, Einstein's formula predicts a quite unexpected behavior. The heat capacity of all solids should approach zero with the absolute temperature.

Unlike cavity radiation, specific heats had long been a standard subject in both physics and chemistry. It is therefore noteworthy that Einstein's surprising formulation remained virtually unmentioned in the published literature of either field during the four years after its publication in the *Annalen der Physik*. For this initial neglect, two factors are presumably responsible. Though Einstein's theory of specific heats now seems a straightforward extension of Planck's work, it did not seem that to black-body theorists at the time. If Einstein were right, then the quantum discontinuity could no longer be associated simply with the interaction between radiation and matter; nor could one still hope that discontinuity might be disarmed by encapsulation within an improved electron theory. Although Einstein had required some reference to radiative phenomena to justify quantizing the energy of ions, his theory of specific heats was thereafter a statistical-mechanical theory, independent of electromagnetic considerations. Equilibrium demanded, as Einstein quickly pointed out, that the energy not only of ions but also of neutral atoms be quantized.[10] The application of classical mechanics to *any* atomic process was thus in jeopardy and with it the whole of kinetic theory. Even to the few physicists persuaded that black-body radiation demanded a break with classical physics, Einstein's theory of specific heats may well have seemed as radical as the light-particle hypothesis with which it was associated by authorship.

More important, the low temperature behavior predicted by Einstein's formula conflicted with the virtually universal interpretation of existing observations on the specific heat of solids.[11] Especially after the careful research of Kopp in the 1860s, experiment had been thought to confirm the Dulong–Petit law. It stated, first, that the specific heat per molecule of any solid is very nearly the sum of the specific heats per atom of the molecule's components. Presently more important, the law also held that, for elements which form solids at normal temperatures, the specific heat is approximately 6.4 calories per mole-degree. The precise figure varied, of course, usually falling in the range 6.1 to 6.5, but occasionally running as high as 6.8 (bromine and iodine) or at low as 5.4 (phosphorus and sulfur). Boron, silicon, and carbon were the only ordinarily solid elements regularly cited for thermal capacities below that range. Their specific heats were, however, known to increase rapidly with temperature to normal levels, and the effect was usually attributed to uncommonly large variations in the fraction of the added heat required to do work against internal forces. Since almost no one thought such variations troublesome, little systematic information was available concerning the temperature dependence of the specific heat of solids at the low temperatures where marked departures from the Dulong-Petit law would regularly have been observed.

The general phenomenon that Einstein claimed to have explained in his 1907 paper was therefore one that, in the eyes of most of the professionals concerned, simply did not exist. Presenting evidence for his new law, Einstein's primary recourse was to a single, frequently reproduced table for the gram specific heat of diamond. Converted to molar specific heat in calories per degree, it read 0.76 at -50.5 °C, 1.84 at 58.3 °C, 5.3 at 606.7 °C, and 5.5 at 985.0 °C, a change of heat capacity far larger than that recorded for any other element.[12] Considering other substances, he showed only that low specific heat at normal temperatures was regularly accompanied by unusually large infrared resonance frequencies ($h\nu/kT > 0.1$). In 1907 the experimental evidence for declaring the Dulong–Petit law a special case was not at all strong.

Four years later that situation had changed decisively, and the status of the Dulong–Petit law had begun a drastic decline. The change was due, however, neither to Einstein nor to the quantum, but rather to a newly proposed solution for a long-standing problem in physical chemistry.[13] Thermodynamic analysis of chemical equilibria or of the direction of spontaneous reactions requires knowledge of two functions of the state variables. The first is the internal energy U of the constituents of the reaction under study, and it is readily determined by experiment. The second is their free energy, usually in this period the so-called Helmholtz free energy A, defined as $A \equiv U - TS$, where T is the absolute temperature and S the entropy. It too can be directly determined by experiment, but only for the limited range within which the state variables may be changed reversibly. (Determination of free energy is in this respect like that of entropy and unlike that of energy.) The resulting limitation on the thermodynamic study of reactions is severe; a number of ad hoc hypotheses, often equivalent to setting $A = U$, had been devised to circumvent it; none of them, however, had for long withstood experimental tests.[14]

This is the lacuna that the physical chemist Walther Nernst (1864–1941) filled in a lecture published early in 1906.[15] The functions A and U are connected by a standard differential equation, which follows at once from the definitions of free energy and of entropy:

$$U = A - T\left(\frac{\partial A}{\partial T}\right)_V.$$

With U known from experiment, graphical integration permits the determination of A, provided some way can be found for selecting the constant of integration. Nernst's original suggestion, admittedly speculative but backed by much indirect and extrapolated evidence from experiments, was that, near absolute zero, $\partial A/\partial T = \partial U/\partial T$, independent of pressure, state of aggregation, and allotropic form. Since U itself includes an arbitrary additive constant, this hypothesis is equivalent to the statement that A and U are mutually tangent functions at $T = 0°$.[16†]

Both confirmation and application of that suggestion depended on an expanded knowledge of the behavior of specific heat near absolute zero. Nernst soon set the staff of his new Berlin Institute to work on the problem and simultaneously searched the literature for data he had previously ignored. Early evidence of the result appeared in the fifth edition of his *Theoretische Chemie*, published in 1906. Though his section on the Dulong–Petit law was an almost verbatim copy of the one he had published in the fourth edition three years before, it was immediately followed by an entirely new section entitled "The Influence of Temperature on Atomic Heat." The latter opened by reminding readers of the three exceptional cases of heat dependence mentioned in the preceding section and at once continued:

> A number of newer measurements have generalized this result, in that almost without exception there is a very notable influence of temperature [on specific heat].... Under these circumstances [the success of] the rule of Dulong and Petit seems almost accidental.[17]

Five of the sixteen measurements he cited were preliminary results from his own laboratory. The others, which recorded molar specific heats ranging only from 4.1 to 6.1, were drawn from a scattered experimental literature that had not previously seemed to challenge the Dulong-Petit law. As Nernst was to remark at the beginning of 1911, the inadequacy of the classical kinetic theory of heat might have been recognized decades earlier by pursuit of clues in the existing literature on the temperature dependence of specific heats. In practice, he continued, that discovery came about in a totally different way.[18†]

It was his new heat theorem and the resulting research on specific heat that first directed Nernst's attention to Einstein's work, and it was Nernst who primarily led others, both chemists and physicists, to it. He first cited Einstein's paper on specific heats in a discussion of his own new heat theorem in 1909.[19] During February of the following year, in the second of a series of papers reporting new experiments on specific heats, Nernst remarked, in his final summary, that "one receives the impression that they [specific heats] converge to zero in accordance with the requirements of Einstein's theory."[20] A month later he pointed out that his heat theorem could, in fact, be derived from Einstein's formula, together with the requirements of "molecular theory."[21] I know no earlier reference to Einstein's work on specific heats, and in these there is still no reference to the quantum, only to "Einstein's theory," the basis for which Nernst appears as yet to have largely ignored. Presumably that theory was for him "essentially a calculation rule," a judgment he pronounced on the quantum theory as a whole at the start of the following year.[22]

Whatever he may have thought of the theory, however, Nernst was, by March 1910, sufficiently impressed with its results to undertake a visit to Zurich for discussions with Einstein. Writing to his friend, Jakob Laub, the latter reported:

> I consider the quantum theory certain. My predictions with respect to specific heats seem to be strikingly confirmed. Nernst, who was just here, and Rubens are eagerly occupied with experimental tests, so that people will soon be informed about the matter.[23]

Years later, that visit of Nernst's was recalled by the physical chemist George Hevesy (1885–1966), who from 1908 to 1910 worked in Zurich as

Assistant at the Eidgenossische Technische Hochschule. In his circles, at least, it had, he recalled, "made Einstein famous. Einstein [in 1909] came as an unknown man to Zurich. Then Nernst came, and people in Zurich said, 'This Einstein must be a clever fellow, if the great Nernst comes so far from Berlin to Zurich to talk to him.'"[24]

No records remain of the conversation, but they are a likely, though by no means the only possible, source of Nernst's apparently new realization of the magnitude of the problems with which his new heat theorem would associate him. The tone of an important letter addressed to the Belgian industrialist Ernest Solvay in July 1910, is very different from that of his previous references to "Einstein's theory."

> It appears that we are currently in the midst of a revolutionary reformulation of the foundation of the hitherto accepted kinetic theory of matter.
>
> On the one hand, its consistent elaboration leads to a radiation formula which conflicts with all experience, a situation that no one disputes; on the other hand, the consequences of that same theory include theorems about specific heats (constancy of the specific heat of gases with changes of temperature, validity of the Dulong–Petit rule to the lowest temperatures), which are completely contradicted by many measurements.
>
> As Planck and Einstein especially have shown, these contradictions vanish if one restricts the motions of electrons and atoms about their rest positions (doctrine of energy quanta); but this conception is so foreign to the previously used equations of motion that its acceptance must doubtless be accompanied by a wide-ranging reform of our fundamental intuition.[25]

That evaluation of Einstein's work on specific heats—now seen, however, as part of quantum theory, the natural result of Einstein's having "generalized Planck's view to all vibrating atoms, not just to vibrating electrons"—was central to the important invited address that Nernst presented at an open meeting of the Berlin Academy on 26 January 1911,[26] and it recurs regularly in his subsequent papers.

Only after Nernst's public, experimentally documented endorsement did other chemists and physicists begin even to mention Einstein's work on specific heats. The suddenness and speed with which the topic was then recognized and taken up are, however, remarkable. Planck says nothing about the quantum and specific heats in his Columbia lectures of 1908 or in his survey of the state of the quantum in 1910; the earliest of his many allusions to a topic on which he thereafter increasingly worked occurs in the first of the papers in which, during 1911, he belatedly began to incorporate discontinuity in his black-body theory.[27] Lorentz makes no mention of specific heats in his Rome lecture of 1908 or, more surprisingly, in his Göttingen lectures, "Old and New Questions of Physics," in 1910; his first recorded references, like those of many other physicists, are in the discussions at the first Solvay Congress late in 1911; and his first contribution to the topic appeared in the following year.[28] Also late in 1911, Arnold Sommerfeld (1868–1951) outlined the relation of the quantum to the theory of specific heat at the Karlsruhe Naturforscherversammlung, a gathering at which both Nernst and the young Austrian physicist, Fritz Hasenöhrl (1874–1915), delivered papers specifically addressed to the topic.[29] Except for Nernst's, these are all first references, and they are by no means the only ones during 1911. As Figures 2a and 2b, again drawn from the *Fortschritte der Physik*, illustrate dramatically, the quantum theory of specific heats entered physics to stay during that year.

Between 1911 and the beginning, around 1916, of systematic work on the quantum theory of spectra, the specific heat problem played a major role in the further spread and development of the quantum theory. Just because specific heats and, more generally, the kinetic theory of matter differed from the black-body problem in being standard subjects for physics and physical chemistry, their transition to quantum topics greatly broadened the audience exposed to the quantum theory. At the same time, unequivocal experimental evidence on the behavior of specific heats at low temperatures provided a new and stronger basis for the conviction that Planck's energy distribution law or something closely resembling it must replace the classical equipartition theorem. Conviction on that point did not, of course, imply commitment to a discontinuous physics. Attempts to derive the Planck distribution classically presumably continued for some years.[30] But the fundamental proofs developed for the black-body problem by Einstein, Lorentz, Ehrenfest, and Poincaré applied equally to the specific heat problem: no Planck-like distribution could be reconciled with a continuous physics. The argument for the necessity of discontinuity could be carried directly from the original quantum topic to the new one; no redesign or fresh start was required.

The emergence of the specific heat problem therefore inevitably spread both acquaintance with and conviction about the quantum. An additional, less obvious effect, proved, however, to be even more important. While restricted to the black-body problem, the mystery of the quantum was closely associated with the properties of an un-

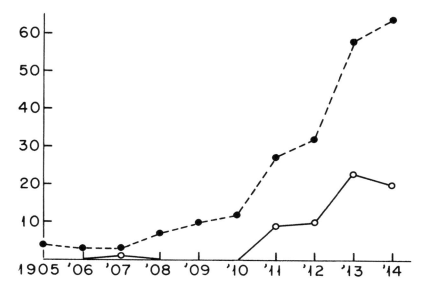

Figures 2a and 2b: Total and Specific-Heat Authors, 1905-14.
The solid circles indicate how many authors published on quantum topics each year. The open circles show how many of them dealt with the theory of specific heats. Figure 2a, above, is plotted on a linear scale; Figure 2b on a semilog scale to display the rate of exponential growth.

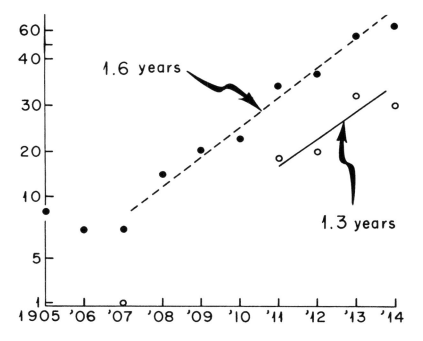

known entity, the Planck resonator. When such resonators were identified with bound electrons, the problems and paradoxes presented by their interaction with radiation were transferred to the rapidly developing field of electron theory. As previously pointed out, however, neither radiation nor electrons were centrally implicated in the treatment of specific heats. Heat capacities were accounted for mechanically in terms of the translational, vibrational, and rotational energies of atoms and molecules. To these micro-mechanical entities, the applicability of the classical equations of motion had been almost universally assumed. Even those who hoped that mechanics could be absorbed into a generalized electromagnetic view of nature did not expect to see changes in the behavior predicted by mechanical theory to result.

Specific heat theory thus identified the locus of quantum effects with a specificity that the black-body problem had not. At least for atomic and molecular processes, a new mechanics would be needed. Einstein, in his 1907 paper, sidestepped the challenge to provide one by restricting himself to "the simplest representation which one can construct of thermal motion in solids," sinusoidal vibrations about an equilibrium position.[31] Since such vibrations could occur in any of three dimensions in a lattice, he simply multiplied Planck's formula for resonator energy by three. When specific heat theory became a central topic in 1911, however, that simplification provided no resting point, and the search began for what were subsequently known as quantum conditions, restrictions more general than Planck's on the permissible mechanical motions of multi-dimensional systems. From 1911 through 1923 their development was a central element in the evolution of a quantum theory applicable to a remarkable range of atomic and molecular phenomena. Two of that theory's main problems and an embryonic form of one of its most important research techniques were first publicly discussed during 1911, a result primarily of the new concern with specific heats. As that concern became central, the direction of the quantum theory's development began to change.

One of the new problems has already been mentioned as implicit in Einstein's 1907 paper, and it was first publicly discussed in the paper presented to the 1911 Naturforscherversammlung by the young Austrian Hasenöhrl. He pointed out that, though "Planck's assumption" seems to provide a promising clue to a satisfactory theory of specific heats, "great difficulties arise when one tries to apply it to systems of several degrees of freedom."[32] A few weeks later the problem

BLACK-BODY THEORY AND STATE OF THE QUANTUM 219

arose again at the first Solvay Congress, initially through Lorentz's criticitism of Nernst's method of decomposing the three-dimensional vibrations of atoms in a solid. Einstein, who had clearly thought about the problem before, entered the discussion to remark that the application of existing methods to vibrations in three dimensions led to a mean energy that was not, as the specific heat formula required, three times the energy for the one-dimensional case. "In its current state," he said, "the theory of quanta leads to a contradiction when one tries to apply it to systems having several degrees of freedom."[33] Increasingly numerous attempts to quantize multi-dimensional systems date from this exchange, but no proposal attracted much attention until the introduction of the Wilson–Sommerfeld phase integral quantum conditions in 1915.

A second major problem, the quantization of rotational energy, emerged earlier in the same year, when Nernst turned his attention from the specific heat of solids to that of gases.[34] The well-known anomalies in the latter, which from 1894 on had led first Boltzmann and then Jeans to consider the interaction between gas molecules and the ether,[35] might also, Nernst pointed out, be accounted for in terms of the quantum theory. Unlike the translational motion of a molecule, rotational and vibrational motions are each characterized by a frequency. If these motions could absorb and emit energy only in whole-quantum units $h\nu$, then they would not begin to contribute to molecular specific heat until the temperature was great enough so that kT became of the order of $h\nu$. Nernst's evidence for this quantum effect was predominantly qualitative, but significant quantitative support was quickly provided in papers by the young Dane, Niels Bjerrum (1879–1958), then working at Nernst's laboratory in Berlin.[36] The data Bjerrum treated first gave information only about vibration spectra, but during 1912 he extended the treatment to rotations in an important paper on the band structure of the infrared absorption spectrum.[37] Even before that quantitative step was taken, however, the subject of quantizing rotations had figured prominently in Einstein's address to the Solvay Congress and in the discussion that followed.

Unlike the frequency of a one-dimensional vibrator, that of a rotator increases with energy and thus, statistically, with temperature. As a result, no Planck-like condition can be used to restrict rotational energy unless the frequency by which h must be multiplied is known from theory in advance. In his Solvay address, Einstein reported that his attempts to find the rotation frequency of a diatomic molecule by existing techniques had failed because of mathematical complexities.[38] Lorentz then pointed out in the discussion that, since the energy of a rotating sphere was given by $q\nu^2$ with q a constant, it would be natural to quantize the rotation by setting $q\nu^2 = nh\nu$, with n an integer. Permissible frequencies would then be given by the formula $\nu = nh/q$.[39] That way of quantizing rotation was quite obvious once a problem was presented to which it could be applied. It had, in fact, been suggested a month earlier at Karlsruhe in the discussion of a first full report on the magneton by Pierre Weiss (1865–1940), and it was reinvented, probably independently, during 1912 by Bjerrum for application to rotation spectra and by J. W. Nicholson (1881–1955) for application to rotating rings of electrons.[40]

Lorentz's suggestion provoked much discussion, some of it skeptical that the actual rotation frequencies of a molecule (as against the statistical distribution of such frequencies) could be restricted by the quantum. That skepticism, in turn, reminded Lorentz of a problem he had previously discussed with Einstein, an especially simple example of a motion in which frequency increases with energy. Imagine, he suggested, a simple pendulum vibrating with an amplitude and frequency such that its energy is just $h\nu$. What happens if one grasps the string of the pendulum between two fingers at its point of support and then lowers one's fingers, thus shortening the pendulum's length? It appears, Lorentz stated, that the pendulum's energy will then be smaller than that corresponding to the new frequency. Einstein, who had made progress with the problem since their previous discussion, at once intervened. If, he said, the string were shortened infinitely slowly, then the energy of the pendulum would increase with frequency so that it continued to equal precisely $h\nu$. "The same holds," Einstein added, "for an electrical circuit without resistance and for free radiation."[41] Two years later, Ehrenfest independently developed the principle of the adiabatic invariance of the quantum conditions from considerations of this sort.[42] That principle, especially as deployed by Bohr, rapidly became a central research tool of the old quantum theory.

Quanta and the structure of radiation

Black-body theory and specific heats were the two quantum topics well established by the end of the period 1911–12. Those who knew them from the inside were by then persuaded that both demanded a distribution law like Planck's, and the theoreticians among them recognized that the result would be some more or less profound modification of classical theory. If they belonged to the small group that had been studying the work of Planck, Einstein, and Lorentz since 1906, they were almost surely also persuaded that the change would involve the introduction of fundamental discontinuity. Detailed studies of the black-body problems had convinced some; transfer of black-body results to specific heats rapidly spread conviction to others. Until the assimilation of the Bohr atom from late 1915, no other quantum topic had a remotely comparable impact. By 1911–12, however, there were numerous other such topics, some since forgotten, others subsequently to become central themes in the development of the quantum theory. A brief survey of the main subjects to which one physicist or another had suggested that that theory might apply will facilitate understanding of the state of the quantum early in the second decade of this century.

Though the coherence that results is somewhat artificial, the heterogeneous assortment of "other quantum" topics must be subdivided for purposes of exposition. A first set, to be considered in this section, consists of those that relate to the radical suggestions made by Einstein in 1905. His arguments for the particulate properties of high frequency radiation had been, as previously indicated, both theoretical and abstract; the observed validity of the Wien distribution law at high frequencies provided their primary experimental basis.[43] In the last pages of his paper, however, Einstein did suggest that further evidence for his heuristic viewpoint was provided by Stokes's rule for fluorescent radiation and perhaps also by experiments on photoelectric and photoionization effects.

Stokes's rule (that the wavelength of fluorescent radiation is always greater than or equal to that of the exciting radiation) was well established. But, in the absence of a quantum theory applicable to spectral frequencies, Einstein's suggestion did little to explain the phenomenon and offered no guidance at all for further research. As to the photoelectric and related effects, little evidence was available. Einstein relied almost exclusively on a paper published in 1902 by Philipp Lenard (1862–1947), one that demonstrated only that the maximum velocity of emitted cathode particles was independent of the intensity but varied with the nature of the incident illumination.[44] In 1907 and 1909, respectively, A. F. Joffé (1880–1960) and Rudolph Ladenburg (1882–1952) pointed out that recent quantitative observations were compatible with the frequency dependence of photoelectron velocity predicted by Einstein's law. Others disagreed, however, and the accuracy of the law was generally acknowledged only after the thorough experiments reported in 1914 to 1916 by R. A. Millikan (1868–1953).[45] Similarly ambiguous quantitative information relating the quantum to ionization processes in gases began to appear only in 1914 with the development of the research on electron collisions undertaken for other reasons by James Franck (1882–1964) and Gustav Hertz (b. 1887).[46] Though all these experiments did much to spread conviction about the quantum, it should be noted that they did not attract physicists to Einstein's light-particle hypothesis. With few exceptions, its acceptance dates only from the discovery of the Compton effect in 1922.

During the years 1907 to 1909, a new series of suggested quantum applications followed those offered by Einstein in 1905. The earliest was due to Wien, who used Planck's theory and his own measurements of the intensity of radiation from moving ions (canal rays) to compute the entropy and temperature of the H_β line of hydrogen. Though Wien took pains to emphasize the numerous special hypotheses his computation demanded, he was apparently pleased by the order of magnitude agreement (a multiplicative factor of nearly five) between his value and Planck's for the "elementary quantum of energy."[47] Later, in 1907, Wien suggested also that the quantum might be used to compute the minimum wavelength of x-rays produced by the rapid deceleration of electrons with energy determined by the voltage across the tube.[48] When first introduced, that method of computing the limit of the continuous x-ray spectrum could be checked only to a rough order of magnitude, but it was closely confirmed after the discovery of x-ray diffraction in 1912.

Wien's application of the quantum to x-rays was typical of many that were to come during much of the following decade: find a phenomenon characterized by a frequency and an energy, and attempt to relate the two through an equation like $E = h\nu$. For those who took the quantum seriously, that technique was obvious. Wien's use of it had, in fact, already been proposed by the man who proved to be the most ardent and fecund of the quantum's early advocates. He was the ambitious and irascible Johannes Stark (1874–1957), a brilliant and imaginative experimentalist, but a highly idiosyncratic theoretician. Stark first utilized the quantum in a strange paper submitted to the *Physikalische Zeitschrift* in late 1907.[49] From an article just published by Planck he took what he called "Planck's relation," $E = Mc^2$. That formula he applied to the rest-mass of the electron, equated the resulting energy with $h\nu$, and emerged with an intrinsic frequency for the stationary electron. If the electron possessed an intrinsic frequency, he pointed out, it must have some structure or at least be anisotropic. By postulating a special structure, he claimed to account for both the negative and positive charge of the atom as well as for the main features of atomic and molecular structure.

As that description may indicate, the main text of Stark's paper displayed him at his very worst, speculating wildly and eclectically in an area he lacked the training or patience to enter. But a long footnote early in the paper shows what the same mentality could do when confined to a region Stark knew from his own experiments. In that note, Stark first anticipated Wien's computation of x-ray wavelength; then, independent of Einstein, whose priority he acknowledged in the paper to be discussed next, he reversed the relationship and computed the maximum velocity of photoelectrons released from a metal by radiation of known frequency. In both cases, he pointed out, his numerical results agreed adequately with the very small amount of relevant experimental information.

One month later Stark submitted the first of two papers that developed a plausible, though since discarded, explanation of a puzzling effect he had himself discovered two years before.[50] The radiation from canal rays, observed along an axis parallel to the direction of ionic motion, is shifted in frequency by the Doppler effect. When he examined that radiation spectroscopically in 1905, Stark had therefore expected to find the lines of the normal spectrum broadened continuously by Doppler shifts corresponding to ionic velocities from zero to some maximum. Instead, he found a pronounced minimum between the radiation from fast-moving ions and that from ions which moved slowly or were at rest. Returning to this puzzling result after his first use of the quantum in 1907, Stark suggested that the significantly displaced radiation was due to ions stimulated by collisions with a stationary atom, so that it could be emitted only by ions with a velocity above a minimum specified by the quantum relation $\frac{1}{2}\alpha m v^2 = h\nu$, where α is a "radiation constant" dependent on the particular type of ion involved. The slow moving ions that produced the undisplaced radiation must, he thought, be stimulated by a different mechanism, presumably by collision with the energetic electrons in cathode rays. As late as 1921 his explanation was still cited as one of the quantum's significant applications.[51]

Stark next, still in 1907, applied the quantum to band spectra, another area in which he had done much previous work.[52] Unlike series spectra, which he attributed to the vibrations of electrons in rotating rings in the interior of the atom, band spectra were, in his view, due to the recapture of valence electrons removed from a molecule by ionization. Setting the ionization energy equal to $h\nu$ yielded a lower limit for the wavelength in the band. Longer wavelengths were generated when an electron returned to the molecule on a spiral path, releasing part of its energy at the perihelion of each loop, and radiating at a frequency again determined by equating $h\nu$ with the energy lost. A long series of papers by Stark and collaborators followed, dealing with band spectra and also with fluorescent radiation.

Even these papers did not exhaust Stark's zeal for the quantum. Late in 1908 he discussed its relevance to photochemistry and suggested three fundamental laws derived from it.[53] Then, in 1909, he announced his support of Einstein's light-particle hypothesis, thus becoming its only established physicist-advocate besides its author.[54] In his first paper on the subject, Stark used his own estimate of x-ray wavelength to compute an approximate breadth for the pulse emitted during the deceleration of an electron. It was very small compared with the mean time between pulses, known from the density of electrons in the cathode ray beam. Since he also believed that the times of successive collisions were uncorrelated, Stark concluded that photo-emission by x-rays must be due to single pulses and that their energy must be concentrated at the site of the electron to be emitted. No such concentration could, he felt, be reconciled with what he called the "aetherwave hypothesis." This much was relatively common, but Stark was not finished. Instead, in a second paper, which inaugurated an extended series, he attempted to prove that x-rays could not be just ordinary radiation by showing experimentally that the angular dependence of x-ray intensity from a normal tube violated the symmetry properties that, in his view, the "aetherwave hypothesis" required.[55]

Once again Stark had overreached himself, a fact pointed out promptly and in detail by Arnold Sommerfeld, the professor of theoretical physics at Munich and one of the foremost experts on the equations of electromagnetic theory.[56] He had, Sommerfeld reported, been considering x-ray emission due to electron deceleration for some time, and Stark's article now led him to report some results. The sorts of asymmetries observed by Stark could, he went on, be accounted for perfectly by electromagnetic theory, without any recourse to the quantum. That was the point Sommerfeld proceeded to demonstrate in an extended article, filled with formulas and with diagrams of the angular distribution of radiation intensity to be expected from existing theory. Stark responded that he had not had electromagnetic theory in mind but rather the "aetherwave hypothesis," to which Sommerfeld replied, not altogether without heat, that the aetherwave hypothesis now appeared to exist only in Stark's own paper and did not therefore require the refutation its author had provided.[57]

That episode would be only amusing except that it marks Sommerfeld's first involvement with quantum theory, a field that engaged his attention at once and of which he became, from 1915, a principal architect. Early in his first response to Stark, Sommerfeld had carefully pointed out that his calculations applied only to that part of x-radiation due to the deceleration of cathode rays, not to the so-called characteristic x-radiation discovered by Barkla in 1906. In the latter case, he pointed out, "It is quite possible that Planck's quantum of action may play a role."[58] Whether or not he was comfortable with that concession, others were not, as Wien made clear in a letter written very shortly after the article appeared. Its opening sentence expressed agreement with Sommerfeld's position, but Wien then continued:

> I think, however, that even the energy element will have to be traced back to electromagnetic processes when we have *one* comprehensible emission mechanism based on the electron. You surely do not yourself maintain that there is both electromagnetic and non-electromagnetic x-radiation.[59]

That Sommerfeld shared these reservations is apparent in an important paper on γ-radiation that he presented to the Munich Academy of Science early in 1911. There he expanded the treatment he had developed for x-rays and also accounted for γ-radiation in terms of the considerable acceleration imparted to an electron (β-ray) as it was expelled from the interior of an atom. If the acceleration were rapid enough, the resulting γ-ray would, he pointed out, have "the character of a projectile, scarcely distinguishable in its energy localization from a corpuscular radiation or the hypothetical light-quantum."[60] Much of Sommerfeld's long article consists of an elaborate mathematical derivation of the consequences of his proposed mechanism at relativistic velocities. Midway through the paper, however, he inserts an explicitly speculative section, to the importance of which he had called special attention in his introductory summary. "We transfer," Sommerfeld wrote, "the basic hypothesis of the Planck radiation theory to radioactive emission and suppose that, in each such emission, exactly one quantum of action is given up. The 'action' of an emission (the time integral of energy) we set equal to the acceleration time τ multiplied by the total emitted energy."[61] Values of h computed from crude observations of the relative γ- and β-ray energies then proved to be in the range 2.4×10^{-27} to 1.4×10^{-28} erg·sec. Applying the same hypothesis to the normal x-radiation from which he had previously excluded the quantum, Sommerfeld found $h = 8 \times 10^{-27}$, remarkable agreement throughout.[62]

Planck was greatly impressed with these results, especially with the fundamental role they gave to the quantum of action as against the quantum of energy,[63†] and Sommerfeld seems to have shared his excitement. Appearing at the Karlsruhe Naturforscherversammlung in September of 1911, he announced that, although he had been invited to report on the state of relativity, he would instead, with the organizers' permission, discuss the theory of quanta. Relativity, he felt, was already a closed field; in quantum theory on the other hand, "the basic concepts are still in flux and the problems innumerable."[64] The first part of his paper was a summary of the results attained by others; the second developed his own quantum hypothesis from the basis supplied by his γ-ray paper of 1911. "In every purely molecular process," Sommerfeld wrote, "a determinate universal quantity of action is taken up or given off, namely the quantity $\int_0^\tau H \, dt$ [where] τ is the time required by the action process, [and] H is to be taken as a mere abbreviation for $T - U$," kinetic minus potential energy.[65] In the remainder of the paper Sommerfeld justified his formulation by its congruence to relativity theory, applied it to the photoelectric effect, and suggested how it might be developed for more standard quantum topics. At the Solvay Congress in the following month—a meeting from which Stark was notably absent—Sommerfeld further elaborated his version of the quantum.

The quantum and atomic structure

One last set of new quantum topics will complete this survey of the main areas of quantum discussion to the years 1911 and 1912: proposed relationships between the constant h and atomic structure. Such proposals were varied, both in substance and in objective. Some physicists postulated particular structures in order to explain h; others took h as given and applied it to the elucidation of atomic structure or of some otherwise surprising aspect of atomic behavior, often spectroscopic.[66] Papers of this sort appeared first in 1910, and their number increased rapidly thereafter. Bohr's 1913 paper "On the Constitution of Atoms and Molecules" is among them.

Work of this general sort is illustrated by a paper which the young physicist-historian of science, A. E. Haas (1884–1941), presented to the Vienna Academy of Science in 1911.[67] Seeking a physical version of the Planck resonator, Haas examined the behavior of a single electron within the diffuse positively charged sphere of the plum pudding model developed by J. J. Thomson (1856–1940). Within that sphere, characterized by radius a and uniformly distributed total charge $+e$, the electron is drawn to the center by a restoring force $e^2 r/a^3$, where r is its distance from the center. In the absence of radiation losses, it can thus move in a stable circular orbit with a characteristic frequency, independent of orbit size, given by $\nu = e/2\pi a\sqrt{ma}$, where m is the electronic mass. As the radius of the electron's orbit increases, the total energy of the atom rises from 0, with the electron at rest at the center, to e^2/a, with the electron moving on the surface, its normal position.

Except for the restriction to the one-electron case,[68†] this much was standard, but Haas's next step was not. A mathematical slip permitted him to conclude that the atom's total energy would continuously decrease as the electron was moved from the surface of the sphere to infinity. The maximum absorbable energy, e^2/a, thus became a second intrinsic characteristic of the atom, and Haas plausibly equated it with h times the other intrinsic characteristic, the frequency ν.[69] Using existing estimates of e, m, and a, he was then able to calculate both Planck's constant and also the characteristic frequency of his one-electron atom model. Within the limits of uncertainty of e and a, the first coincided with Planck's own value for h, the second with the limiting frequency of Balmer's formula for the hydrogen spectrum. That agreement was sufficiently impressive to persuade a few contemporaries that Haas, despite his obvious mistake about maximum energy, might be on the right track.[70] Lorentz who still hoped to explain the quantum as a property of resonators, spoke with interest of Haas's work both in his Wolfskehl lectures of 1910 and at the Solvay Congress a year later.[71] During 1911 both Wertheimer and Schidlof referred to Haas's work in developing related, quantum-producing atom models.[72]

Other attempts to relate the quantum to atomic structure were apparently independent. In 1910 J. J. Thomson examined the properties of circular electronic orbits in the field of an electronic dipole and showed how they could account for Einstein's photoelectric law, including the appearance of a constant with value close to Planck's.[73] In the following year his countryman Nicholson applied h with notable success to restrict the orbital frequencies of an atomic model he had previously developed to account for unexplained lines in the solar spectrum.[74] When Pierre Weiss, later in the year, reported to the 1911 Naturforscherversammlung his discovery of a natural unit of molecular magnetism, two discussants immediately suggested that the effect was likely to be due to the quantum; one of them sketched a model-based theory of the effect, and Langevin developed it further at Solvay.[75]

Even for the year 1911 these examples do not quite exhaust the efforts to relate the quantum to atomic structure.[76] But they should sufficiently indicate the rapid emergence, beginning in 1910, of still another area in which physicists called upon the quantum. As the discussion of Haas's work illustrates, many of their models were ingenious and suggestive. But few could be controlled by existing experiment, and the range of data even potentially relevant to them was minuscule. They were of interest primarily in their invocation of the quantum, and their utility was correspondingly limited. They spoke to no other recognized problems, provided no basis for a program of coherent research, and thus closely resembled the entire body of quantum literature that has occupied us since the close of the discussion of black-body theory and of specific heats. Until late in 1913 those two remained the only established quantum topics. X-rays, the photo-effect, band spectra, atom models, and the rest were matters for occasional speculation. Some of these speculations would prove to have a future; others would shortly be forgotten; as yet there were no criteria to determine which were which.

Prior to that determination, the main importance of this other quantum literature is the evidence it provides of increasing professional consciousness of the existence of the quantum and of its likely place in the future development of physics. Figures 3a and 3b, examined in conjunction with the earlier data from the *Fortschritte der Physik*, provide graphical evidence of that increase. They thus reinforce an impression, drawn primarily from other sources, that by 1911 and 1912 the quantum had entered physics to stay and that a major reconstruction of classical theory had become inevitable.

The state of the quantum

Return briefly to the years 1905 and 1906. Black-body theory was then an esoteric specialty of concern to very few. Among them only

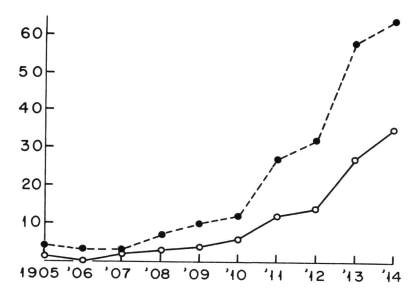

Figures 3a and 3b: Total and "Other Quantum" Authors, 1905–14. The solid circles indicate how many authors published on quantum topics each year. The open circles show how many of them dealt with topics other than black-body theory and specific heats. Figure 3a is plotted on a linear scale; Figure 3b on a semilog scale to display exponential growth.

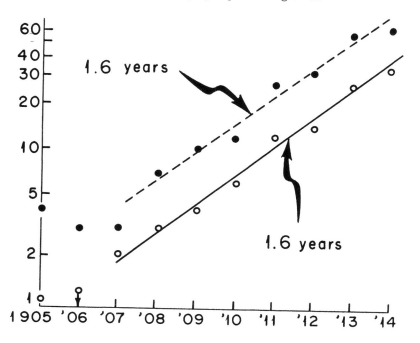

Einstein, Ehrenfest, and perhaps Laue—all very young and virtually unknown—appear to have believed that Planck's success would necessitate a fundamental reconstruction of physics. Planck himself, Lorentz, and presumably Wien—all senior figures of deserved reputations—thought them wrong, and most other physicists aware of the problem would surely have agreed. Except as a puzzling universal constant characteristic of all the main cavity-radiation laws, the quantum was scarcely an object of professional consciousness.

By the middle of 1910 that situation had altered, but not yet decisively. Einstein's reputation had grown. Planck, Lorentz, Wien, and perhaps Jeans had recognized the need for discontinuity. Converts to the quantum now included most of the senior as well as the junior experts on the theory of cavity radiation. That group, unlike its original core, was unlikely either to have made a mistake or to have been swept away by enthusiasm. The quantum had been established, but still only for cavity-radiation where it remained deeply entangled with the more general problems of electron theory. Planck was not being unrealistic when he wrote to Nernst in June 1910 describing the projected Solvay Congress as premature and urging that it be postponed a year or more. "My experience," he wrote, "leads me to the opinion that scarcely half of those you envisage as participants have a sufficiently lively conviction of the pressing need for reform to be motivated to attend the conference.... Of the entire list you name [presumably eighteen in number] I believe that beside ourselves [only] Einstein, Lorentz, W. Wien and Larmor are deeply interested in the topic."[77]

What Planck underestimated, however, was the extent and speed of the change that would be created by Nernst's, and perhaps also by Sommerfeld's, not yet public intervention. The latter presented his paper on γ-rays, the first in which he had developed a quantum hypothesis, to the Munich Academy on 7 January 1911 and to the University Physics Colloquium four days later. For the latter audience, it was the first quantum topic to be discussed, but it was followed by six others before the end of the year and by eight more during 1912.[78] Nernst's far larger role as a quantum publicist has already been described. Though he had stated his new conviction about the quantum in letters proposing the Solvay Congress during the summer of 1910, he did not speak publicly of the need for fundamental reform or of the relation between his new law and radiation theory until the following January.[79] During 1911, however, papers in which he and his coworkers pursued the theme appeared in numerous periodicals, and Nernst himself returned to it in the paper he read to the annual Naturforscherversammlung at Karlsruhe in late September.

That meeting provides other evidence, as well of the extent of the change in awareness of the quantum. Prior to 1911 physicists attending the Naturforscherversammlung had heard just two papers involving the quantum, one by Stark in 1908 and the second by Einstein in 1909.[80] At the Karlsruhe meeting there were four quantum papers, and the subject figured prominently in the discussion of a fifth. None dealt with black-body theory, previously the central topic. Nernst spoke on specific heats as did the young Austrian Hasenöhrl. Sommerfeld presented the version of a quantum theory he had developed from his x-ray work, and Haber reported some new photochemical effects that, he thought, would demand explanation in terms of the quantum. The fifth paper was Weiss's first major report on the magneton, the one discussants promptly related to the quantum. For Germans, at least, that set of papers might have been sufficient, but a month after they were read, a number of Europe's leading physicists assembled at Brussels for the Congress that Nernst had organized to discuss the needed reformulation of physics. Their papers and a full account of the accompanying discussion appeared during 1912, only two years after Planck's sensible advice that the Congress itself be postponed. After that time few physicists at any major center are likely to have remained unaware or disdainful of the quantum.

Only those who had worked on the subject were, of course, yet convinced that fundamental reform and, more particularly, a discontinuous physics would be required, and that group was still largely German or in the German intellectual orbit. But the fundamental proofs of Einstein and Lorentz had been available for several years, and they had so far withstood all critical scrutiny. A month after his return from Brussels, Poincaré presented to the French Academy of Sciences the first version of his own detailed proof of the necessity of discontinuity. The full version followed in January 1912, after which date French publication on the quantum, if initially sparse, is nevertheless continuous.[81] James Jeans, who, whatever his private views, was still widely thought to maintain the adequacy of classical theory, publicly announced his conversion at the September 1913 meeting of the British Association for the Advancement of Science. The occasion was his introductory address—the first quantum talk for a British Association audience—to a scheduled panel, "Discussion on Radiation," in which Larmor, Lorentz, and Pringsheim all participated.[82]

232 THE EMERGENCE OF THE QUANTUM DISCONTINUITY

Thereafter, Jeans at once became the quantum theory's leading British expert and spokesman. His *Report on Radiation and the Quantum Theory*, prepared for the Physical Society of London in the following year, remained for more than a decade the standard English language account of the subject.[83] Though Germany was, at the outbreak of the first World War, the only country in which the quantum was well established and the call for fundamental reform widespread, physicists in other countries had begun to convince themselves of what their German colleagues already knew. Insofar as such events are datable, the first stage of the conceptual transformation leading from classical to modern physics was complete.

Part Three

EPILOGUE

X
PLANCK'S NEW RADIATION THEORY

During the years 1911 and 1912, as the frontier of research on the quantum moved from black-body theory to the theory of specific heats, Planck too had been active. The result of his efforts, since known as his "second theory," was still another way of applying the quantum to cavity radiation and related problems. For the subsequent development of quantum concepts, it was in many respects comparable to the other long-forgotten theories briefly mentioned in Chapter IX, and an incidental effect of discussing it will be to suggest how productive some of these short-lived theories could be. To the concerns of this volume, however, Planck's second theory has a larger importance, for attention to it will return the narrative to themes that were dominant through its early stages, but that have since been lost from view. The papers in which Planck elaborated the second theory are products of his first attempts to incorporate discontinuity and develop a non-classical physics. Inevitably their preparation necessitated his reconsidering and re-evaluating a number of the ideas, including natural radiation, that had been fundamental to his earlier research on cavity radiation. Comparison of the second edition of Planck's *Lectures*, completed late in 1912, with the first edition of 1906 displays the magnitude of the resulting conceptual reorientation. By doing so, it brings this book, now considered as a story about Planck, to a second close.

Planck's "second theory"

When Planck, in correspondence with Lorentz late in 1908, belatedly acknowledged that his black-body theory demanded discontinuity, he attributed it to the resonator and, more specifically, to the excitation process. "There exists," he wrote, "a certain threshold: the resonator does not respond at all to very small excitations; if it responds to larger ones, it does so only in such a way that its energy is an integral multiple of the energy element $h\nu$."[1] Though Planck does not yet say so, he appears to have believed that, once the resonator was excited, its behavior was governed unproblematically by classical theory. That, in any case, is the position he made explicit fifteen months later. Writing to Lorentz in January 1910, he said:

> The discontinuity must enter somehow; otherwise one is irretrievably bound to the Hamiltonian equations and the Jeans theory. Therefore, I have located the discontinuity at the place where it can do the least harm, at the excitation of the oscillator; its decay can then occur continuously with constant damping.[2]

The same conceptions, somewhat more tentatively expressed, are to be found in the published paper with which Planck in 1910 announced his return to quantum problems. There, too, he speaks of "a kind of threshold for the response of an oscillator," and, in summarizing his views on discontinuity, he states that "it appears necessary to take the energy of an elementary oscillator... *at least at the moment of its excitation*, as an integral multiple of $h\nu$."[3] Though discontinuity is unavoidable, it need not affect both emission and absorption.

The culmination of these scattered hints about the asymmetry of emission and absorption was a new version of radiation theory, one which Planck first presented to the German Physical Society early in 1911, developed in further papers during the year that followed, and formulated canonically in the second edition of his *Lectures*, completed late in 1912. By the time he presented the first of these papers, however, Planck had found a reason to reverse his initial localization of discontinuity, and he developed it for his Physical Society audience:

> The absorption of a finite energy quantum from radiation of finite intensity can only occur in a finite interval of time.... Now, with increasing frequency, the energy element $\varepsilon = h\nu$ becomes steadily greater while the intensity J declines so rapidly that for short waves the time required [to absorb a quantum] soon becomes immensely long. This contradicts the [previous threshold] hypothesis; for if the oscillator had once begun to absorb energy, and the incident radiation were then cut off, it would be absolutely blocked from taking up its full energy quantum, which it must do from time to time in order to restore its average energy.[4]

Guided by these considerations, Planck now supposed that absorption occurred continuously, in accordance with classical laws. Though discontinuity was still required, it entered through "the hypothesis that the emission of energy by an oscillator occurs by leaps, in accord with energy quanta and the laws of chance."[5] Discontinuity was thus shifted from the excitation to the emission process.

The hypothesis of discontinuous emission is common to all Planck's papers on the second theory, but they differ on the process itself. In his first paper an oscillator may emit a single quantum at any time, the probability of its doing so in the interval dt being $n \eta$ dt, where η is a probability constant (< 1) and n is the number of whole quanta in the oscillator's energy U just prior to emission ($nh\nu \leq U < (n + 1)h\nu$). In his subsequent papers and in the second edition of the *Lectures*, an oscillator may emit only when it reaches one of the phase-plane boundaries for which $U = nh\nu$, and it must then give up its entire energy, emitting just n quanta. The probability that an oscillator emits as it crosses any boundary is now simply η, regardless of its energy U at that boundary. For present purposes, only this later version of the theory need be considered. I shall largely follow the simplified and mature account that Planck published in the *Annalen der Physik* early in 1912,[6] omitting reference to the two slightly different derivations he had provided during the preceding year.[7†]

After some introductory remarks on the structure of his derivation and the hypotheses it requires, Planck turned directly to the electrodynamic part of his argument. Like all previous derivations of his black-body law, it opens with the standard equation for oscillator energy:

$$U = \tfrac{1}{2}Kf^2 + \tfrac{1}{2}L\dot{f}^2, \qquad (1)$$

where, again, f is the oscillator's moment, and K and L are constants related to the oscillator's frequency by the equation $2\pi\nu = \sqrt{K/L}$. Planck's next step, however, is a drastic departure from all his quasi-classical derivations, and it results in a great abridgement and simplification of the subsequent discussion. Since the mechanism by which oscillators emit is now unknown and since finite line breadth does not result in energy redistribution, he drops the damping term from the oscillator equation, which then becomes simply

$$Kf + L\ddot{f} = E_z. \qquad (2)$$

As before, E_z is the field strength of the incident radiation in the direction of the oscillator's motion.

To investigate absorption, a classical process, Planck assumes that at time $t = 0$ the oscillator is at rest ($f = \dot{f} = 0$), having just emitted all its energy. If the exciting field E_z is specified by a Fourier series, the coefficients in the corresponding series for f are readily determined with the aid of equation (2). Next, using equation (1), oscillator energy and its rate of change can be found as functions of time. Planck shows straightforwardly that, for a stationary exciting field, the rate of absorption is independent of time and proportional to the field intensity J. Quantitatively, he finds that

$$\frac{dU}{dt} = \frac{J}{4L}. \tag{3}$$

That is the only consequence of electromagnetic theory his new derivation will require. Planck produces it in five pages, less than one-fifth the space he had needed in 1900 to develop the electromagnetic requisites of his original theory.

With the rate of absorption determined, Planck turns to the nonclassical emission process. At some instant, he imagines, there are A_n oscillators in the phase-plane ring bounded by the pair of ellipses for which $U = (n-1)h\nu$ and $U = nh\nu$. As these oscillators continue to absorb, they will move across the ring to the exterior boundary, at which $A_n\eta$ of them will give up their entire energy by quantum emission and $A_n(1-\eta)$ will move into the next ring. Since the same time is required to move through each ring, the process can be stationary, i.e., maintain the ring population at all times, only if $A_n(1-\eta) = A_{n+1}$. It follows that, in a stationary radiation field, the total number of oscillators in all rings must be given by

$$N = A_1\{1 + (1-\eta) + (1-\eta)^2 + (1-\eta)^3 + \cdots\} = \frac{A_1}{\eta}.$$

That formula leads directly to the first of the results Planck wishes to obtain. In a stationary field there must be $N\eta$ oscillators in the central ellipse, $N\eta(1-\eta)$ in the first ring surrounding it, $N\eta(1-\eta)^2$ in the next, and so on. More generally, if P_n is the fraction of all oscillators that have energy between $(n-1)h\nu$ and $nh\nu$, then $P_n = \eta(1-\eta)^{n-1}$, and the corresponding numbers supply a distribution function that can be used to compute oscillator entropy. For this purpose Planck again turns to Boltzmann, but this time to his H-theorem definition of entropy, not to his combinatorial definition. For a collection of N oscillators, Planck writes, the total entropy is given by[8]

$$S_N = -kN \sum_1^\infty P_n \log P_n = -kN\left\{\frac{\log \eta}{\eta} + \left(\frac{1}{\eta} - 1\right)\log\left(\frac{1}{\eta} - 1\right)\right\}. \tag{4}$$

To compare that equation with Planck's earlier expressions for entropy, the constant η must be expressed as a function of mean oscillator energy $\overline{U}$. Since oscillators are distributed uniformly within any given ring, the mean energy of the oscillators with energy between $(n-1)h\nu$ and $nh\nu$ must be just $(n-\frac{1}{2})h\nu$. For all oscillators, the corresponding mean is

$$\overline{U} = \sum_{1}^{\infty}\left(n-\frac{1}{2}\right)P_n h\nu = \left(\frac{1}{\eta}-\frac{1}{2}\right)h\nu, \qquad (5)$$

a result which, when inserted into equation (4), yields

$$S_N = kN\left\{\left(\frac{\overline{U}}{h\nu}+\frac{1}{2}\right)\log\left(\frac{\overline{U}}{h\nu}+\frac{1}{2}\right) - \left(\frac{\overline{U}}{h\nu}-\frac{1}{2}\right)\log\left(\frac{\overline{U}}{h\nu}-\frac{1}{2}\right)\right\}. \qquad (6)$$

Finally, the mean entropy of a single oscillator may be written as $S = S_N/N$, and the temperature can then be determined by the standard relation $\partial S/\partial \overline{U} = 1/T$. Straightforward manipulation of eq. (6) thus yields a new distribution law,

$$\overline{U} = \frac{h\nu}{2}\frac{e^{h\nu/kT}+1}{e^{h\nu/kT}-1} = \frac{h\nu}{e^{h\nu/kT}-1} + \frac{h\nu}{2}. \qquad (7)$$

That law differs from Planck's earlier form only in attributing additional energy $h\nu/2$ to all oscillators, a point to which we shall return.

Planck's argument has so far dealt entirely with the distribution of oscillator energy, but what he requires is a distribution law for the field. In the past he had bridged the gap between them by calling on the classical relationship $u = 8\pi\nu^2 U/c^3$, but the introduction of discontinuous emission now made that inappropriate. Faced with the same difficulty, Einstein had suggested that the classical formula could be expected to apply to the average value of the field energy, and Ehrenfest had indicated how the factor $8\pi\nu^2/c^3$ might be obtained by direct quantization of the field. Planck, in 1912, found still another route from oscillator to field energy. Further refined by Bohr in and after 1913, it proved a primary determinant of the development of the old quantum theory.

Planck's first step was doubtless guided by the result he knew he required. Early in his paper he had noted that, physically, the mean energy of an oscillator must increase with the field intensity J. Within his theory that increase with field intensity was brought about by a decrease in the emission probability η, or by the corresponding increase in the probability $(1 - \eta)$ that an oscillator would pass without emission from one ring to the next. A plausible quantitative formulation of that insight he presented in the form: "The ratio of the probability that no emission takes place to the probability that it does take place is proportional to the intensity of the radiation which excites the oscillator."[9] Taking p to be the proportionality constant, that statement is equivalent to

$$\frac{1-\eta}{\eta} = pJ \quad \text{or} \quad \frac{1}{\eta} = 1 + pJ = 1 + \frac{4\pi}{3}pu.$$

Using the right-hand equation to replace $1/\eta$ in equation (5) yields

$$\bar{U} = \left(pJ + \frac{1}{2}\right)h\nu = \left(\frac{4\pi}{3}pu + \frac{1}{2}\right)h\nu, \qquad (8)$$

and inserting that relation in equation (7) gives

$$u = \frac{3}{4\pi p} \frac{1}{e^{h\nu/kT} - 1}. \qquad (9)$$

To complete his derivation, Planck requires only a value for p, and he finds it by employing a second hypothesis, one introduced immediately after the sentence quoted above:

> The value of the proportionality constant [p] we shall determine by application [of equation (9)] to the special case of very large radiation intensity. In that case, as we know, the standard formulas of classical dynamics and the Rayleigh radiation law which follows from them hold for all oscillator periods.[10]

High radiation intensity corresponds, however, to high temperature or low frequency. Under those conditions the classical equation $u = 8\pi\nu^2 U/c^3$ must hold. Applying it to the limiting form of equation (8) yields $p = 3c^3/32\pi^2 h\nu^3$, precisely the value required to transform equation (9) to the standard form of Planck's radiation distribution law. His second theory of black-body radiation has therefore achieved its primary goal. While bringing it to that point, Planck has, as we shall see, invented a technique that, in the very next year, would begin to become the Correspondence Principle.

Revising the *Lectures*

In much the form just sketched, the second theory is the one Planck presented in the second edition of his *Lectures*, published early in 1913.

Comparing it with its predecessor provides further examples of the extent and nature of the changes in Planck's thought during the almost seven years separating the two editions. In both, the first two chapters treat the thermodynamics of radiation without recourse to either Newtonian mechanics or electromagnetic theory, and Planck's text is virtually unchanged. The single significant exception is ten pages added to the second edition at the end of Chapter 2.[11] They are a prelude to a drastically altered treatment of natural radiation throughout the remainder of the book.

In the added pages Planck expands the radiation field in a Fourier series, squares it to evaluate the instantaneous field energy, and then averages the result over a time interval of arbitrary length, long compared with the periods of the significant field components. If the resulting measurable intensity is to be time independent, he points out, the amplitudes C_n of the terms in the Fourier series for the field must be individually small, and both they and the phase constants θ_n must "vary in an entirely irregular manner from one term number to another."[12] There is no mention of prohibited states, and Planck's attitude towards that notion becomes clear shortly afterwards when he reaches the part of his text where, in the first edition, he had emphasized the need for it. The following passage from the second edition is identical with that from the first except in two important respects: the clause in boldface, below, exists only in the first edition, the footnote only in the second.

> The proposition that in nature all states and events composed of countless uncontrollable parts are primitively disordered is the prerequisite **and also the sure guarantee** for the unequivocal determinateness of measurable events, both in mechanics and also in electrodynamics.... *
>
> ---
> * To avoid misunderstanding, I wish particularly to point out that the question of whether or not the hypothesis of primitive disorder is actually fulfilled everywhere in nature is not affected by the preceding considerations. Here it need only be shown that, where the hypothesis does not hold, natural events are not unequivocally determinate from the thermodynamic ("macroscopic") point of view.[13]

These changes have their primary effect in the last chapter, which in both editions deals with irreversible radiation processes, Planck's electromagnetic H-theorem. In the first edition the treatment is the elaborate one Planck had provided in 1899.[14] "Analyzing resonators" are required to make measurements of physically meaningful quantities, and natural radiation, as a prohibition of ordered states, is required to justify defining the coefficients in the series for the radiation intensity J as[15]

$$A_\mu^0 = \frac{2\nu_0^2}{\rho} \int d\nu \, C_{\mu+\nu} C_\nu \frac{\sin^2 \delta_\nu}{\nu^3} \sin(\theta_{\mu+\nu} - \theta_\nu),$$

$$B_\mu^0 = \frac{2\nu_0^2}{\rho} \int d\nu \, C_{\mu+\nu} C_\nu \frac{\sin^2 \delta_\nu}{\nu^3} \cos(\theta_{\mu+\nu} - \theta_\nu). \tag{10}$$

In the second edition the treatment is far briefer and simpler. Cavity resonators are undamped, and there are no analyzing resonators. The increase of entropy with time is no longer a consequence of electrodynamics but of straightforward probabilistic considerations. Natural radiation does not enter the discussion. Instead, equations (10) are replaced by the simpler pair,

$$A_\mu = \overline{C_{\mu+\nu} C_\nu \cos(\theta_{\mu+\nu} - \theta_\nu)},$$

$$B_\mu = \overline{C_{\mu+\nu} C_\nu \sin(\theta_{\mu+\nu} - \theta_\nu)},$$

which Planck explains as follows: "The lines over the expressions on the right denote the mean values over a narrow spectral range [of ν] for a given value of μ. If such mean values do not exist [i.e., if they are dependent on the range over which ν is varied], then there is no determinate spectral intensity."[16] Natural radiation, now called primitive disorder, has been transformed from a physical prohibition on permissible states to something very like a mathematical stipulation of randomness. Planck's subsequent autobiographical explanation of natural radiation is in the latter guise.[17]

Other, more drastic changes in Planck's second edition reflect his new perception of his achievement in the area he had made his own, the electromagnetic theory of black-body radiation. In the 1906 *Lectures*, after exhausting the consequences of thermodynamics for cavity radiation, Planck turned in Chapter 3 to a study of the mechanism that would permit the derivation of more detailed and precise results. That chapter's title was "Emission and Absorption of Electromagnetic Waves by a Linear Oscillator," and it contained a careful derivation, from the field equations and the laws of motion, of the behavior of a resonator in an arbitrary radiation field. Only after those derivations were completed did Planck turn, in Chapter 4, to the statistical arguments required to permit the selection of a unique entropy function. In the second edition the structure of the argument is inverted. Planck proceeds directly from the thermodynamic discussion of the first two chapters to the topic "Entropy and Probability," previously postponed to Chapter 4. Most of his earlier derivations of emission and absorption formulas disappear permanently from the book; those electrodynamic oscillator theorems that are preserved appear only after a full treatment of statistics.

Except for this change in position and logical function, the presentation of statistics in the second edition is very like that in the first, the most significant alteration occurring in Planck's discussion of the ideal gas. In 1906, gas theory had been introduced to illustrate Boltzmann's unfamiliar combinatorial approach. The statistics of oscillators had been treated next, and Planck had emphasized the apparently characteristic novelty—fixed cell size—that entered with it. In the second edition, a fixed cell size g is introduced at the start, thus subjecting the ideal gas to the quantum. When computing gas entropy, g remained an unspecified parameter "which must relate closely to the laws—at this time entirely unknown—of microscopic molecular interaction."[18] Meanwhile, it's value could be calculated from experiments on the absolute value of entropy (Nernst's third law). In the radiation case, g is simply set equal to h by "the quantum hypothesis"; the energy equation for an oscillator, which was not needed at this place in the first edition, is introduced to determine the shape of cells; and the mean energy and entropy of a collection of oscillators are then calculated as above. There is no longer even a pretence of supplying a mechanism for oscillator interaction. The hypothesis $g = h$ is explicitly the novelty upon which all else depends.

Only after the discussion of statistics does Planck, in his second edition, take up the interaction of oscillator and field, the subject with which he had begun his research on black-body theory almost twenty years before. As he does so, his text ceases, for the first time, to be modeled on that of his first edition, and it does not again return to that model except in brief discussions of natural units and of the Rayleigh–Jeans cavity. Absorption is handled by the much simplified electrodynamic argument sketched above, emission by the statistical hypotheses presented immediately after it. The electromagnetic H-theorem, now without the electrodynamic mechanism on which its resemblance to Boltzmann's theorem had depended, is greatly abridged, and it disappears entirely from the next revised edition of the book.[19] These drastic simplifications account for another striking sort of difference between Planck's first two editions. Unlike all other pioneering works known to me, the second edition of his *Lectures* is, despite some additional material, shorter than the first. Simultaneously, deprived of the original and consequential electrodynamic theorems that Planck had developed between 1894 and 1900, it is focused upon the quantum in a way that the first edition was not.

Some uses of the second theory

In view of what quantum concepts were to become, Planck's second theory has inevitably appeared a blind alley to both historians and physicists. Usually it enters their discussion as an index of Planck's conservatism, his inability to accept the more radical restriction on permissible energy levels he had himself introduced at the end of 1900. But, though both Planck and his second theory were more conservative than a number of their contemporaries,[20†] there are several difficulties with those evaluations. For Planck himself the second theory was not a retreat but a radical step, the first theory from his pen to allow any place at all for discontinuity. More to the point, before Bohr's application of restricted energy levels to *absorption* spectra in 1913, none of the phenomena thought to demand the quantum had a bearing on the choice between energy quantization and Planck's new probabilistic-emission approach. Aesthetic considerations may have seemed to demand that emission and absorption be treated symmetrically as in classical theory. But Planck's theory worked, and there was, as we shall see at once, qualitative evidence elsewhere for its primary oddity, the statistical emission process. Finally, whatever its ultimate fate, the second theory did explain some puzzling phenomena with sufficient precision to provide a base for exploring others. During the five years after it appeared, several longer-lasting contributions to the quantum theory were based upon it. Though even brief exploration of these points will carry us beyond the temporal bounds of this book, it will nevertheless provide an appropriate point at which to close. The reception and the use of Planck's second theory illustrate the state of the quantum on the eve of the first World War with a concreteness impossible to achieve within the survey format of the preceding chapter.

One source of the second theory's appeal, regularly mentioned in subsequent discussions,[21] was that it reduced the magnitude of the break with classical theory. The point is not that discontinuity was restricted to emission, but rather that vibrators, whether Planck oscillators or massy atoms, could possess all possible energies. To preserve the energy continuum was not, however, to rescue classical theory, and the primary attractions of the second theory were probably of another sort. Like many of the other new formulations sketched in the last chapter, Planck's second theory suggested possible new applications for the quantum. In his very first paper on the subject, he mentioned several, two of which seemed to demand something very like his probabilistic-emission mechanism.

One, of course, was radioactivity, which may conceivably even have been the source of his probabilistic decay constant η. Planck attempted no theory, pointing out only that "radioactivity seems effortlessly to fit the hypothesis of 'quantum emission'" and that that hypothesis might explain why α- and β-rays could apparently be emitted only with certain predetermined velocities.[22] Concerning a second topic, the interaction of electrons with matter and fields, Planck had more to say. Oscillators must, he suggested, be stimulated not only by radiation but also by collisions with free electrons. Whatever the velocity with which an electron struck an oscillator, its recoil might well be governed by a process like quantum emission. If so, the electron would recede from impact with a velocity governed only by the oscillator's frequency, energy, and charge. Having their velocities and energies determined by the oscillator's characteristics would, Planck pointed out, deprive "free" electrons of their own degrees of freedom, a fact that might account for their well-known failure to contribute to specific heats. Lacking a frequency, free electrons could not be directly subjected to the quantum, a fact that had led Lorentz to distinguish them categorically from oscillators and to insist that they were governed in full by classical laws. Planck, who had repeatedly rejected that distinction in correspondence with Lorentz,[23] was pointing out that the quantum could affect free electrons through interactions in which the intrinsic frequency of oscillators played a role.

Of even more obvious promise was Planck's application of probabilistic quantum emission to the photoelectric effect. Triggered by incident light or x-rays, electron emission would nevertheless be governed only by the oscillator's characteristics, not by the temperature or the intensity of the incident radiation. As in the radiation case, furthermore, the energy of the emitted electron would be drawn all at once from the emitting oscillator, to which it would be gradually restored by the field.[24] Returning to the problem at the Solvay Congress late in 1911, Planck further elaborated this view. Even at absolute zero, he pointed out, oscillators governed by his new theory would possess finite energies randomly distributed between zero and $h\nu$. A very weak field would therefore quickly raise the energy of oscillators near the top of the range to $h\nu$, at which point it would, with specified probability, be transformed to the translational energy of an emitted electron. An increase in the strength of the incident field would, according to this theory, increase only the number of emitted electrons, not their energy.[25] Even if Einstein's photoelectric law proved correct, there would be no reason to accept its derivation from light-particles. Two days after the Congress closed, Wien attempted by letter to persuade Stark that Planck's new theory provided a full explanation of secondary emission.[26] Thereafter, it figured repeatedly in attempts to explain the success of Einstein's law.[27]

The aspect of Planck's second theory that attracted the most and the longest lasting attention was, however, the extra term $h\nu/2$ in his new formula for oscillator energy, equation (7) above. Planck himself had not expected it to be experimentally consequential: it did not appear at all in the distribution law for radiant energy, and it would also, he thought, disappear when the oscillator energy was differentiated by temperature in the computation of specific heats.[28] But, as Einstein and Otto Stern (1888–1969) pointed out early in 1913, the latter conclusion depended on the assumption that the frequency of all motions contributing to specific heat were independent of temperature. Doubtless that was the case for Planck oscillators and probably also for the vibrations of an atom or ion in a solid lattice, but it could not be presumed to apply, for example, to the rotational frequency of a diatomic molecule of a gas. Writing the rotational energy of a molecule as $E_r = \frac{1}{2}J(2\pi\nu)^2$ where J is the molecule's moment of inertia, Einstein and Stern derived expressions for rotational specific heat both from the standard and from the new form of Planck's distribution law. Comparing their results with recent measurements on the specific heat of hydrogen at low temperatures, they concluded that "the existence of a zero-point energy of size $h\nu/2$ [is] probable."[29]

That conclusion was immediately noted and explored by the rapidly growing group of physical chemists concerned with low temperature research. A few of them, furthermore, quickly applied it also to the search for quantum effects on the low-temperature specific heats of monatomic ideal gases. Because the energy in such gases is entirely translational, the choice of a frequency to appear in the distribution law was inevitably somewhat arbitrary. But the obvious candidates (for example, number of mean free paths traversed per unit time) all gave similar results, and none were in satisfying agreement with experiment during the period when only Planck's original distribution law was applied. By April 1913, however, it was recognized that the discrepancies already known would be greatly reduced by the introduction of a zero-point energy.[30] A year later, in the light of additional research, Arnold Eucken (1884–1950) felt able to report "with full certainty that, of the existing formulas [for translational energy at low temperatures], those without zero-point energy are certainly not satisfactory, while those which include a zero-point energy at least lead to no contradiction."[31] After this report, more and more research made use of a zero-point energy; the concept has a continuous history to this day. Its further pursuit is, however, irrelevant here, for by the end of the decade in which it was first announced, the concept was gradually becoming independent of Planck's second theory, within which it had originated.[32]

Other aspects of that theory had meanwhile made possible significant contributions to the analysis of atomic and molecular spectra. Planck's introduction of multi-quantum emission provides an important example: if an oscillator crossing the nth elliptical boundary in the phase plane emitted at all, the second theory required that it radiate just n quanta and that emission terminate with the oscillator in its ground state of zero energy. Early in 1913 that conception of emission was central to the first of the viewpoints developed by Niels Bohr (1885–1962) in his efforts to understand spectral emission and to quantize the multiple stationary states which that understanding appeared to require.

Some six months before he conceived of applying Rutherford's planetary atom model to spectra, Bohr had used the quantum to eliminate that model's mechanical instability and to determine such parameters as the sizes, the rotational frequencies, and the ionization and binding energies of various atoms and molecules.[33†] Only around Christmas 1912 do his letters show signs of a concern with spectra and of a consequent realization that atoms might possess not just one, but a succession of stationary states. Not until February 1913 was his attention for the first time called to the likely relevance of the Balmer formula. Bohr immediately recognized how equations he had previously derived for the stable ground state might, by the introduction of a running index n, be manipulated to yield that formula, including the numerical value of the Rydberg constant. What he still required was a conception of the emission process that could justify the steps in his derivation, and for that, in his first paper on the subject, he turned to Planck:

> Now the essential point of Planck's theory of radiation is that the energy radiation from an atomic system does not take place in the continuous way assumed in the ordinary electrodynamics, but that it, on the contrary, takes place in distinctly separated emissions, the amount of energy radiated out from an atomic vibrator of frequency ν in a single emission being equal to $\tau h\nu$, where τ is an entire number, and h is a universal constant.[34†]

A footnote at the end of that passage cites three of Planck's four published papers on the second theory (only his Solvay address is omitted). At the time they were a unique source for the multi-quantum emission process of which Bohr quickly made two applications. Not quite compatible with one another or with the more developed analysis supplied later in his paper, they display an essential formative stage in the emergence of his model. (The paper was written at great speed: a draft was mailed to Rutherford less than a month after Bohr first recognized the relevance of Balmer's formula.) Qualitatively, Bohr appears to have supposed that an entire spectral series was emitted when a single electron at rest outside the atom was captured through a multi-quantum transition to the ground state. To formulate that conception quantitatively, Bohr next imagined that, when an electron made the transition from rest to the nth quantized orbit, it emitted just n quanta of frequency $\tfrac{1}{2}\omega_n$ where ω_n was the rotation frequency of the electron in that orbit. (The factor of $\tfrac{1}{2}$ was, Bohr suggested, required to average the electron's initial and final frequencies, 0 and ω_n.) The total emitted energy would then be just $\tfrac{1}{2}nh\omega_n$, and it was set equal to the difference between the electron's energy at infinity and in the nth orbit. Straightforward manipulation led to what have since been the standard energy levels for the Bohr hydrogen model.[35]

For Bohr, both that emission mechanism and that method of quantizing were preliminary. No signs of them appear in his later papers, and even in this one they are followed by a second discussion in which no signs of the multi-quantum emission mechanism appear. In replacing it, however, Bohr seems again to have drawn on Planck, borrowing a technique introduced by the latter in his 1912 article on the second theory. Further elaborated by Bohr with an important addition by Einstein, it became the Correspondence Principle, a constant guide in the development of the quantum theory until 1926.

Immediately after concluding the derivation just sketched, Bohr pointed out that there was something wrong with it. The frequency at which a planetary electron rotates is, he remarked, a function of its energy. An electron making the transition from rest to the nth orbit could not, therefore, emit n quanta at the same frequency, for its own mechanical frequency would change with the emission of each successive quantum. In a transition between energy levels, he concluded, only a single quantum could be emitted. If the transition were between levels with energy W_n and W_m, then the radiated frequency would be governed by the equation $h\nu_{nm} = W_n - W_m$.

This is, of course, the emission process for which Bohr is known, but to employ it he needed a new quantization technique. By analogy to Planck's phase-plane ellipses, he determined energy levels by a relation of the form $W_n = f(n)h\omega_n$. To derive a formula like Balmer's, he pointed out, $f(n)$ would have to be proportional to n, so that the energy W_n must be $cnh\omega_n$. What remained was to determine the constant c, and Bohr did that by the since standard elementary argument that compares the classical and quantum frequencies in the high-energy, low-frequency limit, where the two should be identical. That is just the method Planck had introduced in 1912 to determine the parameter p required to complete his second theory. The paper in which he did so is one Bohr cited, and this part of Bohr's approach was likely drawn from it.[36]

By 1916 Planck's formulation of his second theory had made an even more decisive contribution to the evolving quantum theory of atoms, molecules, and their spectra. Because many people and papers were involved, not even an historical sketch of this development can be provided here, but its beginning and initial fruits are worth noting. As pointed out in Chapter V, the first edition of Planck's *Lectures* included a two-page discussion of an alternative derivation of his distribution law for resonator energy. If f were a resonator's moment and g the corresponding conjugate momentum, then the f-g phase plane could be divided into elliptical rings of area h, and the probability of obtaining any particular energy distribution could be determined.[37]

In the 1906 *Lectures*, that discussion was very much an aside, interesting because it provided a possible bridge between the phase-space treatment demanded by Planck's previous discussion of gases and the quantized energy-continuum treatment Planck next developed for resonators. But in the second edition, built around Planck's second theory, a generalization of the phase-plane treatment was fundamental. Phase space entered early in the discussion of probability and entropy, and the "quantum hypothesis," which now applied to molecules as well as resonators, was at once introduced in phase-space terms:

> If $\phi_1, \phi_2, \phi_3, \ldots$ are the "generalized coordinates" and $\psi_1, \psi_2, \psi_3, \ldots$ the corresponding "momentum coordinates" or "moments" which determine the instantaneous state of a single molecule, then the phase space ...[required for probabilistic computations] contains as many dimensions as there are coordinates ϕ and momenta ψ for each molecule. The elementary region of [equal] probability is, in classical statistical mechanics identical with the (macroscopically) infinitely small element of phase space:
>
> $$d\phi_1\,d\phi_2\,d\phi_3\cdots d\psi_1\,d\psi_2\,d\psi_3\cdots.$$
>
> According to the quantum hypothesis, on the other hand, every elementary region of probability possesses a definite finite magnitude:
>
> $$G = \int d\phi_1\,d\phi_2\,d\phi_3\cdots d\psi_1\,d\psi_2\,d\psi_3\cdots, \qquad [11]$$
>
> the size of which is the same for all regions but which in other respects depends upon the nature of the molecule or system in question. The form and position of the individual elementary regions are determined by the limits of the integral and must be worked out specially in each individual case.[38]

Dealing with the molecules of a gas, Planck felt able to compute only the magnitude of the elementary cell, determined from experiments on the absolute value of entropy. To do more would, be said, require access "to the laws—at this time entirely unknown—of microscopic molecular interaction."[39] But for the one-dimensional oscillator Planck went farther. Equation [11] was explicitly rewritten as[40]

$$\iint d\phi\,d\psi = h, \qquad (12)$$

which fixed the size of cells. The oscillator equation, which determined curves of constant energy in the phase plane, fixed their elliptical shape. During the three years after that treatment appeared, Planck tried to extend it for application to the rotator and to molecules with several degrees of freedom. In 1915, for example, he pointed out that the elementary phase-space cells must be bounded by surfaces of constant energy, $u_0, u_1, u_2, \ldots$, and that the content of equation [11] above, might therefore be further specified by the relationship[41]

$$\int\!\!\int_{u=0}^{u=u_n}\!\!\cdots\int d\phi_1\,d\phi_2\cdots d\phi_f\,d\psi_1\,d\psi_2\cdots d\psi_f = (nh)^f.$$

By the end of the year, that effort culminated in a paper called "The Physical Structure of Phase Space," a title that clearly shows where and how Planck was trying to localize the problem of the quantum.[42]

However other physicists may have felt about Planck's second theory, the second edition of the *Lectures* and the papers that followed it were widely and closely read. In the event, they provided important clues for the growing group that, since 1911, had been seeking a coherent formulation of quantum laws, one that would cover Planck oscillators, rotators, three-dimensional vibrators, sometimes the photoeffect, and from 1913, the Bohr atom. During 1915, there appeared three independent versions of a solution that was to prove canonical for over a decade. All three authors—Arnold Sommerfeld, William Wilson (1875–1965), and Jun Ishiwara (1881–1947)—explicitly cited either the second edition of Planck's *Lectures* or his phase-plane approach, and at least two of them modeled essential elements of their approach on his.[43] Though they shared other sources (especially the Bohr atom) and adopted different approaches, what they took from Planck is clear. The elementary cells of equation [11] are bounded between surfaces of constant energy, and the total area contained by the nth such curve is

$$\int\int_{u=0}^{u=u_n} d\phi\, d\psi = nh. \qquad (13)$$

If a frequency can be attributed to the corresponding system, then that system must possess a natural cycle after which the coordinate ϕ either returns to its initial value ϕ_0 or else increases to $\phi_0 + 2\pi$. Under these circumstances, eq. (13) may be rewritten

$$\int \psi\, d\phi = nh,$$

where the integral is to be taken over a single cycle. All three authors introduced this easy reformulation of Planck's way of determining cells in the phase plane, and all but Ishiwara suggested that it could be applied separately to each of the degrees of freedom of multi-dimensional systems. What resulted was, of course, the famous Wilson–Sommerfeld phase-integral quantum conditions, important both as a highly general means of quantization and for the clues they provided to future quantum-theoretical advance. Within a year of their announcement, mathematical physicists concerned with the quantum had discovered what would soon become their basic tool. It was Hamilton–Jacobi theory, long available to them but only in unfamiliar advanced texts on celestial mechanics.

The fate of the second theory

When Planck addressed the Solvay Congress late in 1911, he showed first how his black-body law could be obtained by restricting the energy of oscillators to integral multiples of $h\nu$ and then how that restriction could be avoided by means of his second theory. In the discussion that followed, both derivations were treated with respect, and at the meeting's close, the French physicist Marcel Brillouin (1854–1948) voiced a reaction that is likely to have been widespread:

> It seems certain that we must from now on include among our physical and chemical conceptions that of discontinuity, of an element varying by jumps of which we had no notion a few years ago. How is it to be included? That is what I see less well. Will it be in the first form proposed by M. Planck, despite the difficulties which that raises, or will it be in the second? Will it be in the form proposed by M. Sommerfeld, or in some other still to be sought? I do not yet know; each of these forms is well adapted to one group of phenomena, less well to others.[44]

Through Brillouin prefaced those remarks with the observation that they might seem conservative to the youngest members of his audience, only Einstein is likely to have disagreed categorically, and Einstein took Planck's second theory seriously enough to have joined Stern in publishing the first concrete evidence that favored it over its predecessor.[45†] Only discontinuity was certain; about the manner of its entry nothing final could yet be said; very likely several of the proposals currently available captured aspects of some more definitive, still not quite visible, formulation. Such attitudes are what assured the second theory a life-span sufficient to permit the contributions just sketched.

What changed the status of the second theory was, of course, the application of the quantum to spectra. Planck recognized the difficulty early. Bjerrum's quantization of the diatomic rotator in 1911 had quickly led to successful analyses of the infrared rotation spectrum. In 1914 Planck pointed out that it was "scarcely possible to understand," in terms of his second theory, how emission from a uniformly rotating dipole in a stationary radiation field "could proceed by quanta." That remark prefaced his short-lived attempt to eliminate the quantum entirely from the interaction between radiation and matter, and to locate it instead in a non-classical mechanism for collisions between material particles.[46] Similar difficulties for the second theory followed from Bohr's treatment of spectra, and they were especially urgent because the Bohr theory displayed the identical discontinuous process at work in absorption and in emission. As Bohr's conception of atomic structure and spectral emission was accepted and became the primary tool for quantum research, the second theory gradually disappeared from the literature of physics.

The first to abandon it were, of course, those primarily involved with spectra. Planck's second theory is mentioned neither in Bohr's "Quantum Theory of Line Spectra" (1918) nor in Sommerfeld's *Atomic Structure and Spectral Lines* (1st ed., 1919), the two main texts for the generation of quantum physicists trained during the half-dozen years after they appeared.[47] Among physical chemists, on the other hand, given their special concern with the zero-point energy, the second theory seems to have remained a topic for discussion into the middle 1920s.[48] Elsewhere there was room for disagreement, and its existence is amusingly illustrated in a card Planck sent to Ehrenfest in the spring of 1915. "Naturally," Planck, began, "I would be glad to send you a copy of the proof of my publication on the rotating dipole." Anticipating his correspondent's reaction, however, Planck then continued:

> Of course my project can't be carried through without resort to hypotheses, and I fear that your hatred of the zero-point energy extends to the electrodynamic emission hypothesis that I introduced and that leads to it. But what's to be done? For my part, I hate discontinuity of energy even more than discontinuity of emission. Warm greetings to you and your wife.[49]

That Planck was not merely being stubborn is indicated by the continuing discussion of his second theory during the next half-dozen years. In 1919 the second enlarged edition of *An Elementary Account of the Foundations of the Quantum Theory* by Siegfried Valentiner (1876–1958) treated the second theory as the fundamental quantum formulation, though recognizing that it was incompatible with the views of Bohr and Einstein, which it also discussed.[50] A standard American monograph, "The Quantum Theory," published in 1920 by E. P. Adams (1878–1956), derived the distribution law for oscillator energy both from what its author called "Bohr's theory of stationary states" and from what he named "Planck's 'Cell' Theory." Only from the second enlarged and revised edition of that monograph, published in 1923, did the second theory disappear, as it did also from the second edition of Jeans's *Report*, published the following year.[51]

Planck's own evaluation of his second theory followed a similar course. The second and third editions of his *Lectures*, the latter a verbatim reprint published in 1919, developed only the hypothesis of quantum emission. The fourth, published in 1921, qualified his conviction by presenting parallel accounts of two alternative formulations: the first in terms of quantized energy levels, the second his quantum emission treatment. But the fifth edition, published only two years later, made energy quantization the standard form and reserved only a few passing remarks for the alternative he had invented twelve years before. During 1922 the Stern–Gerlach experiment had persuaded him that, "at least in a special case, precisely these states [the discrete energy levels demanded by Bohr's theory] are realized predominantly in nature."[52] Faced with such concrete and direct experimental evidence of the existence of stationary states, Planck, too, abandoned the second theory.

Planck, of course, had been through all this before. Though the theory presented in the first edition of his *Lectures* proved far more consequential than the one described in its successor, the two had similar histories and played similar roles in the development of physics. Neither called for the existence of a discrete energy spectrum, the characteristic that in restrospect has seemed the essential characteristic of any quantum theory. Each nevertheless accounted for one or more esoteric aspect of nature with a simplicity and precision that gave its author and some of his contemporaries the confidence required to attempt its further development. Though these attempts, in each case, soon resulted in fundamental modification of the theory that had permitted their design, they had meanwhile assisted in the identification of additional quantum phenomena and in a deeper understanding of the nature of a still emerging new physics. While they retained that role, Planck remained a primary contributor to the development of a theory he never came quite to believe. Among the group concerned with the quantum, his research continued to command attention and respect throughout the first two decades of this century. By their close, Max Planck was in his sixties.

NOTES

Abbreviations and format

The abbreviations used in citing the periodical literature below are from the convenient list in Karl Scheel, "Physikalische Literature," *Handbuch der Physik*, ed. H. Geiger and K. Scheel, Vol. I (Berlin, 1926), pp. 180–186. These abbreviations are generally identical with those listed annually by the German abstracting journal, *Physikalische Berichte*, from its origin in 1920. Titles of a few journals no longer current when these lists were prepared are given in full except for the especially cumbersome titles of the annual reports of the meetings of the British Association for the Advancement of Science, which are cited as *Report of the British Association*, followed by the year of the meeting. Because no ambiguity is thereby created, specification of a journal by series number or, in the case of *Ann. d. Phys.* by editor's name, is omitted.

Many of the articles cited are also available in an edition of the author's collected works, and this form will often be more readily accessible to readers. Ordinarily, therefore, articles are cited below in both forms, and, for the two authors most frequently cited, references to collected works are by volume number and page or pages, alone. Thus, reference to the original publication of a paper by Planck will regularly be followed by a notation like, I 493–504. It indicates that the article in question is to be found on pp. 493–504 of the first volume of Max Planck, *Physikalische Abhandlungen und Vorträge*, 3 vols. (Braunschweig, 1958). A similar notation following citation of a paper by Boltzmann is a reference to the *Wissenschaftliche Abhandlungen von Ludwig Boltzmann*, ed. Fritz Hasenöhrl, 3 vols. (Leipzig, 1909, and New York, 1968). A few other similar abbreviations, used less frequently (usually only within a single chapter), will be introduced here and there below. Occasionally an article to be cited will also have appeared in one or more other places, a particular likelihood in the case of an author like H. A. Lorentz who worked in a country where the native language was neither English, French, nor German. Except in cases where some point of direct historical interest depends upon it, however, references to such articles are to their first appearance in a major scientific journal. Some additional information about other places of publication will be found in the bibliography that follows, but no attempt has been made at a complete listing.

All articles are cited in full the first time they are referred to in each chapter. Thereafter they are cited by abridged title together with a parenthetical reference which permits their location within the bibliography. Thus, (Planck, 1900b) refers to the second of the listed articles published by Planck in 1900, a notation also used when citing articles for the first time within footnotes. The bibliography also provides available information about the date on which a published article was read, submitted, or received.

Most of the notes which follow consist entirely of citations, acknowledgments, brief comments on available secondary sources, or transcriptions in the original language of unpublished manuscript material quoted in translation in the main text. A few, however, include substantive discussion that significantly amplifies or qualifies the text to which it is attached. Notes of the latter sort have been indicated in the text by a dagger (†) after the footnote number and below by the same mark preceding the number. Readers should realize, however, that the distinction between substantive and non-substantive notes is, inevitably, sometimes subjective.

Notes to Preface

1. See especially, Hans Kangro, *Vorgeschichte des Planckschen Strahlungsgesetzes* (Wiesbaden, 1970); Martin J. Klein, "Max Planck and the Beginning of the Quantum Theory," *Archive for History of Exact Sciences*, 1 (1962), 459–479, and, "Planck, Entropy, and Quanta, 1901–1906," *The Natural Philosopher*, 1 (1963), 83–108. Other relevant articles by these and other authors are cited here and there below and are assembled in the first section of the bibliography. For a more exhaustive survey of existing secondary literature see, however, the bibliography in Kangro's volume.

2. One result of my own earlier misreading should be acknowledged at once. In rejecting the suggestion of T. Hirosige and S. Nisio that Planck's papers on his second theory had a special role in the development of Bohr's atom model, I once wrote that Bohr was "unlikely to have drawn anything from them he could not have taken as well, or ever better, from Planck's original formulation" (J. L. Heilbron and T. S. Kuhn, "The Genesis of the Bohr Atom," *Historical Studies in the Physical Sciences*, 1 (1969), 211–290, quotation from p. 268n.). That remark, to which I return in Chapter X, now seems to me clearly mistaken.

3. The project is described and its findings catalogued in: T. S. Kuhn, J. L. Heilbron, P. L. Forman, and Lini Allen, *Sources for History of Quantum Physics: An Inventory and Report* (Philadelphia, 1967).
4. That grant also supported much earlier background study as well as part of my work on the publications cited in notes 2 and 3, above, and on a review article, "The Turn to Recent Physics," *Isis*, *58* (1967), 409–419.

Notes to Chapter I

1. There is no full and balanced study of the history of black-body theory before Planck. But Daniel Siegel of the University of Wisconsin has been working on the subject for some time, and I am indebted to him for comments on an earlier and considerably fuller version of the sketch that follows. Hans Kangro, *Vorgeschichte des Planckschen Strahlungsgesetzes* (Wiesbaden, 1970), includes an excellent account, simultaneously pioneering and authoritative, of the development of the relevant experimental techniques and measurements. My remarks on those subjects, both in this chapter and in Chapter IV, are largely derived from his far more detailed account, now available in English as *History of Planck's Radiation Law* (London, 1976). A brief account of the development of relevant theory, including sketches of the derivations of the main black-body laws, is included in Chap. 12, "Classical Radiation Theory," of E. T. Whittaker's *History of the Theories of Aether and Electricity: The Classical Phase*, revised and enlarged ed. (Edinburgh, London, etc., 1951). More conceptually oriented accounts of some aspects of nineteenth-century theories of radiant heat will be found in Ernst Mach, *Die Principien der Wärmelehre* 2nd ed. (Leipzig, 1900), pp. 125–148, and S. G. Brush, "The Wave Theory of Heat," *British Journal for the History of Science*, *5* (1970), 145–167. Since systematic accounts of classical black-body theory have almost disappeared from the literature of physics, the first two chapters of Max Planck, *The Theory of Heat Radiation*, trans. Morton Masius (Philadelphia, 1914, and New York, 1959), still provide a useful account. The same material appears in all the many German editions (often very different in other respects) of Planck's *Vorlesungen über die Theorie der Wärmestrahlung*, from the second edition of which the preceding is translated.
2. G. R. Kirchhoff, "Über den Zusammenhang zwischen Emission und Absorption von Licht und Wärme," *Monatsberichte der Akademie der Wissenschaften zu Berlin*, 1859, pp. 783–787, and "Über das Verhältnis zwischen dem Emissionsvermögen und dem Absorptionsvermögen der Körper für Wärme und Licht," *Ann. d. Phys.*, *109* (1860), 275–301; also available in Gustav Kirchhoff, *Gesammelte Abhandlungen* (Leipzig, 1882), 566–597. A somewhat less general formulation of the same results had been published in 1858 by the Scottish physicist Balfour Stewart. Kirchhoff stated his law in terms of the intensity of plane waves; intensity of spherical waves is used in the text that follows. The latter will be required later, and an attempt to preserve both forms would introduce irrelevant confusion. On this subject see Daniel

Seigel, "Balfour Stewart and Gustav Robert Kirchhoff: Two Independent Approaches to 'Kirchhoff's Radiation Law'," *Isis*, *67* (1976), 565–600.
3. Ludwig Boltzmann, "Über eine von Hrn. Bartoli entdeckte Beziehung der Wärmestrahlung zum zweiten Hauptsatze," and "Ableitung des Stefan'schen Gesetzes betreffend die Abhängigkeit der Wärmestrahlung von der Temperatur aus der Elektromagnetischen Lichttheorie," *Ann. d. Phys.*, *22* (1884), 31–39, 291–294; III, 110–121. For the radiometer controversy see S. G. Brush and C. W. F. Everitt, "Maxwell, Osborne Reynolds, and the Radiometer," *Historical Studies in the Physical Sciences*, *1* (1969), 105–125, and A. E. Woodruff, "William Crookes and the Radiometer," *Isis*, *57* (1966), 188–198.
4. J. C. Maxwell, *A Treatise on Electricity and Magnetism* (Oxford, 1873), §793.
5. Josef Stefan, "Über die Beziehung zwischen der Wärmestrahlung und der Temperatur," *Wiener Ber. II*, *79* (1879), 391–428.
6. Wilhelm Wien, "Eine neue Beziehung der Strahlung schwarzer Körper zum zweiten Hauptsatz der Wärmetheorie," *Berl. Ber.*, 1893, pp. 55–62.
7. The remainder of this section closely follows *Vorgeschichte* (Kangro, 1970), Chapts. 1–4. References to the very numerous original papers will be found there and in Kangro's very useful bibliography. Only the ones discussed here are cited below.
8. It is Plate IV, Fig. 3, from S. P. Langley, "Observations on Invisible Heat-Spectra and the Recognition of Hitherto Unmeasured Wave-lengths, Made at the Allegheny Observatory," *Phil. Mag.*, *21* (1886), 394–409, also reproduced by Kangro.
9. W. A. Michelson, "Essai théorique sur la distribution de l'énergie dans les spectres des solides," *Journ. de Phys. et le Radium*, *6* (1887), 467–479.
10. H. F. Weber, "Untersuchungen über die Strahlung fester Körper," *Berl. Ber.*, 1888, pp. 933–957.
11. Friedrich Paschen, "Über Gesetzmässigkeiten in Spectren fester Körper und über eine neue Bestimmung der Sonnentemperatur," *Göttinger Nachr.*, 1885, pp. 294–305, and "Über Gesetzmässigkeiten in den Spectren fester Körper, erste Mittheilung," *Ann. d. Phys.*, *58* (1896), 455–492.
12. Wilhelm Wien, "Über die Energievertheilung im Emissionsspectrum eines schwarzen Körpers," *Ann. d. Phys.*, *58* (1896), 662–669.
13. For these later measurements, about which a bit more will be said in Chapter IV, below, see *Vorgeschichte* (Kangro, 1970), Chapt. 7. Planck's derivation of Wien's distribution law is discussed in Chapter III.
14. Much information about the development of thermodynamics before about 1855 is included in D. S. L. Cardwell, *From Watt to Clausius* (Ithaca, 1971), and Robert Fox, *The Caloric Theory of Gases from Lavoisier to Regnault* (Oxford, New York, etc., 1971). Unfortunately, very little has been written about the subject's subsequent evolution.
15. T. S. Kuhn, "Conservation of Energy as an Example of Simultaneous Discovery," *Critical Problems in the History of Science*, ed. Marshall Clagett (Madison, 1959), pp. 321–356, and *Watt to Clausius* (Cardwell, 1971), pp. 235, 241.
16. *Ibid.*, pp. 247–249.

17. R. J. E. Clausius, *Abhandlungen über die mechanische Wärmetheorie*, 2 vols. (Braunschweig, 1864–67). The second volume deals with the mechanical theory of electrical phenomena, especially with the application of thermodynamic arguments to electricity. The English translation is of the first volume only: *The Mechanical Theory of Heat*, ed. T. Archer Hirst (London, 1867). The French translated both volumes: *Théorie méchanique de la chaleur*, trans. F. Folie, 2 vols. (Paris, 1868–69).
18. R. J. E. Clausius, *Die mechanische Wärmetheorie*, 2nd rev. and enlarged ed., 3 vols. (Braunschweig, 1876–89), quotation from Vol. I, pp. v–vi. The first volume appeared in 1876, the second in 1879. The third volume, of which the subtitle is *Die kinetische Theorie der Gase*, was a posthumous publication edited by Max Planck and Carl Pulfrich. Bibliographical information about it is inconsistent, but it seems to have appeared in two installments, the first 48 pp. in 1889 as part of the second edition, and the full work, xvi + 264 pp., with the third edition in 1891. The first volume of the third edition was published in 1887, the second in 1879. Again there were English and French translations: the English, by Walter R. Browne, was from the first volume of the 2nd ed. (London, 1879), the French, by F. Folie and E. Ronker was from the first two volumes of the third edition (Paris, 1888–93).
19. In an autobiographical fragment to be quoted more fully below, Planck spoke of coming upon "the papers of Rudolph Clausius...[and working his] way deeply into them" in Berlin in 1877. It is likely that what Planck referred to as "the papers" included Clausius's second edition.
20. The biographical material here and elsewhere in this chapter is familiar. A convenient source, which includes some new information, is Armin Hermann, *Max Planck in Selbstzeugnissen und Bilddokumenten* (Hamburg, 1973). Pages 114–125 of *Vorgeschichte* (Kangro, 1970) are a rich source with respect to Planck's scientific education and early teaching experience. They include (pp. 116–119) material from notes Planck made during 1878 when reading the German edition of Tyndall's *Heat as a Mode of Motion*, thus recording what is likely to have been his first exposure to problems of radiant heat and light.
†21. Max Planck, *Wissenschaftliche Selbstbiographie* (Leipzig, 1948), p. 7; III, 374; *Scientific Autobiography and Other Papers*, trans. F. Gaynor (New York, 1949), p. 14. What Planck meant when he referred to the laws of thermodynamics as absolute is a problem still to be worked through in detail, but much very useful information and discussion are included in E. N. Hiebert, *The Conception of Thermodynamics in the Scientific Thought of Mach and Planck*, Wissenschaftlicher Bericht Nr. 5/68, Ernst Mach Institut (Freiburg i. Br., [1968]). Though Planck may once have thought that the first and second laws had the status of a priori generalizations, he had concluded by the mid-1880s that "the entropy principle, like that for energy, arises from the observation of certain cyclic processes" (Hiebert, p. 25; Planck (1887a), p. 503; I, 197). For him, as for Ernst Mach whose influence he acknowledged, the relevant observations were the minimal and readily accessible experiences which appear to bar the occurrence of perpetual motions of the first and second kinds. Thermodynamics was thus fundamentally empirical in the same sense as, say, mechanics, but the former was a great deal more securely

based. Even that situation, Planck indicated, need not be permanent: "It is by no means impossible that, if science reaches a higher level of development, another empirical law—perhaps the mechanical world view—will have a better claim to serve as the basis for [scientific] deductions" (Hiebert, p. 22, from Planck (1887b), p. 142). Additional useful information about Planck's philosophy of science is to be found in Stanley Goldberg, "Max Planck's Philosophy of Nature and His Elaboration of the Special Theory of Relativity," *Historical Studies in the Physical Sciences*, 7 (1976), 125–160, of which an advanced draft was kindly supplied to me by the author.

22. *Vorgeschichte* (Kangro, 1970), p. 115.
23. *Selbstbiographie* (Planck, 1948), pp. 8f.; III, 375f.; Gaynor trans., pp. 15f.
24. R. J. E. Clausius, "Über eine veränderte Form des zweiten Hauptsatzes der mechanischen Wärmetheorie," *Ann. d. Phys.*, 93 (1854), 481–506. This paper constitutes the fourth chapter of the first edition of Clausius's *Mechanische Wärmetheorie*.
25. R. J. E. Clausius, "Über verschiedenen für die Anwendung bequeme Formen der Hauptgleichungen der mechanischen Wärmetheorie," *Ann. d. Phys.*, 125 (1865), 353–400. This paper was published too late for inclusion in the first edition of Clausius's book, but it was included as the ninth chapter in the English translation.
26. See, in particular, *Mechanische Wärmetheorie*, 2nd ed. (Clausius, 1876), I, pp. 94, 224, and 3rd ed. (Clausius, 1879), I, pp. 94, 222. In both places Clausius, with the aid of equation (4), gives $dQ \leq T\,dS$ as the form equivalent to equation (3).
†27. Max Planck, *Über den zweiten Hauptsatz der mechanischen Wärmetheorie* (Munich, 1879); I, 1–61. See especially the introduction and the prefatory paragraphs to Section I. These differences between Planck's and Clausius's approaches to the second law are also discussed in *Mach and Planck* (Hiebert, 1968), pp. 10–16. Their importance should not, however, disguise their subtlety. What was at stake was a difference in emphasis or in identification of essentials. At the end of his "bequeme Formen" (Clausius, 1865), Clausius wrote: "The second fundamental theorem, in the form which I have given it, asserts that all transformations occurring in nature may take place in a certain direction...by themselves, that is, without compensation; but that in the opposite...direction they can only take place in such a manner as to be compensated by simultaneously occurring...transformations. The application of this theorem to the universe leads to a conclusion to which W. Thomson first drew attention...: *The entropy of the universe tends to a maximum.*" That was not Clausius's usual version of the second law, but it was likely sufficient to prevent his seeing anything new in the aspects of Planck's thesis emphasized here. Perhaps that is why Planck was unable to establish communication with him (*Selbstbiographie* (Planck, 1948), p. 11; III, 378; Gaynor trans., p. 19). Work done in one of my seminars by Dr. Yung Sik Kim has helped me recognize how hard it is to retrieve more than hints of Planck's ultimately standard version of the second law from Clausius.

28. The quoted phrase is the title of an important three-part paper which Planck published in 1887: "Über das Princip der Vermehrung der Entropie. Erste Abhandlung," *Ann. d. Phys.*, *30* (1887), 562–582; "...Zweite Abhandlung," *ibid.*, *31* (1887), 189–203; "...Dritte Abhandlung," *ibid.*, *32* (1887), 462–503; I, 196–273. But the conception is at least six years older, as the next quotation in the text indicates.
29. The clause occurs in the very first paragraph of the introduction to Planck's thesis, *Zweiten Hauptsatz* (Planck, 1879).
30. Max Planck, "Verdampfen, Schmelzen und Sublimieren," *Ann. d. Phys.*, *15* (1882), 446–475; I, 134–163. Quotation from pp. 472f.; I, 160f. A footnote refers readers to Planck's thesis.
31. Max Planck, *Das Princip der Erhaltung der Energie* (Leipzig, 1887).
32. E. N. Hiebert, "The Energetics Controversy and the New Thermodynamics," *Perspectives in the History of Science and Technology*, ed., D. H. D. Roller (Norman, Okla., 1971), pp. 67–86, provides a useful general discussion of the state of thermodynamics and of Planck's role in its development during the late nineteenth century. Additional information is scattered through his *Mach and Planck* (Hiebert, 1968).
33. Quoted above, p. 14.
34. *Selbstbiographie* (Planck, 1948), pp. 11, 18–22; III, 378, 385–389; Gaynor, trans., pp. 20–21, 29–33.
35. In the early pages of his *Scientific Autobiography* (Planck, 1948), the author-subject suggests that the close friendship between his father and the professor of physics at Kiel played a significant role in his first appointment. It cannot, however, have influenced the call to Berlin, which followed quickly.
†36. Though widely scattered (retrievable only through the index) the remarks on eighteenth- and nineteenth-century kinetic theories in *Caloric Theory of Gases* (Fox, 1971) are particularly cogent, and they provide useful entry to the rich literature on the subject. Here it should be noted that, though gas particles were often seen as literally space-filling in the seventeenth century, eighteenth-century scientists more often conceived them to be held in place (almost as in a lattice) by repulsive interparticulate forces. Except for the direction of the force, the models for solids and liquids were the same. Particularly well-known examples of the resulting vibratory and rotatory theories of heat are provided in the work of Daniel Bernoulli, Count Rumford, and Humphrey Davy. The dictionary legitimates describing their theories as "kinetic," but their model is not the one which that term now brings to mind.
37. The earliest, dating from about 1820, was due to John Herapath, for whom see S. G. Brush, "The Development of the Kinetic Theory of Gases. I. Herapath," *Annals of Science*, *13* (1957), 188–198. Other useful papers by Brush on early proponents of the presently relevant gas model will be found in the bibliography of *Caloric Theory of Gases* (Fox, 1971).
38. R. J. E. Clausius, "Über die Art der Bewegung welche wir Wärme nennen," *Ann. d. Phys.*, *100* (1857), 353–380, and "Über die mittlere Länge der Wege, welche bei Molecularbewegung gasförmigen Körper von den einzelnen Molecülen zurückgelegt werden, nebst einigen anderen Bemerkungen über

die mechanischen Wärmetheorie," *ibid.*, *105* (1858), 239–258. These papers were soon translated in the *Philosophical Magazine* and are conveniently available in S. G. Brush, *Kinetic Theory*, 2 vols. (Oxford, New York, etc., 1965), I, 111–147.

39. This and a number of other points in Clausius's paper had been anticipated in a manuscript submitted by J. J. Waterston to the Royal Society in 1846 and rejected by them. On this topic see: S. G. Brush, "The Development of the Kinetic Theory of Gases. II. Waterston," *Annals of Science*, *13* (1957), 273–282, and E. E. Daub, "Waterston, Rankine, and Clausius on the Kinetic Theory of Gases," *Isis*, *61* (1970), 105–106.

40. J. C. Maxwell, "Illustrations of the Dynamical Theory of Gases," *Phil. Mag.*, *19* (1860), 19–32, and *ibid.*, *20* (1860), 21–37. Reprinted in *The Scientific Papers of James Clerk Maxwell*, ed. W. D. Niven, 2 vols. (Cambridge, 1890, and New York, 1952), I, 377–409, and also (in part) in *Kinetic Theory* (Brush, 1965), I, 148–171.

41. On the background to Maxwell's introduction of statistics into physics see C. W. F. Everitt, *James Clerk Maxwell* (New York, 1975), pp. 135–137, and the papers there cited. Everitt's book is for the most part a reprint of his article on Maxwell in the *Dictionary of Scientific Biography*, ed. C. C. Gillispie, Vol. 9 (New York, 1974), pp. 198–230, where the relevant discussion occurs on pp. 218f.

42. J. C. Maxwell, "On the Dynamical Theory of Gases," *Phil. Mag.*, *32* (1866), 390–393, and *ibid.*, *35* (1868), 129–145, 185–217; and *Phil. Trans.*, *157* (1867), 49–88. Reprinted in Maxwell's *Scientific Papers*, II, 26–78, and *Kinetic Theory* (Brush, 1965), II, 24–87.

43. For other very similar titles, see the books by Burbury and Jeans listed in the bibliography. The only other genre which may include extended accounts of molecular statistics is represented by Clausius's *Mechanische Wärmetheorie*, where the statistical material is restricted to volume 3 subtitled *Die kinetische Theorie der Gase*. The volumes by O. E. Meyer and by Kirchhoff listed in the bibliography have similar titles and similarly restricted subject matters. The lack of other sorts of extended treatments of molecular statistics before the beginning of the twentieth century provides my reason for identifying the nineteenth-century subject as gas theory rather than statistical mechanics.

44. For example, H. W. Watson, *A Treatise on the Kinetic Theory of Gases* (Oxford, 1876), pp. 46–51.

†45. J. Willard Gibbs, *Elementary Principles in Statistical Mechanics* (New York, 1902; reprinted 1960). A. Einstein, "Kinetische Theorie des Wärmegleichgewichtes und des zweiten Hauptsatzes der Thermodynamik," *Ann. d. Phys.*, *9* (1902), 417–433. Einstein's statistical papers will be discussed in Chapter VII, below. Contrasting these titles with those given in the text above and in note 45 suggests the sudden change in subject matter these two works introduce. Both begin by specifying models that can be developed without approximations suitable only to gases or some other special sort of system, a characteristic found, to my knowledge, in only one earlier paper (to be discussed in the next paragraph of this note). Both are centrally

concerned with the statistical treatment of entropy, a subject previously developed only by Boltzmann (and, after 1898, Planck). The transition to a subject matter in which these concerns were primary requires far more study, but a useful early index of what occurred is provided by the course of lectures H. A. Lorentz delivered in Paris in 1912: *Les Théories statistiques en thermodynamique*, ed. L. Dunoyer (Leipzig and Berlin, 1916). Again, the title is revealing.

It is worth nothing, however, that the development of this whole subject might have been very different if Maxwell had lived a few more years. The transition from gas theory to statistical mechanics is closely associated with the adoption of the so-called ensemble approach, which examines the statistical distribution at an instant of time of N identical systems uniformly distributed in phase. The main previous approach had instead examined the statistical distribution over time of the states of a single system, which generally had, for simplicity, to be a gas. Maxwell had followed this standard approach in his early papers, but he brilliantly developed a full ensemble treatment in 1878, the year before his death: "On Boltzmann's Theorem on the Average Distribution of Energy in a System of Material Points," *Trans. Cambridge Phil. Soc.*, 12 (1871–79), 547–570, presented 6 May 1878; *Scientific Papers*, II, 713–741. Unfortunately, however, that paper was widely misunderstood among Maxwell's British followers, many of whom seem to have supposed that each of Maxwell's N systems was a molecule and that all of them together constituted a gas or other aggregate.

That reading, which would surely have been corrected if Maxwell had lived, was doubtless facilitated by the close parallels it created between Maxwell's paper and (Boltzmann, 1868), of which it was rederiving and generalizing the results. Boltzmann had there considered N complex molecules, each with n degrees of freedom, and he had treated the collection as something very like an ensemble during the interval between collisions. Watson had further elaborated that approach and given it wide currency in his *Treatise* (Watson, 1876). For a later survey that clearly displays the relevant misunderstandings see (Bryan, 1894); for a late, but not the final, stage in its elimination see (Rayleigh, 1900a). Note also that Boltzmann was not among those who misunderstood. He reproduced Maxwell's proof in his (Boltzmann, 1881), and he used the ensemble approach for special purposes in (Boltzmann, 1885 and 1887). But he never made it central. Historically, that was the contribution of Gibbs and Einstein.

46. This is that subtitle of Vol. 3 of Clausius's *Mechanische Wärmetheorie*, for which see note 18, above.
47. G. R. Kirchhoff, *Vorlesungen über die Theorie der Wärme*, which is Vol. 4 of Kirchhoff's *Vorlesungen über mathematische Physik* (Leipzig, 1877–94).
48. Quoted in *Mach and Planck* (Hiebert, 1968), p. 21, from (Planck, 1887b).
49. Max Planck, *Gleichgewichtszustände isotroper Körper in verschiedenen Temperaturen* (Munich, 1880); I, 62–124. The relevant discussion occurs on the first page of the introduction.
50. See, for examples, *Vorgeschichte* (Kangro, 1970), pp. 124f., and *Mach and Planck* (Hiebert, 1968), p. 28n.

†51. Max Planck, "Allgemeines zur neueren Entwicklung der Wärmetheorie," *Verhandlungen der Gesellschaft deutscher Naturforscher und Ärzte*, 1891, Pt. 2, pp. 56–61; *ZS. f. phys. Chem.*, 8 (1891), 647–656; I, 372–381; quotation from the second paragraph. The rest of this article is of special interest, for it suggests that Planck was beginning to recognize differences in the status of the first and second laws. He first asks how the phenomenological approach to thermodynamics can permit "one to look more deeply into the world of molecules than even the kinetic theory" (*ibid.*). That power, he insists, must derive from something more than the laws of thermodynamics by themselves. To isolate the required additional element, he points out that, though an observed violation of the first law could at once be exploited to produce perpetual motion, a violation of the second law would not necessarily permit heat to be transported without compensation from a colder to a warmer body. Applications of the second law depend, Planck continues, on the introduction of idealizations, many of which (for example, van der Waals's equation or a semi-permeable membrane able to separate N_2O_4 from NO_2) cannot be approximated in the laboratory. It is the presently inexplicable success of these idealizations, he suggests, which gives the second law its special power. Conversely, any given idealization might fail, an event which "would for the first time open the prospect of determining the boundaries, for so long sought in vain, which limit the validity of ideal processes and perhaps also of the second law" (p. 656; I, 381). Planck would never, I think, have written a similar passage about the first law. That he could have written this one at a significantly earlier date is doubtful, but his choice of the term "natural" for irreversible processes (see below) suggests his early recognition of the intimate relation between ideal processes and the second law.

52. Ludwig Boltzmann, "Über den Beweis des Maxwellschen Geschwindigkeitsverteilungsgesetzes unter Gasmolekülen," *Münchener Ber.*, 24 (1894), 207–210; *Ann. d. Phys.*, 53 (1894), 955–958; III, 528–531.

53. "Verdampfen" (Planck, 1882), 474f.; I, 162f.

54. J. C. Maxwell, *Theory of Heat* (London and New York, 1871). There were two more English editions in 1872 and a fourth edition in 1875 from which both German translations cited by Planck were made. In the ninth English edition (1888), from which I have copied the passage, the quotation occurs on pp. 328f. It appears in substantially identical form, however, in earlier versions.

55. On Maxwell's Demon, see E. E. Daub, "Maxwell's Demon," *Studies in History and Philosophy of Science*, 1 (1970), 213–227, and M. J. Klein, "Maxwell, His Demon, and the Second Law of Thermodynamics," *American Scientist*, 58 (1970), 84–97.

56. P. G. Tait, *Sketch of Thermodynamics*, 2nd ed. (Edinburgh, 1877), pp. xviif., 36f.

57. J. C. Maxwell, "Tait's 'Thermodynamics'," *Nature*, 17 (1877–78), 257–259, 278–280, with quotation from pp. 279f.; *Scientific Papers*, II, 660–671, with quotation from p. 670.

58. M. J. Klein, "Mechanical Explanation at the End of the Nineteenth Century,"

Centaurus, *17* (1972), 58–82, has isolated and illustrated contemporary interest in mechanical systems that displayed analogies to the behaviors both of the second law and of the electromagnetic ether. The systems primarily discussed were of the so-called monocyclic type developed by Helmholtz; Clausius was an active participant in the discussion.
59. *Selbstbiographie* (Planck, 1948), p. 21; III, 388; Gaynor trans., p. 32.
60. The earliest reference I know to the possibility of actually observing deviations from average behavior are in Boltzmann's "Entgegnung auf die wärmetheoretischen Betrachtungen des Hrn. E. Zermelo," *Ann. d. Phys.*, *57* (1896), 773–784; III, 567–578; *Kinetic Theory* (Brush, 1965), 218–228. The references, one of which is to Brownian motion, occur on pp. 778, 572, and 223, respectively. A quantitative theory of fluctuations first appears in *Statistical Mechanics* (Gibbs, 1902) and, more centrally and consequentially, in A. Einstein, "Zur allgemeinen molekularen Theorie der Wärme," *Ann. d. Phys.*, *14* (1904), 354–362.
61. *Mechanische Wärmetheorie*, 3rd ed. (Clausius, 1879), I, 386.
62. The terminology is introduced in the introductory pages of *Zweiten Hauptsatz* (Planck, 1879).
63. "Verdampfen" (Planck, 1882), last paragraph, italics added.
64. *Mechanische Wärmetheorie*, 2nd ed. (Clausius, 1876), I, 224, and 3rd ed. (Clausius, 1879), I, 222. The phrases given in quotation marks are Clausius's chapter and section titles.
65. William Thomson, "The Kinetic Theory of the Dissipation of Energy," *Proc. Edinburgh*, *8* (1874), 325–334; *Mathematical and Physical Papers*, 6 vols. (Cambridge, England, 1882–1911), V, 11–20; (Brush, 1965), II, 176–187. Boltzmann's views will be considered in the next chapter.
66. E. F. F. Zermelo, "Über einen Satz der Dynamik und die mechanische Wärmetheorie," *Ann. d. Phys.*, *57* (1896), 485–494; *Kinetic Theory* (Brush, 1965), 208–217. The sentence quoted is in the first paragraph. The paper by Boltzmann cited in note 60 is a reply, and there are two additional items in the exchange.
67. Max Planck to Leo Graetz, 23 May 1897: "... ich halte [die Fragen, die den Anlass zu den Diskussion zwischen Boltzmann u. Zermelo geben,]... für das Wichtigste, was die theoretische Physik gegenwärtig beschäftigt. Mit Zermelo, der mein Assistent ist, habe ich oft u. eingehend darüber verhandelt; sonst ist hier in dem grossen Berlin kaum einer, der sich wirklich lebhaft für diese Dinge interessiert, soweit mir bekannt ist, besonders seit Willy Wien's Fortgang nach Aachen. In dem Hauptpunkt der Frage stehe ich auf Zermelo's Seite, indem auch ich der Ansicht bin, dass es prinzipiell ganz aussichtslos ist, die Geschwindigkeit irreversibler Prozesse, z. B. der Reibung oder der Wärmeleitung, in Gasen, auf wirklich strengem Wege aus der gegenwärtigen Gastheorie abzuleiten. Denn da Boltzmann selber zugibt, dass sogar die *Richtung*, in der Reibung u. Wärmeleitung wirkt, nur aus Wahrscheinlichkeitsbetrachtungen zu folgern ist, so war völlig unverständlich, woher es dann kommt, dass unter allen Umständen auch die *Grösse* dieser Wirkungen einen ganz bestimmten Betrag darstellt. Die Wahrscheinlichkeitsrechnung kann wohl dazu dienen, wenn man vorher garnichts weiss, einen Zustand als

den wahrscheinlichsten zu finden; sie kann aber nicht dazu dienen, wenn ein unwahrscheinlicher Zustand gegeben ist, nun den daraus folgenden zu berechnen; denn dieses ist dann nicht mehr durch Wahrscheinlichkeit, sondern durch die Mechanik bestimmt, u. es wäre vollkommen unbegründet vorzunehmen, dass die Veränderungen in der Natur immer in der Richtung von geringerer zu grösserer Wahrscheinlichkeit erfolgen.

"In jedem Falle scheint mir eine definitive Entscheidung der Frage nur auf dem Wege möglich, dass man sich erst einmal von vornherein auf einen der beiden Standpunkte stellt u. nun zusieht, wie weit man damit kommt, ad lucem oder ad absurdum. Und da ist die Arbeit sicherlich leichter u. aussichtsvoller, wenn man den zweiten Hauptsatz als wirklich strenggültiges Naturgesetz adaptiert (wobei dann allerdings die kinetische Gastheorie in ihrer jetzigen Form nicht mehr ausreicht) als wenn man sich, um die Gastheorie zu *retten*, mit mathematisch kaum formulirbaren Voraussetzungen über den Anfangzustand der Welt behilft, mit denen man weiter garnicht machen kann als eben nur das, was ihre Einführung veranlasst. Das ist ein Verzicht auf jede tiefere Einsicht. Zermelo geht aber weiter, u. das halte ich für unrichtig; er meint, der zweite Hauptsatz, als Naturgesetz, sei überhaupt unverträglich mit jeder mechanistischen Naturauffassung. Denn die Sache ändert sich wesentlich, wenn man von diskreten Massenpunkten (wie den Molekülen in der Gastheorie) zu continuierlicher Materie übergeht. Ich glaube u. hoffe sogar, dass sich auf diesen Wege eine streng mechanische Deutung des zweiten Hauptsatzes finden lassen wird; aber diese Sache ist offenbar sehr schwierig und verlangt Zeit."

The original is in the Deutsches Museum, Munich, and its importance was first noticed by Hans Kangro, who quotes significant fragments on pp. 128–130 of the monograph previously cited. I am much indebted to him for volunteering a copy.

68. Max Planck, "Über irreversible Strahlungsvorgänge. Erste Mittheilung," *Berl. Ber.*, 1897, pp. 57–68; I, 493–504.
69. *Ibid.*, pp. 58f.; I, 494f.
70. *Ibid.*, p. 60; I, 496.
71. With respect to the continuum, the exception is B. G. Doran, "Origin and Consolidation of Field Theory in Nineteenth-Century Britain: From the Mechanical to the Electromagnetic View of Nature," *Historical Studies in the Physical Sciences*, 6 (1975), 133–260. It traces a putative British conception of a physically continuous ether with irreducibly non-mechanical properties from Faraday and Kelvin in the 1840s to Larmor at the end of the century. If such a tradition had existed in Germany, Planck's position would appear less strange, but the article provides no clue to its export. More to the point, though Dr. Doran's article deserves sympathetic critical scrutiny, I do not myself believe that it establishes, even for Britain, the existence before the 1890s of the tradition it describes.
72. For Planck's musical interest, which continued throughout his life, see, for example, *Max Planck* (Hermann, 1973b), pp. 7ff. It is also worth recording that Planck, on his arrival in Berlin, undertook to study the properties of a mathematically pure tone system on a special harmonium recently presented

to the Institute for Theoretical Physics. His report on the subject is, "Ein neues Harmonium in natürlicher Stimmung nach dem System von C. Eitz," *Verh. d. D. Phys. Ges.*, *12* (1893), 8–9; I, 435–436. See also Planck's remarks on the subject in his *Selbstbiographie* (Planck, 1948), p. 16; III, 383; Gaynor, trans., pp. 26f. Resonance might well have been on his mind.

73. H. von Helmholtz, *Vorlesungen über theoretische Physik*, Vol. 5: *Dynamik continuirlich verbreiteter Massen*, ed. O. Krigar-Menzel (Leipzig, 1902), p. 1. Italics added.
74. *Ibid.*, p. 3.
75. *Ibid.*, pp. 4–9, and, for Planck's notebook reference, *Vorgeschichte* (Kangro, 1970), p. 115.
76. Max Planck, "Absorption und Emission electrischer Wellen durch Resonanz," *Berl. Ber.*, 1895, pp. 289–301; *Ann. d. Phys.*, *57* (1896), 1–14; I, 445–458, where the discussion of the acoustic case is in §3. "Erste Mittheilung" (Planck, 1897a), §3. The analogy is also discussed in Planck's intervening paper, "Über electrische Schwingungen, welche durch Resonanz erregt und durch Strahlung gedämpft werden," *Berl. Ber.*, 1896, pp. 151–170; *Ann. d. Phys.*, *60* (1897), 577–599; I, 466–488. The relevant discussion occurs late in §1.
77. *Ibid.*
78. *Treatise* (Maxwell, 1873), II, Chapts. 5–9.
79. On these men see "Mechanical Explanation" (Klein, 1972).
80. I owe these remarks about dispersion theory to conversations and correspondence with Jed Z. Buchwald, who is currently preparing a paper on the subject.
81. On post-Maxwellian electromagnetic theory see: Tetu Hirosige, "Origins of Lorentz' Theory of Electrons and the Concept of the Electromagnetic Field," *Historical Studies in the Physical Sciences*, *1* (1969), 151–209; Salvo D'Agostino, "Hertz's Researches on Electromagnetic Waves," *ibid.*, *6* (1975), 261–323; and Russell McCormmach, "H. A. Lorentz and the Electromagnetic View of Nature," *Isis*, *61* (1970), 459–497.
82. "Absorption und Emission" (Planck, 1895) and "Über electrische Schwingungen" (Planck, 1896). The terms "conservative damping" and "consumptive damping" are introduced only in the latter, but the emphasis on an analysis which involves no frictional or electrical resistance is common to both.
83. To avoid misleading readers, it should be emphasized that this previously quoted phrase first occurs in Planck's "Erste Mittheilung" in 1897, and one critic has suggested to me that Planck's concern with a demonstration of irreversibility did not arise until then. I can produce no categoric counterevidence, but the suggestion seems to me clearly wrong. The problem is a product of Planck's version of thermodynamics; the emphasis on a blackbody theory that deals with "conservative processes" (in Planck's rather special sense of the term) appears in his first electromagnetic paper, published in 1895.
84. The paper which Planck cites is H. Hertz, "Die Kräfte electrischer Schwingungen, behandlet nach der Maxwell'schen Theorie," *Ann. d. Phys.*, *36*

(1889), 1–22. It was reprinted as Chap. 9 of Hertz's *Untersuchungen über die Ausbreitung der elektrischen Kraft* (Leipzig, 1892) and several later editions. The book is available in English as *Electric Waves*, trans. D. E. Jones (London and New York, 1893).
85. "Absorption und Emission" (Planck, 1895), p. 14; I, 458.

Notes to Chapter II

1. The first hint of a concern with irreversible change in gases appears in (Maxwell, 1860): "Prop. VI. Two systems of particles move in the same vessel; to prove that the mean *vis viva* of each particle will become the same in the two systems." The demonstration depends upon some very special assumptions, and Maxwell does not again attempt to prove that a gas initially in a non-equilibrium state must move towards one. He does claim to have shown by other techniques (Maxwell, 1866) that his law is the only possible equilibrium distribution, but Boltzmann, in the paper cited in the next note, offers cogent criticism of his proof.
2. Ludwig Boltzmann, "Weitere Studien über die Wärmegleichgewicht unter Gasmolekülen," *Wiener Ber. II*, *66* (1872), 275–370; I, 316–402; S. G. Brush, *Kinetic Theory*, 2 vols. (Oxford, New York, etc., 1965–66), II, 88–175.
3. For the development of statistical mechanics in general, including much additional information about Boltzmann, see the voluminous writings of S. G. Brush. Especially helpful are his "Foundations of Statistical Mechanics, 1845–1915," *Archive for History of Exact Sciences*, *4* (1967), 145–183, and his introductions to Ludwig Boltzmann, *Lectures on Gas Theory*, trans. S. G. Brush (Berkeley, 1964), cited below as BB, and to the two volumes of *Kinetic Theory*. On topics of special relevance to this chapter, see also his "The Development of the Kinetic Theory of Gases, VIII. Randomness and Irreversibility," *Archive for History of Exact Sciences*, *12* (1974), 1–88. For Boltzmann, see also, René Dugas, *La Théorie physique au sens de Boltzmann* (Neuchatel, 1959), and M. J. Klein, "The Development of Boltzmann's Statistical Ideas," in *The Boltzmann Equation*, ed. E. G. D. Cohen and W. Thirring, *Acta Physica Austraica*, Suppl. X (Vienna and New York, 1973), pp. 53–106.
4. Ludwig Boltzmann, *Vorlesungen über Gastheorie. I. Theil: Theorie der Gase mit einatomigen Molekülen, deren Dimensionen gegen die mittlere Weglänge verschwinden* (Leipzig, 1896). The second volume, which differs considerably from the first with respect to the issues important here, was published in 1898.
5. In this discussion Boltzmann actually uses the symbols ε, η, ζ for the velocity components. In a paper to be discussed below he uses u, v, w instead. The same sorts of purely notational changes occur repeatedly in Planck's successive papers and elsewhere. To avoid needless confusion I have, throughout this volume, therefore often altered an author's original symbols. Occasionally those altered symbols have, without comment, been substituted for the originals within the text of quotations. Note also that Boltzmann's text in the *Gas Theory* deals with the case of two mixed gases, one with distribution

function f, the other with distribution F. For simplicity, I have restricted attention to the behavior of a single gas. Nothing of present relevance is lost in the process.

6. *Gastheorie*, I (Boltzmann, 1896b), 17; BB, 37. Brush drops the troublesome phrase, "on the average," as well as a full sentence in parentheses immediately before the sentences quoted above. Similar difficulties arise elsewhere, so that Brush's translation, though a valuable guide to Boltzmann's text, provides no foundation for close analysis. Sometimes it eliminates conceptual difficulties from the original; on other occasions, for example the last full sentence on p. 59, it creates difficulties of sense where the original contained none.
7. *Ibid.*, I, 100; BB, 111. A more elaborate discussion of this requirement occurs at I, 45–47; BB, 61f.
8. *Ibid.*, I, 18; BB, 38.
9. *Ibid.*, I, 23; BB, 42. This equation, numbered (18) in Boltzmann's text, is an expanded version of an earlier equation (17) which, for brevity, uses special notation of no present relevance. To avoid introducing the notation here, I have occasionally, below, referred to this equation in places where Boltzmann refers to his equation (17).
10. *Ibid.*, I, 27; BB, 46.
11. *Ibid.*, I, 30; BB, 48.
12. *Ibid.*, I, 32; BB, 49.
†13. Boltzmann's use of the symbol H for $\int f \log f \, d\omega$ is quite new in the *Gas Theory* (Boltzmann, 1896b). When he originally developed the H-theorem in "Weitere Studien" (Boltzmann, 1872) and for some years thereafter, he had called his function E. That selection of symbol and an oddity about the choice of sign associated with it provide evidence about the limits to understanding of thermodynamics in the early 1870s. Presumably E was selected to suggest that Boltzmann's function was intimately related to entropy. But entropy should increase rather than decrease as the gas passes to equilibrium. One expects Boltzmann to reverse the sign in his definition. Apparently his failure to do so results from a misunderstanding of Clausius's version of the second law. As late as (Boltzmann, 1877a) he writes the second law as $\int dQ/T \leq 0$, a form that applies only when the integral is taken over a full cycle. Boltzmann, however, applies it to isolated systems traversing open paths, a situation for which the inequality sign should be reversed. That more than an eccentric choice of sign is involved emerges at the very end of "Weitere Studien" (Boltzmann, 1872). There Boltzmann produces an explicit form for $E_{\min}$ and says that it differs only by a multiplicative and an additive constant from the expression for entropy supplied in a previous paper. That statement is literally correct, but the required multiplicative constant is negative; more than a difference in normalization is involved. The difficulty is first eliminated in (Boltzmann, 1877b), where the second law is given as the tendency of entropy to increase for isolated systems traversing open paths. That treatment is repeated in the *Gas Theory*, where H replaces E thus eliminating the implied relation to entropy.
14. *Gastheorie*, I (Boltzmann, 1896b), 38; BB, 55.

15. *Ibid.*, I, 47–59; BB, 62–74. Actually, in these pages Boltzmann makes entropy proportional to log W a formulation to be considered below. But he had previously shown that H is proportional to $-\log W$, so that his result is equivalent to equation (5), which is, except for the sign mistake in the original, the form he had produced in "Weitere Studien" (Boltzmann, 1872).

†16. It is likely but by no means certain that Planck had not, before the mid-1890s, paid much attention to the details of Boltzmann's treatment of the second law. Given his central concerns, he can scarcely have been unaware of the existence and general nature of Boltzmann's argument, but nothing more than a nodding acquaintance need underlie such criticisms as those quoted in the preceding chapter from his 1891 address. Zermelo's example is in this respect instructive. In the first paragraph of his reply to Boltzmann's criticism of the recurrence paradox, Zermelo remarks without apparent embarrassment that he had not known Boltzmann's "gas-theoretical investigations" when his initial paper was prepared ("Über mechanische Erklärungen irreversibler Vorgänge," *Ann. d. Phys.*, *59* (1896), 793–801; *Kinetic Theory*, II, 229–237). That many others were in a similar state is suggested by a sentence in Planck's 1897 letter to Graetz. Immediately before the first of the long passages quoted above, Planck remarks: "I have frequently discussed these matters deeply with Zermelo, my Assistant; otherwise in the whole of Berlin there is scarcely anyone with a really lively interest in these questions, especially since Willy Wien's departure [in 1896] for Aachen." (For the source of this letter and the original text see n. 67, Chapter I.) The extent of Zermelo's acquaintance with Boltzmann has, however, already been indicated, and Wien, when statistical arguments got deep, had to appeal to others for assistance in judging them (see Chapter VIII).

Presumably Planck knew more about Boltzmann's work than they, but probably not very much. Before 1900, in any case, he very rarely cites Boltzmann's papers, and only one citation leads to a paper employing statistical techniques. Planck's concern with it is only to point out that his methods lead to a result identical with that gained in "the theoretical investigations of Gibbs, Boltzmann, and van der Waals by very different means." ("Über das Prinzip der Vermehrung der Entropie. Dritte Abhandlung," *Ann. d. Phys.*, *32* (1887), 462–503, p. 484n.; I, 232–273, p. 254n.) Though five of Boltzmann's statistical papers are cited in Kirchhoff's *Theorie der Wärme* (pp. 171, 210), the citations are Kirchhoff's, not Planck's. All refer to articles published prior to the presentation of the last relevant manuscript available to Planck (for the lecture series in 1884), and none carries the initials "D. H." (*Der Herausgeber*) which Planck carefully attached to his own editorial interventions.

It seems likely, therefore, that Planck's detailed acquaintance with Boltzmann's statistical theories began only after the appearance of the *Gas Theory* of 1896. That book is the only work by Boltzmann which Planck cites in his *Annalen* article, and all the parallels to be developed in the next chapter are to matters covered in its first chapter. Planck may well, however, have begun to read the *Gas Theory* before his radiation theory demanded it. His own confrontation with Boltzmann in 1894–95 and his involvement in

Zermelo's confrontation during 1896–97 would likely have drawn him to Boltzmann's first systematic account of the subject.
17. "Weitere Studien" (Boltzmann, 1872), pp. 295, 307; I, 334, 345; *Kinetic Theory*, II, 106, 117. Italics added.
†18. The attempt at a strictly mechanical proof is in Boltzmann's "Über die mechanische Bedeutung des zweiten Hauptsatzes der Wärmetheorie," *Wiener Ber. II, 53* (1866), 195–220; I, 9–33. It is discussed in "Boltzmann's Statistical Ideas" (Klein, 1973). Though conceivable, it is unlikely that Boltzmann in 1872 thought his H-theorem mechanical (and thus deterministic) in some sense very close to that of his first attempt to derive the second law. More likely, as Norton Wise has persuaded me, he thought of f as the limiting distribution function for an infinite number of molecules and simply failed to note that fluctuations significant for his proof would occur in relatively short periods of time if the number were large but finite.
19. Ludwig Boltzmann, "Bemerkungen über einige Probleme der mechanischen Wärmetheorie," *Wiener Ber. II, 75* (1877), 62–100; II, 112–138; partially translated in *Kinetic Theory*, II, 188–193. The quotation is on p. 72; II, 121; *Kinetic Theory*, II, 192f.
20. *Gastheorie*, I (Boltzmann, 1896b), 43; BB, 59.
21. Josef Loschmidt, "Über den Zustand des Wärmegleichgewichtes eines Systems von Körpern mit Rücksicht auf die Schwerkraft. I," *Wiener Ber. II, 73* (1876), 128–142. The relevant paragraph is on p. 139.
22. "Bermekungen" (Boltzmann, 1877a), p. 71; II, 120f.; *Kinetic Theory*, II, 192. Italics added.
23. Ludwig Boltzmann, "Über die Beziehung zwischen dem zweiten Hauptsatze der mechanischen Wärmetheorie und der Wahrscheinlichkeitsrechnung respektive den Sätzen über das Wärmegleichgewicht," *Wiener Ber. II, 76* (1877), 373–435; II, 164–223. Cited below simply as (Boltzmann, 1877b).
24. In the second of the mathematical appendices added to an elementary book on kinetic theory, Meyer had in 1877 derived the Maxwell distribution law by computing the most probable distribution of molecular velocities. The derivation was imperfect, as Boltzmann carefully pointed out, but it introduced a new and very powerful sort of argument, one which Boltzmann arrived at independently. See, O. E. Meyer, *Die kinetische Theorie der Gase* (Breslau, 1877), pp. 259–269. A revised edition was published in 1899.
25. (Boltzmann, 1877b), pp. 376–396; II, 167–186. Boltzmann's resort to an energy rather than a velocity distribution is characteristic. His first derivation of the H-function had also proceeded from an energy distribution function.
26. *Ibid.*, p. 378; II, 170. Boltzmann uses $\mathfrak{P}$ for the permutability in 1877, switching to Z in the *Gas Theory*.
27. Boltzmann (1877b), pp. 396–401; II, 186–190.
28. *Ibid.*, pp. 401–403; II, 190–193.
29. Boltzmann's actual reference is not to his 1872 paper but to a more general version of the H-theorem he had published in 1875.
30. (Boltzmann, 1877b), p. 403; II, 192. Actually, the quantity with the properties Boltzmann attributes to the probability measure is not Ω but rather the logarithm of the permutability Z, defined in equation (6). The two

differ by a constant which may not be suppressed if the permutability measure for two bodies is to be the sum of their individual measures. Boltzmann corrects the mistake without comment in the first volume of the *Gas Theory*.
31. (Boltzmann, 1877b), p. 426; II, 215.
32. *Ibid.*, p. 429; II, 217f.
33. *Ibid.*, p. 430; II, 218. The entire passage is italicized in the original.
34. *Gastheorie*, I (Boltzmann, 1896b), 38–47; BB, 55–62. For the relation between probability and entropy see pp. 58–61; BB, 73–75. Readers using the translation should note the substitution of "maximum" for "minimum" in BB, p. 58, line 5.
35. *Ibid.*, pp. 60, 61; BB, 74, 75.
36. Quoted above on p. 27.
37. (Boltzmann, 1877b), p. 404; II, 193f. The two pages following are also relevant.
38. H. W. Watson, *A Treatise on the Kinetic Theory of Gases* (Oxford, 1876), p. 12. The relevant Boltzmann paper is, "Studien über das Gleichgewicht der lebendigen Kraft zwischen bewegten materiellen Punkten," *Wiener Ber.* II, *58* (1868), 517–560; I, 49–96. This paper was written before Boltzmann invented the H-theorem, and it employs a different approach to the problems of gas theory, facts which may help account for his incorrectly identifying the theorem. Maxwell's version was presented only in 1878, but Watson acknowledges access to Maxwell's notes (*op. cit.*, p. iv). The article is, "On Boltzmann's Theorem on the Average Distribution of Energy in a System of Material Points," *Trans. Cambridge Phil. Soc.*, *12* (1871–79), 547–570; *Scientific Papers of James Clerk Maxwell*, ed. W. D. Niven, 2 vols. (Cambridge, 1890), II, 713–741.
39. *Gastheorie*, I (Boltzmann, 1896b), 40; BB, 56. The theorem is developed on p. 27; BB, 46.
40. (Boltzmann, 1877b), p. 374; II, 165. The first sentence, like much else in Boltzmann's introduction, is intended to demonstrate that he had conceived the idea of computing the probability of states before its publication by O. E. Meyer.
41. *Ibid.*, pp. 406–408; II, 196–198.
42. *Ibid.*, p. 408; II, 197.
43. On this subject see *La Théorie physique* (Dugas, 1959) and other sources listed in footnote 40 of "Boltzmann's Statistical Ideas" (Klein, 1973). The relationship between the continuous and the discrete in Boltzmann's physical thought needs more study, but not in this place. Those concerned with it should, however, notice one notational simplification introduced without comment above. Though Boltzmann writes integrals involving the distribution function with infinite limits, he employs only finite sums. I have here allowed the constraint on total energy to eliminate the extra terms.
44. G. H. Bryan, "Report on the Present State of our Knowledge of Thermodynamics. Part II.—The Laws of Distribution of Energy and their Limitations," *Report of the British Association*, 1894, pp. 64–102.
45. For Boltzmann's acknowledgement of his debt, see, for example, *Gastheorie*, I (Boltzmann, 1896b), vi, 20n; BB, 22, 40n.
46. In addition to Boltzmann, the participants in the discussion in *Nature*

were: Bryan, Burbury, Culverwell, Fitzgerald, Larmor, Schuster, and Watson. Their letters are readily traced through the indices to volumes 50–52. "Randomness and Irreversibility" (Brush, 1974) provides a fuller description of the debate.

47. E. P. Culverwell, "Dr. Watson's Proof of Boltzmann's Theorem on Permanence of Distributions," *Nature*, 50 (1894), 617.
48. S. H. Burbury, "Boltzmann's Minimum Function" and "The Kinetic Theory of Gases," *Nature*, 51 (1894–95), 78, 175f. Burbury's objective is to disprove Loschmidt's paradox. He therefore emphasizes that Condition A does not apply to the reversed motion.
49. Ludwig Boltzmann, "On Certain Questions of the Theory of Gases," *Nature*, 51 (1894–95), 413–415; III, 535–544. The quotation, within which I have inverted the order of the two nearby sentences, is on p. 415.
50. G. R. Kirchhoff, *Vorlesungen über die Theorie der Wärme*, Vol. 4 of Kirchhoff's *Vorlesungen über mathematische Physik* (Leipzig, 1877–94).
51. Ludwig Boltzmann, "Über den Beweis des Maxwellschen Geschwindigkeitsverteilungsgesetzes unter Gasmolekülen," *Münchener Ber.*, 24 (1894), 207–210; *Ann. d. Phys.*, 53 (1894), 955–958; III, 528–531.
52. Note that Boltzmann's criticism is not quite correct. Even when treating direct collisions Kirchhoff selects a special pair of molecules, i.e., one which will shortly undergo collision. But the problem of the correlation between the coordinates of molecules which have just collided is nonetheless real.
53. Max Planck, "Über den Beweis des Maxwellschen Geschwindigkeitsverteilungsgestezes unter Gasmolekülen," *Münchener Ber.*, 24 (1894), 391–394; *Ann. d. Phys.*, 55 (1895), 220–222; I, 442–444.
54. Ludwig Boltzmann, "Nochmals das Maxwellsche Verteilungsgesetz der Geschwindigkeiten," *Münchener Ber.*, 25 (1895), 25–26; III, 532–534. Boltzmann also discusses Planck's argument with interest in *Gastheorie*, I (Boltzmann, 1896b), 44–45; BB, 59–60.
55. The likelihood that Boltzmann would have seen this point is increased by Burbury's emphasis on Condition A's inapplicability to the reversed motion (above, n. 48).
56. Ludwig Boltzmann, "Nochmals das Maxwellsche Verteilungsgesetz der Geschwindigkeiten," *Ann. d. Phys.*, 55 (1895), 223f.; III, 532–534. The latter reproduces the text of the original version, placing the revised portion of the text in a footnote.
†57. Ludwig Boltzmann, "On the Minimum Theorem in the Theory of Gases," *Nature*, 52 (1895), 221; III, 546. The importance of the mean molecular path's being very long compared with intermolecular distance is also recognized in the closing pages of (Boltzmann, 1878). There, Boltzmann is replying to Loschmidt's remark that collision number will not be proportional to ff_1 unless the energy of molecules in $d\omega$ and $d\omega_1$ is independent. The functions f and f_1 refer to two different sorts of molecules and are dependent on both position and velocity. The question of independence is whether, within a given cell in position space, the energy of one sort of molecule can be especially large without necessitating (for example, by thermal conductivity) a correspondingly high or low value of the energy of the other sort of molecule.

The discussion of 1878 can be distinguished from that just quoted by the absence of any reference to probabilities; "independence" does not seem to be statistical independence at all. Instead, the question appears to be whether the actual value of the energy of one sort of molecule is a function of the actual value of the energy of the other. Under these circumstances, the special status of inverse collisions and related probabilistic considerations do not arise.

58. It is possible, of course, that the preceding quotation was written after the corresponding passages in the *Gas Theory* and represents a step beyond them. Nothing seems, in this case, to depend upon establishing the order of composition.
59. *Gastheorie*, I (Boltzmann, 1896b), 20; BB, 40.
60. Max Planck, "Über irreversible Strahlungsvorgänge," *Ann. d. Phys.*, 1 (1900), 69–122; I, 614–667. The citation occurs on p. 75; I, 620.
61. *Gastheorie*, I (Boltzmann, 1896b), 21; BB, 40f.
62. *Ibid.*, p. 21–22; BB, 41. Because of its importance and obscurity, I have rendered the last sentence literally: "Auch Kirchhoff steckt die Annahme, dass der Zustand molekular-ungeordnet sei, schon in die Definition des Wahrscheinlichkeitsbegriffs." I take Boltzmann to mean that for Kirchhoff molecular disorder is a consequence of the very nature of probability. Note also that, having interpreted Kirchhoff in this way, Boltzmann now appears to approve his derivation.
63. *Ibid.*, p. 23; BB, 42. Italics added. The phrase replaced by an ellipsis prohibits molar order, but that is the part of the hypothesis for which Boltzmann will have no later need.
64. See, for example, the remarks on molecular disorder in "Randomness and Irreversibility" (Brush, 1974).
65. See the last major quotation above as well as the quotations from the *Gas Theory* in the first section of this chapter.
66. Ludwig Boltzmann, *Vorlesungen über Gastheorie. II. Theil: Theorie van der Waals'; Gase mit zusammengesetzten Molekülen; Gasdissociation; Schlussbemerkungen* (Leipzig, 1898), p. 259; BB, 448. The entire last chapter of the volume displays Boltzmann's change of position. Doubtless that change is due not only to the events described above but also to his encounter with Zermelo and the recurrence paradox in 1896.
67. S. H. Burbury, *A Treatise on the Kinetic Theory of Gases* (Cambridge, 1899), p. 33, and compare p. 10n.
68. S. H. Burbury, "On the Conditions necessary for Equipartition of Energy. (Note on Mr. Jeans's Paper, *Phil. Mag. November 1902*.)," *Phil. Mag.*, 5 (1903), 134f. Burbury's reference is to J. H. Jeans, "On the Conditions Necessary for Equipartition of Energy," *Phil. Mag.*, 4 (1902), 585–596. Jeans first calls upon molecular disorder on p. 591, and he directs attention to the need for the hypothesis repeatedly thereafter.
69. J. H. Jeans, "The Kinetic Theory of Gases Developed from a New Standpoint," *Phil. Mag.*, 5 (1903), 597–620. The discussion of molecular disorder occurs primarily on pp. 598f.; the quotation is from pp. 605f.
70. I have found no useful up-to-date survey of the state of these problems,

but a particularly relevant discussion of the situation almost two decades ago is provided by Chap. 3 and Appendix 1 (the latter by G. E. Uhlenbeck) of Mark Kac, *Probability and Related Topics in Physical Science* (London and New York, 1959). Note especially the discussion of the propagation in time of the "'Boltzmann property'" on pp. 112f. The early pages of R. Brout, "Statistical Mechanics of Irreversible Processes. Part VIII: Boltzmann Equation," *Physica, 22* (1956), 509–524, supplies an entrée to the immediately preceding literature and leads indirectly to an article that displays a significant earlier state: R. Peierls, "Zur kinetischen Theorie der Wärmeleitung in Kristallen," *Ann. d. Phys., 3* (1929), 1055–1101.

†71. The following quotation from a Bakerian Lecture of 1909 is relevant: "The motive of this present discussion is the conviction expressed at the beginning, that the statistical method, in Boltzmann's [combinatorial] form, must in some way hold the key to the position, no other mode of treatment sufficiently general being available. The writer has held to this belief, with only partial means of justification, ever since the appearance in 1902 [*sic*] of Planck's early paper extending that method to radiation." J. Larmor, "The Statistical and Thermodynamical Relations of Radiant Energy," *Proc. Roy. Soc. London, 83* (1909–10), 82–95; *Mathematical and Physical Papers*, 2 vols. (Cambridge, 1929), II, 396–412. For the quotation see p. 95; 412.

Doubtless more than an interest in Planck's work was responsible for the increasingly central position assumed by the probability calculus in statistical physics from about 1910. Jeans's influential *Dynamical Theory of Gases* (Cambridge, 1904), though it does not deal with entropy in combinatorial terms, does make much use of probability theory. The attempt to understand the effect of Gibbs's resort to error theory in his *Elementary Principles in Statistical Mechanics* (New York and London, 1902) may also have played a role. This whole subject needs additional research.

72. (Boltzmann, 1878, 1879, 1880, 1881, 1883.) These papers can be located through the bibliography.

73. "Report on... Our Knowledge of Thermodynamics" (Bryan, 1894), pp. 91–95. Bryan does relate Boltzmann's combinatorial discussion to the H-function, but for him and other British physicists H is "Boltzmann's minimum function," not entropy. They view it, that is, as a mathematical component of a proof that the Maxwell distribution is the only possible equilibrium state.

The extent of the separation between the mechanical approach to gas theory and that through the probability calculus is indicated by the context within which Bryan takes up Boltzmann's combinatorial work. It enters in the Report's closing Section, entitled "The Boltzmann–Maxwell Law Considered in Relation to Other Theories." The first sub-section is "The Connection with the Theory of Probability," and Boltzmann's 1877 paper is the first item there discussed.

74. William Thomson, "The Kinetic Theory of the Dissipation of Energy," *Proc. Edinburgh, 8* (1872–74), 325–334; *Mathematical and Physical Papers*, Vol. 5 (Cambridge, 1911), pp. 11–20; *Kinetic Theory*, II, 176–187.

75. S. H. Burbury, "On the Law of Distribution of Energy," *Phil. Mag., 37* (1894), 143–158.

76. "Kinetic Theory...from a New Standpoint" (Jeans, 1903). That the resort to probability theory is what Jeans thought new is indicated by a remark in his *Dynamical Theory of Gases*, published the following year. In the preface he qualifies his claim, pointing out that, though Kirchhoff and Meyer have anticipated the method, the proof is his own. Curiously, Jeans does not here cite Boltzmann.

Notes to Chapter III

1. Max Planck, "Über irreversible Strahlungsvorgänge. Erste Mittheilung," *Berl. Ber.*, 1897, pp. 57–68; I, 493–504. "...Zweite Mittheilung," *ibid.*, 715–717; I, 505–507. "...Dritte Mittheilung," *ibid.*, pp. 1122–1145; I, 508–531. "...Vierte Mittheilung." *Berl. Ber.*, 1898, pp. 449–476; I, 532–559. "...Fünfte Mittheilung (Schluss)," *Berl. Ber.*, 1899, pp. 440–480; I, 560–600. These will be referred to below by their position in the series, e.g., "Dritte Mittheilung," etc. Early in 1900 Planck published a summary article (actually much of it was a verbatim reprint of his "Fünfte Mittheilung") under the same title: "Über irreversible Strahlungsvorgänge," *Ann. d. Phys.*, 1 (1900), 69–122; I, 614–667. It will be cited below as "Strahlungsvorgänge." These papers are also discussed in Hans Kangro, *Vorgeschichte des Planckschen Strahlungsgesetzes* (Wiesbaden, 1970), pp. 125–148.
2. For Planck's explicit references to these objectives see pp. 27–29, above.
3. "Erste Mittheilung" (Planck, 1897a), 57f.; I, 493f.
4. Above, p. 28.
5. The material which follows is drawn entirely from the "Dritte Mittheilung" (Planck, 1897d).
6. *Ibid.*, p. 1131; I, 517.
7. *Ibid.*, p. 1132; I, 518. See also, p. 1145; I, 531.
8. *Ibid.*, p. 1145; I, 531.
9. *Ibid.*
10. Ludwig Boltzmann, "Über irreversible Strahlungsvorgänge," *Berl. Ber.*, 1897, pp. 660–662; III, 615–617.
11. "Zweite Mittheilung" (Planck, 1897b), p. 715; I, 505.
12. For Planck's acknowledgment, see the end of this chapter.
13. "Vierte Mittheilung" (Planck, 1898), pp. 449f.; I, 532f.
14. *Ibid.*, p. 451; I, 534.
15. Though the argument to follow is all contained in Planck's fourth installment, my description of it follows the more systematic presentation in his fifth. Note particularly that the transition from Fourier series to Fourier integrals occurs for the first time in the latter.
16. Well-behaved intensities are, for Planck, those which vary with time only during periods long compared to the natural period of the resonator. If this condition is met, then the integrands of equations (6) and (8) may contribute significantly only for $\mu \ll \nu_0$. This restriction, shortly to be replaced by a stronger one (natural radiation), is required to permit the averaging process which intervenes between equation (5) and equations (6) and (7).
†17. Planck here adopts without mention a very considerable specialization of his treatment. A likely by-product is his failure to recognize a fundamental

shortcoming to be discussed early in Part Two below. The field and intensity defined by equations (6) and (7) contain a large number of frequencies. By virtue of the condition specified in the preceding note, the C_v's are required to be negligible for very low frequencies, but they are otherwise unrestricted. A treatment more general than Planck's would therefore employ an analyzing resonator tuned to a frequency v' which could be varied independent of the frequency v_0 of the cavity resonator. Planck, however, aims to determine only the manner in which his cavity resonator exchanges energy with the field (not with resonators at neighboring frequencies through the field). Since that exchange is mediated entirely by frequencies very near v_0 and since an analyzing resonator with $\rho \gg \sigma$ and $v = v_0$ will interact with a frequency range wider than that which affects the cavity resonator, he considers only this case. It follows that equations (13), below, which Planck employs to define natural radiation, can be used to determine the C_v's only for v near v_0.

18. This definition of δ_v is actually from Planck's 1900 summary in the *Annalen* (Planck, 1900a), p. 87; I, 570, and it differs slightly from the definition Planck had provided in the fifth installment. In an accompanying footnote, Planck thanks Boltzmann for pointing out the mathematical slip, an indication of how closely Boltzmann was following Planck's work. From this definition it follows that $(2/\rho v_0) \int \sin^2 \delta_v \, dv = 1$, a relationship which eliminates most of the dependence of J_0 on the bandwidth of the analyzing resonator. That is why the parameter ρ regularly disappears from the end-products of Planck's derivations, below.

19. "Fünfte Mittheilung" (Planck, 1899), pp. 452f.; I, 572f. See also the "Vierte Mittheilung" (Planck, 1898), pp. 468f., 473f.; I, 551f., 556f. The closing statement appears to suggest that natural radiation is defined by equation (14). But that restriction is not strong enough to permit the manipulations which lead from equation (17), below, to the one immediately following.

20. "Fünfte Mittheilung" (Planck, 1899), pp. 453f.; I, 573f. The sketch to follow is drawn primarily from this closing installment of 1899. Planck had also proved irreversibility in the "Vierte Mittheilung" of the preceding year, but by a far clumsier means. With one exception, to be considered below, the differences between the two treatments are of no present importance.

21. When the various parts of Planck's scattered argument are juxtaposed in this way, it is apparent that the earlier ones have no logical function. Planck need not, that is, have developed a formula for $\overline{Ef}$, the rate of energy absorption. He could instead have evaluated directly the form $(dU_0/dt) + 2v_0\sigma U_0$, without making any reference to its physical significance. Presumably, however, physical significance is what led him to experiment with that form.

22. "Vierte Mittheilung" (Planck, 1898), p. 471; I, 554.

23. Division of the radiation into incoming and outgoing spherical waves plays a fundamental role in Planck's treatment of radiation in his fourth installment, but is not otherwise relevant here.

24. "Fünfte Mittheilung" (Planck, 1899), p. 465; I, 585.

25. *Ibid.*, p. 473; I, 593, italics added. Wien's earlier comments on the special problem of relating temperature and entropy in the radiation case are recorded in (Kangro, 1970), p. 107.

26. The assumption seems entirely gratuitous, and Planck's making it provides additional plausible evidence of the extent to which he is following Boltzmann. If S_t could be shown to be unique, then its maximum would necessarily determine the equilibrium state. But thermodynamic entropy, which is defined only in equilibrium, might be another function altogether.
27. "Fünfte Mittheilung" (Planck, 1899), pp. 476f.; I, 596f. The paragraph in square brackets paraphrases the relevant parts of Planck's intervening text.
28. Max Planck, "Entropie und Temperatur strahlender Wärme," *Ann. d. Phys.*, *1* (1900), 719–737; I, 668–686. The displacement law applied to resonators is introduced casually in a footnote on the last page.
29. "Fünfte Mittheilung" (Planck, 1899), p. 479; I, 599. By 1901 Planck recognized that the constant a is not, in fact, absolute, but depends on the units chosen to measure temperature. See his "Über die Verteilung der Energie zwischen Aether und Materie," *Arch. Néerland.*, *6* (1901), pp. 55–66; *Ann. d. Phys.*, *9* (1902), 629–641; I, 731–757.
†30. The probable importance to Planck of his discovery of a system of absolute natural units was first pointed out in (Klein, 1965, n. 13; and also 1966, pp, 26f.) with thanks to Joseph Agassi. More recently the point has, without acknowledgment, been decisively overemphasized in (Hermann, 1969, pp. 19, 28, 30; and also 1973b, pp. 29f.). Hermann, who describes the two radiation constants in Planck's fifth installment as h and k, suggests that 18 May 1899 might be described as "the 'Birthday of the Quantum Theory,'" and argues that the emergence of h in 1899 may be what Planck is said to have described to his son as "the greatest discovery since Copernicus." But it is Kirchhoff's work, not Planck's, which underlies the latter's claim that the constants a and b are absolute. Exactly the same claim might, with equal justice, have been made by Wien for the constants in the law he suggested in 1896. (The fact that Wien's law is identical to Planck's makes the point obvious, but many other forms involving two disposable parameters would have done as well.)

Planck was doubtless pleased when he noted the special characteristic of the Wien law constants. That his own theory permitted the derivation of a law with such constants presumably increased his confidence in what he had done. But the constants a and b in his formula were old, and the special characteristic, which he was the first to note, was in no way dependent on the theory he had developed. His very important contribution to the status of the constants appeared only in December 1900 when he rewrote a as h/k and related k to Boltzmann's combinatorial definition of entropy. Before that date, though the amplitude constant b of his 1899 paper was in many ways the same as the amplitude constant h of December 1900 (he reports the value $6.885 \cdot 10^{-27}$ erg·sec. for the first, $6.55 \cdot 10^{-27}$ erg·sec. for the second, the difference being due to the difference between the distribution formulas used in curve fitting), Planck's a was not the Boltzmann constant k.
31. "Strahlungsvorgänge" (Planck, 1900a), p. 75; I, 620.
†32. Since Planck had, from 1881, repeatedly and publicly rejected Boltzmann's approach, he cannot have found public capitulation easy. Only a man of his considerable integrity could have handled the situation so nearly straight-

forwardly as he did. But recantation does not appear to have eliminated his sense that he had behaved badly towards Boltzmann, and there are signs that the resulting distress continued until the end of his life. By late 1901 Planck appears to have recognized that his reputation would be permanently associated with black-body theory, his contributions to which depended on Boltzmann's work, both globally and in significant detail. Boltzmann's suicide in 1906 can only have reinforced his sense of the complexity of his position (to him it would have been moral complexity). John T. Blackmore (*Ernst Mach* (Berkeley, 1972), pp. 217–222) has recently suggested that Planck's famous and uncharacteristic attack on Mach in 1908 must have arisen in part from the former's need for a scapegoat; the timing of Planck's attack, fifteen months after Boltzmann's death, increases the already considerable plausibility of that hypothesis. Probably Planck's often quoted remarks (for which see Blackmore, *loc. cit.*) about the extent to which he had been a follower of Mach's during his early career should be read against the same background, for passages like those quoted on pp. 21–23, above, are not easily reconciled with his having taken a positivist position which he later abandoned. Though Planck's autobiographical recollections prove exceptionally reliable on most topics, the remarks which bear directly or indirectly on his relation to Boltzmann must be read with special caution. Planck's emphasis, for example, on the totality of his failure to persuade physicists of the independence and importance of the second law seems excessive; that he then, for no apparent reason, gives Boltzmann credit for establishing these convictions may suggest a strong but contorted sense of obligation (*Wissenschaftliche Selbstbiographie* (Planck, 1948), pp. 192f.; III, 386f; Gaynor trans., pp. 30f).

Notes to Chapter IV

1. The pioneering and still standard account of Planck's work during and immediately after 1900 is M. J. Klein, "Max Planck and the Beginnings of Quantum Theory," *Archive for History of Exact Sciences*, 1 (1962), 459–479, supplemented by his, "Thermodynamics and Quanta in Planck's Work," *Physics Today*, 19, No. 11 (1966), 23–32. Much additional significant detail is contained in Hans Kangro, *Vorgeschichte des Planckschen Strahlungsgesetzes* (Wiesbaden, 1970), Chapter 8. My account of this crucial period owes much to these works, from which it departs, matters of emphasis and detail aside, only in the analysis of Planck's first combinatorial derivations.

2. Planck's remarks about his forthcoming paper were made during an extended contribution to the discussion of papers by Thiesen and by Lummer and Pringsheim, to be considered below. The report of his contribution suggests that it was relatively formal and thus that Planck had had advance notice of the substance of one or both of the papers to be discused. Planck's published paper is, "Entropie und Temperatur strahlender Wärme," *Ann. d. Phys.*, 1 (1900), 719–737; I, 668–686. For the report of his earlier remarks about it, see (Planck, 1900b).

3. The membership list of the Physical Society appeared annually in its *Verhandlungen*.

4. O. Lummer and E. Pringsheim, "Die Vertheilung der Energie im Spectrum des schwarzen Körpers," *Verh. d. D. Phys. Ges.*, 1 (1899), 23–41, quotation from p. 36. (Kangro, 1970) supplies much additional information about the development and transmission of the experimental results which led to the abandonment of the Wien law.
5. O. Lummer and E. Pringsheim, "Die Vertheilung der Energie im Spectrum des schwarzen Körpers und des blanken Platins," *Verh. d. D. Phys. Ges.*, 1 (1899), 215–235, quotations from pp. 223, 225.
6. M. F. Thiesen, "Über das Gesetz der schwarzen Strahlung," *Verh. d. D. Phys. Ges.*, 2 (1900), 65–70.
7. "Entropie und Temperatur" (Planck, 1900c), p. 730n.; I, 679n. Note that Planck's resort to arguments based on local maxima permits him to introduce a wider class of entropy functions than the one he had considered in the previous year.
†8. Planck's derivation of equation (3) depends on aspects of his earlier papers not considered above. To grasp its structure suppose, as Planck does not, that the field surrounding the resonator is spherically symmetric and that polarization may be neglected, which it may not in a more rigorous treatment. Then, by the argument immediately following equation (III-19), $u_0 = 3J_0/4\pi$, where u_0 is the density of radiant energy at frequencies near ν_0 and where J_0 is the mean square value of the electric field parallel to the resonator's axis. In addition, by equation (I-1), $u_0 = (4\pi/c)K_0$, where K_0 is the intensity, in ergs/cm^2, of radiation of frequency ν_0 incident on the resonator. Dropping the subscript zero, equation (III-19) can therefore be rewritten in the form: $(dU/dt) + 2\sigma\nu U = (c^2\sigma/\nu)K$. In that equation the second term on the left is the rate of energy emission by the resonator; the term on the right is its rate of absorption.

Restricting attention to the field incident on the resonator in a cone $d\Omega$ at angle θ to its axis, the rate of energy absorption is $(3c^2\sigma/8\pi\nu)K \sin^2\theta \, d\Omega$. The resonator may therefore be imagined as a vertical absorbing surface with area δs and bandwidth $\delta\nu$, where $\delta s \, \delta\nu = 3c^2\sigma/8\pi\nu$. Since energy is emitted by this surface into $d\Omega$ at a rate $(3\sigma\nu U/4\pi)\sin^2\theta \, d\Omega$, the intensity of the field reradiated normal to the axis of the cone must be $(2\nu^2/c^2)U \sin^2\theta$. To this must be added $K\cos^2\theta \, d\Omega$, the unabsorbed intensity of the incident field, thus yielding the net radiation intensity moving away from the resonator. The difference between the outgoing and incoming intensities is therefore given by, $\Delta K = [(2\nu^2/c^2)U - K]\sin^2\theta$. For the present problem—an equilibrium field and a resonator energy differing from equilibrium by ΔU— the corresponding relation is $\Delta K = (2\nu^2/c^2)\Delta U \sin^2\theta$. These relationships are all special cases of the more general ones in "Fünfte Mittheilung" (Planck, 1899), pp. 455–467; I, 575–587.

In the paper currently under discussion, Planck assumes that the rate L at which entropy is carried by radiation of frequency ν across a unit surface perpendicular to the direction of radiation is a function only of K. The change in L corresponding to the change ΔK in K is thus: $(\partial L/\partial K)(2\nu^2/c^2)\sin^2\theta \, \Delta U + \frac{1}{2}(\partial^2 L/\partial K^2)(4\nu^4/c^4)\sin^4\theta \Delta U^2$. If, as Planck takes for granted, the resonator's cross section for radiant entropy is the same as that for energy, then the rate at which it contributes entropy to the field while returning to equilibrium is given by $(3c^2\sigma/8\pi\nu)\int \Delta L\, d\Omega$. Similarly, the rate at which the resonator's entropy changes during the return to equilibrium is given by $(\partial S/\partial U)(dU/dt)$. The sum of these expressions is the rate of change of total entropy S_t, and it may be evaluated by expanding L and $(\partial S/\partial U)$ about their equilibrium values in powers of ΔU. Remembering that dS_t must be positive for all dU/dt and applying equation (2) plus the readily derived equilibrium condition $L = (2\nu^2/c^2)S$, Planck's equation (3), follows directly.

9. "Entropie und Temperatur" (Planck, 1900c), p. 720; I, 669.
10. O. Lummer and E. Pringsheim, Über die Strahlung des schwarzen Körpers für lange Wellen," *Verh. d. D. Phys. Ges.*, 2 (1900), 163–180, quotations from pp. 163n., 171. There is significant uncertainty about the date at which the information in this paper was first circulated. The printed version states that the paper was "Presented at the meeting of 2 February 1900." That is the meeting at which Thiesen's paper was read and at which Planck described his forthcoming derivation of the entropy function he had previously assumed. The printed description of that meeting (note 2, above) indicates that Lummer and Pringsheim did present a paper there under a title like the above but with "und des Platins" inserted immediately after "Körpers." But that paper had presumably been considerably changed before it was printed later in the year. In a note on the last page of the paper to be discussed next, Planck points out that both his own and Thiesen's contributions had been made before Lummer and Pringsheim "extended their measurements to greater wavelengths." The printed version of the paper reporting those measurements is paginated for binding after the proceedings of the last spring meeting of the Physical Society on 19 June but before those of its first fall meeting on 19 October, and a printer's end note suggests that it was separately published. Probably it did not appear until after 18 September 1900, the date at which its lead footnote states that its contents were presented to the Naturforscherversammlung.
11. Max Planck, "Über eine Verbesserung der Wien'schen Spektralgleichung," *Verh. d. D. Phys. Ges.*, 2 (1900), 202–204; I, 687–689. Because the paper is so short, the brief quotations which follow are not further localized by citation.
†12. Italics added. The sentence following this one is accompanied by a footnote pointing out that the displacement law, applied to resonator entropy, requires $S = \phi(U/\nu)$. The forms taken by S at high frequencies and at low energies must therefore be the same, and the former was already known to be the Wien law. Planck's emphasis on that limit and on the simplicity of his new form suggests that these criteria provided all the guidance he needed to guess successfully the formula for $f(U)$. But there is another possibility. On 7 October 1900, twelve days before his new distribution law was first presented to the Physical Society, Planck learned from Rubens of still unpublished measurements indicating that above 100 °C the intensity of the long wavelength components of the black-body spectrum increased linearly with temperature (Kangro, 1970), pp. 200–206. But, as

previously demonstrated, the radiation intensity is proportional to the energy of the corresponding resonator, so that $U \propto T$ for high temperature and energy. Since $\partial S/\partial U = 1/T$, from thermodynamics, $\partial^2 S/\partial T^2 \propto -1/U^2$ for high energies. That relation could also have led Planck to the new form of $f(U)$, and Planck later attributed to it an essential role (*Selbstbiographie* (Planck, 1948), p. 26f.; III, 393f.; Gaynor, trans., p. 39f.).

13. Max Planck, "Die Entstehung und bisherige Entwicklung der Quantentheorie," *Les Prix Nobel en 1919–1920* (Stockholm, 1922), pp. 1–14; III, 121–134. The quotation occurs on pp. 5 and 125, respectively.
14. Max Planck, "Über das Gesetz der Energieverteilung im Normalspectrum," *Ann. d. Phys.*, *4* (1901), 553–563; I, 717–727. Quotation from p. 555; I, 719.
15. "Wien'schen Spektralgleichung" (Planck, 1900d), p. 203; I, 688.
16. Otto Lummer, "Le rayonnement des corps noirs," *Rapports présentés au Congrès international de physique réuni à Paris en 1900* (Paris, 1900), Vol. 2, pp. 41–99; quotation from p. 92.
17. Wilhelm Wien, "Les lois théorique du rayonnement," *ibid.*, pp. 23–40; quotation from p. 40. As pointed out in (Kangro, 1970), p. 220, Planck cites both this remark and the preceding in his (1901a), p. 555; I, 719.
18. Léon Rosenfeld, "La première phase de l'évolution de la théorie des quanta," *Osiris*, *2* (1936), 149–196. In addition to its intrinsic plausibility, Rosenfeld's suggestion is indirectly supported by one of Planck's autobiographical accounts "Zur Geschichte der Auffindung des physikalischen Wirkungsquantums," *Naturwissensch.*, *31* (1943), 153–159; III, 255–267. See especially the beginning of Part III.
†19. With few exceptions, historians discussing Planck's early derivations of his law employ the symbol W where Planck regularly uses R. Since they simultaneously follow Planck in writing Boltzmann's relation between entropy and probability in the form $S = k \log W$, they somewhat disguise the fact that Planck's R is intended to function like Boltzmann's permutation number Z. Neither yields a probability until renormalized. Since the requisite normalizing factor is fixed in situations where the value of Z or R may vary and since it therefore contributes only an additive constant to the entropy, both Planck and Boltzmann generally ignore it. Boltzmann did, however, once mention its value in passing (see his 1877b), p. 391; II, 181, and it proves to be just the combinatorial form on the right-hand side of equation (8), below. Planck, on the other hand, never did evaluate the normalizing factor relevant to his significantly different problem (see below), but simply assumed its existence. In fact, the normalizing factor required by his problem is not in general expressible in a simple closed form. Its value for special cases (resonators at a small number of commensurable periods) can, however, be computed straightforwardly.
20. In his two early papers on the derivation of his distribution law (see below), Planck simply introduces equation (8) with the words, "From combinatorial theory [*Combinationslehre*] the number of all possible complexions is...," thus treating it as a standard formula. Presumably it was, since he mentions no specific source, but I have not had access to the textbooks required to confirm its accessibility to Planck and his readers. Look, however, at E.

Netto, "Kombinatorik," *Encyklopädie der mathematischen Wissenschaften*, Vol. 1, Pt. 1, ed. W. F. Meyer (Leipzig, 1898–1904), pp. 29–46. Completed in 1898, it is the second article in the first volume of the *Encyklopädie*, and its opening page lists three fundamental operations (permutation, combination, and variation) together with four standard formulas. The last ("combination with repetition") is $(n + k - 1)!/k!(n - 1)!$. Though Planck is likely to have had some other source, the formula's position in Netto's paper suggests that it would not have been hard to find. Planck could, of course, also have found the formula in Boltzmann's (1877b), which he cites in another connection, but his manner of introducing it makes that unlikely. Besides, Boltzmann presents it, not as itself a combinatorial, but as a sum of permutation numbers.

21. Max Planck, "Zur Theorie des Gesetzes der Energieverteilung im Normalspectrum," *Verh. d. D. Phys. Ges.*, 2 (1900), 237–245; I, 698–706. The reference to the alternate method is on p. 242; 703.

22. "Über das Gesetz" (Planck, 1901a).

23. H. A. Lorentz, "Alte und neue Fragen der Physik," *Phys. ZS.*, 11 (1910), 1234–1257; *Collected Papers*, Vol. 7 (The Hague, 1934), pp. 205–257. The derivation occurs on pp. 1253–1257; pp. 248–257. I am grateful to Mr. Allan Needell of Yale University for an exchange of letters which has helped me clarify the discussion which follows.

†24. Lorentz's fuller version of the argument considers the manner in which given total energy E_t may be distributed over N resonators *and* a collection of M molecules interacting with them. That problem permits the determination of the manner in which E_t is divided between the resonators and the molecules, and it is then readily shown that the resonator energy must itself be distributed in the most probable manner. It follows that, if only the distribution over resonators is in question, one may neglect the molecules which permit them to exchange energy. If the trial energy E distributed over the resonators proves to be the wrong fraction of E_t, one simply adjusts the number of molecules or resonators. Planck, in his first derivation, makes a similar remark about adjusting the trial energy E until, it constitutes the proper fraction of E_t ("Zur Theorie des Gesetzes" (Planck, 1900e), p. 241; I, 702.

25. Note that the novelty is the fixed size of the elements ε, not its dependence on frequency. Boltzmann himself had, as noted in Chapter II, pointed out that the size of the energy element must vary with the square root of molecular energy if the Maxwell distribution were to result from his combinatorial derivation. Dimensional considerations alone could therefore have suggested to Planck that the dependence on frequency must be linear, though no such suggestion seems to have been required.

26. "Zur Theorie des Gesetzes" (Planck, 1900e), pp. 239f.; I, 700f.

27. *Ibid.*, italics added.

28. *Ibid.*, pp. 240f.; I, 701f. This is Planck's first use in print of the symbol R. Note that it is here equal to a product of expressions like the one on the right-hand side of equation (8). In Planck's next derivation paper it will be used for a single one of those expressions.

NOTES TO PP. 105-112

29. *Ibid.*, p. 241; I, 702. Planck states the value of k without comment when he first introduces it, then comments at length at the end of his paper (see below).
30. *Ibid.*, pp. 241f.; I, 702f. Italics added.
31. Ludwig Boltzmann, "Über die Beziehung zwischen dem zweiten Hauptsatze der mechanischen Wärmetheorie und der Wahrscheinlichkeitsrechnung respektive den Sätzen über das Wärmegleichgewicht," *Wiener Ber. II*, 76 (1877), pp. 373–435; II, 164–223. For the maximization technique, see, pp. 386–396; II, 177–186. This is, of course, the one paper in which Boltzmann develops his combinatorial techniques at length, and Planck cites it repeatedly after his own involvement with them.
32. From equations (I-1) the displacement law may be written in the form $u_\nu = \nu^3 f(\nu/T)$ or, with the aid of equation (III-19), $U_\nu = \nu g(\nu/T)$. From the last equation and the standard thermodynamic definition of temperature, it follows that $1/T = h(U_\nu/\nu)/\nu = \partial S/\partial U$. Integrating with respect to the single argument U_ν/ν yields the relationship in the text below.
33. Max Planck, *Vorlesungen über die Theorie der Wärmestrahlung* (Leipzig, 1906), pp. 148–153.
34. "Zur Theorie des Gesetzes" (Planck, 1900e), p. 244; I, 705.
35. Max Planck, "Über die Elementarquanta der Materie und der Elektricität," *Ann. d. Phys.*, 4 (1901), 564–566; I, 728–730.
36. Max Planck, "Über der Verteilung der Energie zwischen Aether und Materie," *Arch. Néerland.*, 6 (1901), 55–66; *Ann. d. Phys.*, 9 (1902), 629–641; I, 731–743.
37. Max Planck, "Über irreversible Strahlungsvorgänge. Fünfte Mittheilung (Schluss)," *Berl. Ber.*, 1899, pp. 440–480; I, 560–600. The numerical values are given on the last page.
38. "Über das Gesetz der schwarzen Strahlung" (Thiesen, 1900), p. 67.
39. (Kangro, 1970), pp. 144–148, analyzes the source of these changes in value with special care.
40. Boltzmann (1877b), p. 428; II, 216. A similar and somewhat more convenient formulation is given in the first chapter of Boltzmann's *Gas Theory*, but Planck cites the earlier paper. In his (1901b) Planck shows how Boltzmann's formulas can be transformed to his own.
†41. In 1908 Rutherford and Geiger measured the electronic charge by determining the rate at which charge was transported by α-particle decay. Their value was not quite 1 percent less than Planck's, to which their attention was drawn by Sir Joseph Larmor while their results were being prepared for publication. (See, "Note by Professor E. Rutherford," *Naturwissensch.*, 17 (1929), 483, a reference for which I am indebted to J. L. Heilbron.) The physicist R. W. Pohl (b. 1884), who was a student in Berlin from 1904 and an Assistant from 1906, recorded what is likely to have been the more typical reaction during an interview on 25 June 1963. Asked for his recollection of discussions about the quantum during his first years in Berlin, Pohl said he had been aware of Planck's determination of a new value for the electronic charge and of a new natural constant. (These aspects of Planck's early work were the only ones he mentioned.) But Nernst, he said, had thought the whole computation too hypothetical and abstract to be taken seriously, and he had

accepted that evaluation. (See, pp. 9–10 of the interview in the Archive for History of Quantum Physics.)

42. On this whole subject see, Russell McCormmach, "H. A. Lorentz and the Electromagnetic View of Nature," *Isis, 61* (1970), 459–497.

43. Planck to Lorentz, 6 December 1898: "Ich muss nun gestehen, dass mir die letzten Annahmen [a gravitating ether which is carried with the earth] sehr wenig plausibel erscheinen, da ich gar keine Veranlassung sehe, dem Lichtäther Eigenschaften beizulegen, welche die ponderable Materie besitzt, da er sich doch eben von der ponderablen Materie in den wesentlichsten Punkten unterscheidet." The letter is in the manuscript collection of the Algemeen Rijksarchief, The Hague, Netherlands. Its parts, somewhat scattered, can be found on the first reel of the microfilm of the Lorentz correspondence deposited with the Archive for History of Quantum Physics. I am grateful to Russell McCormmach for bringing this early correspondence to my attention. Some context for it is provided in E. T. Whittaker, *History of the Theories of Aether and Electricity: The Classical Phase*, revised and enlarged ed. (Edinburgh, London, etc., 1951), p. 387, a reference for which I am indebted to John Stachel.

†44. This widely circulated story (including the comparison with Newton) was told by Erwin Planck to the philosopher Bernard Bavink. The latter repeated it to Arnold Sommerfeld, who included it in a memorial address for Planck at the Physical Society meeting in November 1947 (*Ann. d. Phys., 3* (1948), 3–6). More impressive evidence that some such episode occurred has recently been provided by Armin Hermann in his *Max Planck in Selbstzeugnissen und Bilddokumenten* (Hamburg, 1973), p. 29. Having expressed skepticism about the story in previous publications, Hermann received a letter in June 1972 from R. W. Pohl, a good friend of Erwin's, who wrote: "On one of our boat rides, Erwin spontaneously said to me, 'Father knew, in his own words, that his discovery of the new natural constant had the same significance as that of Copernicus.' Therefore, after Planck's death, I took the trouble to see that the constant with its numerical value was placed on his gravestone." Ironically, though the appropriateness of associating Planck's memory with h is not in doubt, his excitement at the turn of the century may rather have derived from the constant k.

Notes to Chapter V

1. The main experimental tests of Planck's law are cited in Max Jammer, *The Conceptual Development of Quantum Mechanics* (New York, St. Louis, etc., 1966), p. 23.
2. Max Planck "Über irreversible Strahlungsvorgänge (Nachtrag)," *Ann. d. Phys., 6* (1901), 818–831; I, 744–757.
3. *Ibid.*, pp. 820f.; I, 746f.
4. See above, p. 99.
5. Max Planck, "Zur Theorie des Gesetzes der Energieverteilung im Normalspectrum," *Verh. d. D. Phys. Ges., 2* (1900), 237–245; I, 698–706. The quotation, to which italics have been added, is on the first two pages.

NOTES TO PP. 119-123

6. Max Planck, "Über das Gesetz der Energieverteilung im Normalspectrum," *Ann. d. Phys.*, 4 (1901), 553–563; I, 717–727. Quotation from p. 556; I, 720.

7. *Vorlesungen über die Theorie der Wärmestrahlung* (Planck, 1906a), pp. 149f.

†8. Planck's argument does not in principle depend on his knowing how time averages are to be taken. But he does know, and he takes the trouble to show his readers. Furthermore, as we shall see, the passages in which he does so are among those eliminated from all later editions of his book. Note also that Planck makes no attempt to show the equivalence of space and time averages. That equivalence problem has a long previous history in statistical mechanics, especially in connection with the equipartition theorem. (See also note 16, below.)

9. "Zur Theorie des Gesetzes" (Planck, 1900e), p. 238; I, 699.

10. *Ibid.*, pp. 242f.; I, 703f.

11. "Über das Gesetz" (Planck, 1901a), p. 558; I, 722. For the context of the quotation from Kries, see note 15, below.

12. See pp. 65f., above.

13. See pp. 81f., above.

14. See above, pp. 54–57.

†15. That information about "the special nature of the resonator vibrations" must be information about the relative probability of various types of vibration is indicated by Planck's quotation from Kries. In his text the quotation is excessively cryptic, but an accompanying footnote leads directly to p. 36 of J. von Kries, *Die Principien der Wahrscheinlichkeitsrechnung* (Freiburg, 1886). Planck's quotation is taken from a chapter called "Die Aufstellung gleichberechtigter Annahmen," roughly "The Representation of Equally Justified Hypotheses," and the phrase he quotes is taken from a theorem stating that: "the relative numerical probability of hypotheses is specifiable when they embrace indistinguishable elementary regions comparable in their magnitudes." The German—"Annahmen in einem zahlenmässig angebbaren Wahrscheinlichleitsverhältniss stehen, wenn sie indifferent und ihrer Grösse nach vergleichbare ursprüngliche Spielräume umfassen"—is scarcely clearer, but its reference to the evaluation of relative probabilities is unequivocal.

†16. Planck's abandonment of all reference to a theoretical justification of the choice of equiprobable elements in the case of gases is somewhat puzzling, but the following considerations provide at least a partial explanation. In his *Lectures*, where Planck spells out in detail those arguments he does use, any reference to Liouville's theorem would have added considerably, in ways not otherwise relevant, to the length and complexity of his text. A fully responsible treatment would, in any case, have had to consider two sets of doubts about the relevance of Boltzmann's argument. First, as seen in Chapter II, Boltzmann had at best shown that equal volumes of phase space are equiprobable for molecules interacting with fixed scattering centers but not with each other. Proof of even that limited theorem, furthermore, could not be carried through without resort to some form of ergodic hypothesis or equivalent, for example the hypothesis that energy is the only constant of a

gas's motion. But the role of ergodic hypotheses, though they had been introduced in the 1870s by Boltzmann and Maxwell, both of whom had reservations about their validity, was neither well worked out nor generally recognized before publication of the Ehrenfests' famous encyclopedia article, "Begriffliche Grundlagen der statistischen Auffassung in der Mechanik," in 1912 (Ehrenfest, 1912). In the first decade of this century the subject was especially controversial because of the close association of ergodicity with equipartition and because of the doubts cast on the latter by measurements of the specific heat of gases. (For citations and additional information on this complex subject see, S. G. Brush, "Foundations of Statistical Mechanics, 1845–1915," *Archive for History of Exact Sciences*, 4 (1967), 145–183, especially §§8–10.) I am aware of only three attempts before 1906 to justify the choice of equiprobable elements: (Boltzmann, 1877b and 1896b), discussed in Chapter II, above; (Einstein, 1902b and 1903), discussed in Chapter VII, below; and Chapter 3 of J. H. Jeans, *Dynamical Theory of Gases* (Cambridge, 1904). Jeans fails to note that, in the absence of an ergodic hypothesis, the selection of a stationary distribution is arbitrary.

17. *Wärmestrahlung* (Planck, 1906a), p. 129.
18. *Ibid.*, p. 131. Planck works out the special boundary conditions required in his Chapter 5.
19. *Ibid.*, p. 132.
20. *Ibid.*, p. 134.
21. *Ibid.*, p. 135, italics added.
22. *Ibid.*, pp. 137f., italics added.
23. *Ibid.*, p. 132.
24. For Planck's autobiographical accounts and for one published passage that may seem to conflict with them, see below, n. 36.
25. *Ibid.*, p. 133.
26. I am, nevertheless, most grateful to Hans Kangro for showing me that it may seem to suggest energy quantization.
27. "Zur Theorie des Gesetzes" (Planck, 1900e), p. 240; I, 701. "Über das Gesetz" (Planck, 1901a), p. 557; I, 721.
28. *Wärmestrahlung* (Planck, 1906a), p. 151. The diagram illustrating complexions is on p. 153.
29. *Ibid.*, p. 154.
30. *Ibid.*, p. 155, italics added. I am especially indebted to Hans Kangro who, though then skeptical of my position, pointed out to me the difficulties in reconciling this part of Planck's argument with more traditional interpretations.
31. Since Planck was "defining" probability, without recourse to Liouville's theorem or an equivalent, nothing depends upon his demonstrating the equivalence or even being aware of it. He very likely was, however, at least before he wrote his *Lectures*.
32. *Wärmestrahlung* (Planck, 1906a), p. 151f. For Boltzmann's and Planck's methods of determining complexions see above, pp. 48f., 104f.
33. Further exploration of Boltzmann's attitude towards the relation between the continuous and the discrete (compare Chapter II, note 38) may require

qualification of this description of Boltzmann's view, but it will not affect the point now at issue.

34. "Zur Theorie des Gesetzes" (Planck, 1900e), p. 239; I, 700.
35. *Wärmestrahlung* (Planck, 1906a), pp. 153f.
36. Max Planck, "Die Entstehung und bisherige Entwicklung der Quantentheorie," *Les prix Nobel en 1919–1920* (Stockholm, 1922), pp. 1–14; III, 121–134, especially pp. 1, 7; III, 121, 127; "Zur Geschichte der Auffindung des physikalischen Wirkungsquantums," *Naturwissensch.*, *31* (1943), 153–159; III, 255–267, especially, p. 159; III, 267; *Wissenschaftliche Selbstbiographie* (Leipzig, 1948); III, 374–401; Gaynor, trans., pp. 13–51, especially pp. 29f.; III, 396f.; Gaynor, pp. 43–45.

In his "Zur Theorie der Wärmestrahlung," *Ann. d. Phys.*, *31* (1910), 758–768; II, 237–247, Planck may seem to be making a more radical claim involving discontinuity. Examining the prerequisites for a derivation of his radiation formula, he says (p. 766; II, 245): "First of all, it seems to me certain that one will not succeed [in the derivation] with the hypothesis of the general continuity of U. For, if one wishes to treat the energy of oscillators as everywhere continuous (i.e., assume h infinitely small), one is led at once to the Jeans radiation formula. Of that I was convinced long before Jeans provided a rigorous proof, indeed ten years ago when Lord Rayleigh was led to the same formula." That passage, however, does *not* say that Planck recognized the difficulties in assuming U continuous at an early date. What he recognized from the start, he says, was simply that allowing h to approach zero would produce the Rayleigh–Jeans law (for which see the next chapter). The passage does not at all supply a date for Planck's recognition of the need for discontinuity, except that it must have occurred before 1910, a fact for which there is other evidence.

37. *Wärmestrahlung* (Planck, 1906a), p. 108n.
38. Planck to Ehrenfest, 6 July 1905: "Auf Ihren werthen Brief vom l.d.M. will ich Ihnen gerne meine Meinung über die von Ihnen angeregte Frage mittheilen. Vor allem stimme ich Ihnen in der Hauptsache vollkommen bei, nämlich dass zur Ableitung des Gesetzes der Energieverteilung im Normalspektrum die Resonatorentheorie (einschliesslich der Hypothese der natürlichen Strahlung) nicht ausreicht, und dass die Einführung des *endlichen* Energiequantums $\varepsilon = h\nu$ eine neue, der Resonatorentheorie an sich fremden, Hypothese bedeutet. Es kommt also durch diese Hypothese ein neues Element in die Theorie hinein, das sich keinesfalls auf rein logischem Wege deduzieren lässt.

"Aber vielleicht ist es nicht ganz ausgeschlossen, auf folgendem Wege vorwärts zu kommen. Wenn man die Annahme einführt, dass die Resonatorschwingungen aus Bewegungen von Elektronen bestehen, so bringt man ebenfalls ein neues Element in die Theorie hinein. Denn weil die Ladung eines Elektrons mit div E proportional ist, so kann E nicht im ganzen Felde um m^2 vergrössert werden, ohne dass die Ladung eines Elektrons ebenfalls im Verhältnis $1:m^2$ wächst. Sind also die Ladungen der Elektronen konstant, so ist der von Ihnen betrachtete Vorgang $E' = m^2 E$, $H' = m^2 H$, $f' = m^2 f$ unmöglich.

NOTES TO PP. 132–134 289

"Es scheint mir nun nicht ganz ausgeschlossen, dass es von dieser Annahme (Existenz eines elektrischen Elementarquantums) eine Brücke gibt zu der Existenz eines energetischen Elementarquantums h, zumal da h von der gleichen Dimension und auch von der gleichen Grössenordnung ist wie e^2/c (e, elektr. El. quantum im elektrostatischen Mass; c, Lichtgeschwindigkeit). Aber ich bin nicht imstande, darüber eine bestimmte Vermutung zu äussern."

The original of this letter is in the Museum Boerhaave, Leyden, and the text is also available on microfilm with the collections of the Archive for History of Quantum Physics.

39. Above, pp. 110f.
40. H. A. Lorentz, "The Theory of Radiation and the Second Law of Thermodynamics," *Proc. Amsterdam*, 3 (1901), 436–450, quotation from p. 442; reprinted in *Collected Papers*, Vol. 6 (The Hague, 1938), pp. 265–279, quotation on p. 271. For further development of the same approach see also H. A. Lorentz, "Boltzmann's and Wien's Laws of Radiation," *Proc. Amsterdam*, 3 (1901), 607–620; *Collected Papers*, Vol. 6, pp. 280–292. The first of these papers was presented on 29 December 1900, the second on 23 February 1901.
41. J. H. Jeans, "On the Laws of Radiation," *Proc. Roy. Soc. London*, 76 (1905), 545–552.
†42. Planck maintained an active scientific correspondence with Lorentz and Wien, both deeply concerned with radiation theory, from about the turn of the century. His contributions have been largely preserved with the Lorentz papers at the Algemeen Rijksarchief, The Hague, and in a collection of Wien's papers recently deposited at the Staatsbibliothek preussischer Kulturbesitz in Berlin. Both are also available on microfilm in some of the depositories of the Archive for History of Quantum Physics. Interestingly enough, neither contains significant references to Planck's theory before April 1908, though from the end of that year discussions of the subject are dense in both. We shall discover in Chapter VIII, however, that 1908 is the year in which Planck first gives evidence of recognizing that his theory breaks with classical physics, and the perplexities surrounding that break are often the subject of these letters thereafter. The contrasting silence in the earlier letters suggests that, though deeply puzzled about the source of the constant h, Planck then thought his theory generally unproblematic.
43. Planck to Nernst, 11 June 1910: "Denn ich kann ohne Übertreibung sagen dass mich seit 10 Jahren nichts in der Physik so ununterbrochen an-, er-, und aufregt wie diese Wirkungsquanten." The letter is quoted in full on pp. 6f. of an unpublished manuscript by Jean Pelseneer, "Historique des Instituts Internationaux de Physique et de Chimie Solvay depuis leur fondation jusqu'à la deuxieme guèrre mondiale," available on Microfilm 58 of the Archive for History of Quantum Physics. I am indebted to Martin Klein for insisting that I face the problems this statement might present to my reconstruction.
†44. See above, pp. 59f. Planck's failure to consider the physical significance of the size of the energy element had a second effect. The mean energy of a resonator with U in the region between $nh\nu$ and $(n + 1)h\nu$ is $(n + \frac{1}{2})h\nu$. Planck's formula for the distribution of resonator energy therefore lacks an

additive term $\frac{1}{2}h\nu$, later to be called the zero-point energy. It is, of course, negligible for $h\nu \ll kT$.

45. The subject of this section has recently been considered in Elizabeth Garber, "Some Reactions to Planck's Law, 1900–1914," *Studies in History and Philosophy of Science*, 7 (1976), 89–126. In going over similar ground, Max Jammer, *The Conceptual Development of Quantum Mechnics* (New York, 1966), p. 23, n. 98, cites A. L. Day, "Measurement of High Temperature," *Science*, *15* (1902), 429–433, for an additional early report on Planck's work. I can find no mention of Planck in that place, but early references other than those below will doubtless be found in time. The pages that follow should, however, suggest both their nature and their low density.

46. *Die Fortschritte der Physik im Jahre 1900, zweite Abteilung enthaltend Physik des Aethers* (Braunschweig, 1901), and the same for the year 1901. *Science Abstracts: Physics and Electrical Engineering*, 4 (1901). Abstracts of Planck's articles are scattered, but readily traced through the indexes.

47. H. G. J. Kayser, *Handbuch der Spectroscopie*. Vol. 2 (Leipzig, 1902), p. 110. The article to which Kayser refers is (Planck, 1900b).

48. *Ibid.*, p. 120.

49. *Ibid.*, p. 68.

50. Woldemar Voigt, *Thermodynamik*, Vol. 2 (Leipzig, 1904), p. 355.

51. Paul Drude, *Lehrbuch der Optik*, 2nd ed. (Leipzig, 1906), p. 517.

52. *Ibid.*, p. 515–517, 519.

53. S. H. Burbury, "On Irreversible Processes and Planck's Theory in Relation Thereto," *Phil. Mag.*, *3* (1902), 225–240; quotation from p. 239 with italics added.

54. Joseph Larmor, "Radiation, Theory of," in *The [Eighth of the] New Volumes of the Encyclopaedia Britannica,...being Volume XXXII of the Complete Work* (London, 1902), pp. 120–128, discussion on pp. 124f. I am indebted to Elizabeth Garber for calling my attention to this article, which is not included in Larmor's *Mathematical and Physical Papers*.

55. Joseph Larmor, "On the Application of the Method of Entropy to Radiant Energy," *Report of the British Association*, 1902, p. 546. *Mathematical and Physical Papers by Sir Joseph Larmor*, Vol. 2 (Cambridge, Eng., 1929), p. 699.

56. Joseph Larmor, "The Statistical and Thermodynamical Relations of Radiant Energy," *Proc. Roy. Soc. London*, *83* (1909–10), 82–95; *Mathematical and Physical Papers*, Vol. 2, pp. 396–411. The information about Larmor's lectures on Planck's theory is drawn from this source as is the quotation in Chapter II, n. 71, above.

57. Joseph Larmor, "On the Statistical Theory of Radiation," *Phil. Mag.*, *20* (1910), 350–353; *Mathematical and Physical Papers*, Vol. 2, pp. 413–415.

58. Lord Rayleigh, "The Dynamical Theory of Gases and Radiation," *Nature*, *72* (1905), 54–55; John William Strutt, Baron Rayleigh, *Scientific Papers*, Vol. 5 (Cambridge, Eng., 1912), pp. 248–252. Note that, if Rayleigh had supposed Planck restricted the value of energy available to a resonator, he would have had no problem in understanding why the result he gained was different from that due to equipartition.

59. J. H. Jeans, "A Comparison between Two Theories of Radiation," *Nature*, 72 (1905), 293–294.
60. J. H. Jeans, "On Non-Newtonian Mechanical Systems, and Planck's Theory of Radiation," *Phil. Mag.*, 20 (1910), 943–954.
61. H. A. Lorentz, "On the Emission and Absorption by Metals of Rays of Heat of Great Wavelength," *Proc. Amsterdam*, 5 (1903), 666–685; *Collected Papers*, Vol. 3 (The Hague, 1936), pp. 155–176. Quotation from pp. 668f., 157f. In referring to alternate ways of evaluating probability, Lorentz may have had any one of several things in mind: the choice of equiprobable elements, the distribution of energy over resonators as against the distribution of resonators over energy, or Planck's justifiable but obscure equation of probability with the total number of ways of distributing a given amount of energy.
62. Paul Ehrenfest, "Über die physikalischen Voraussetzungen der Planck'schen Theorie der irreversiblen Strahlungsvorgänge," *Wiener Ber.*, II, 114 (1905), 1301–1314; *Collected Scientific Papers* (Amsterdam and New York, 1959), pp. 88–101. Quotation on penultimate page. Ehrenfest's attributing restricted values of energy to the field rather than to resonators is probably due to the fact that his article nowhere introduces resonators at all.
63. M. J. Klein, *Paul Ehrenfest, Vol. 1: The Making of a Theoretical Physicist* (Amsterdam, London, New York, 1970), p. 46.
64. Lorentz to Wien, 6 June 1908. This letter is discussed more fully in Chapter VIII, below. Its location is described and the German text quoted in nn. 10 and 19 to that chapter.
65. H. A. Lorentz, ["Discussion"], *Report of the British Association*, 1913, p. 385, italics added.
66. Albert Einstein, ["Review of Plank's *Wärmestrahlung*,"] *Annalen der Physik, Beiblätter*, 30 (1906), 764–766.
67. G. H. Bryan, ["Review of Planck's *Wärmestrahlung*"], *Nature*, 74 (1906), supplement to issue of October 11, pp. iii–iv.
68. Clemens Schaefer, ["Review of Planck's *Wärmestrahlung*"], *Phys. ZS.*, 8 (1907), 224.

Notes to Chapter VI

1. Lord Rayleigh, "Remarks upon the Law of Complete Radiation," *Phil. Mag.*, 49 (1900), 539–540; reprinted in John William Strutt, Baron Rayleigh, *Scientific Papers*, Vol. 4 (Cambridge, Eng., 1903), pp. 483–485. This paper has been carefully discussed by Hans Kangro in his *Vorgeschichte des Planckschen Strahlungsgesetzes* (Wiesbaden, 1970), pp. 189–192, and also by M. J. Klein in his "Max Planck and the Beginnings of the Quantum Theory," *Archive for History of Exact Sciences*, 1 (1962), 459–479, esp. 465–468. Both emphasize the importance of recognizing that it does *not* contain the Rayleigh–Jeans law.
2. Lord Rayleigh, "The Dynamical Theory of Gases and Radiation," *Nature*, 72, (1905), 54–55; reprinted in Rayleigh's *Scientific Papers*, Vol. 5 (Cambridge,

Eng., 1912), pp. 248–252. Because of the paper's brevity, the fragmentary quotations which follow are not further localized.
3. The error resulted from Rayleigh's failure to consider only the first octant of the sphere when counting modes. Jeans's correction was published as a postscript to his paper "On the Partition of Energy between Matter and Aether," *Phil. Mag.*, *10* (1905), 91–98, and it was acknowledged by Rayleigh in "The Constant of Radiation as Calculated from Molecular Data," *Nature*, *72* (1905), 243–244; reprinted in his *Scientific Papers*, V, p. 253.
4. For the response of experimentalists to Rayleigh's suggestion see Kangro, *Vorgeschichte*, pp. 191f. For Planck's knowledge of the temperature dependence of the intensity of long wavelength radiation, see Chapter IV, note 12, above.
5. O. Lummer and P. R. E. Jahnke "Über die Spectralgleichung des schwarzen Körpers und des blanken Platins," *Ann. d. Phys.*, *3* (1900), 283–297.
6. H. Rubens and F. Kurlbaum, "Über die Emission langwelliger Wärmestrahlen durch den schwarzen Körper bei verschiedenen Temperaturen," *Berl. Ber.*, 1900, pp. 929–941, where Rayleigh's formula is rejected on p. 940. A fuller version of the same report is, "Anwendung der Methode der Reststrahlen zur Prüfung des Strahlungsgesetzes," *Ann. d. Phys.*, *4* (1901), 649–666, where Rayleigh's proposal is discussed on pp. 651–653.
7. On this subject see S. G. Brush, "Foundations of Statistical Mechanics, 1845–1915," *Archive for History of Exact Sciences*, *4* (1967), 145–183, esp. 160–162.
8. Ludwig Boltzmann, "On Certain Questions of the Theory of Gases," *Nature*, *51* (1894–95), 413–415, quotation on p. 414; reprinted in Boltzmann's *Wissenschaftliche Abhandlungen* (Leipzig, 1909; New York, 1968), III, 535–544, quotation on p. 538.
9. J. H. Jeans, "The Distribution of Molecular Energy," *Phil. Trans.*, *196* (1901), 397–430, quotation from p. 398.
10. "Partition of Energy" (Jeans, 1905a) was quickly followed by "On the Application of Statistical Mechanics to the General Dynamics of Matter and Aether," *Proc. Roy. Soc. London*, *76* (1905), 296–311.
11. Lord Rayleigh, "The Dynamical Theory of Gases," *Nature*, *71* (1904–05), 559; reprinted in *Scientific Papers*, V, p. 248. This is the first letter in an important series of which the items in *Nature* cited in notes 2 and 3 above are parts. Jean's first response appears on p. 607 under the same title.
12. "Theory of Gases and Radiation" (Rayleigh, 1905b).
13. Compare p. 137, above.
14. J. H. Jeans, "The Dynamical Theory of Gases and of Radiation," *Nature*, *72* (1905), 101–102.
15. J. H. Jeans, "On the Laws of Radiation," *Proc. Roy. Soc. London*, *76* (1905), 545–552.
16. For Jeans's acknowledgment of the need for a new derivation of the Stefan–Boltzmann law, see "Dynamics of Matter and Aether" (1905d), pp. 309–311. For an early attempt to supply one, see his "Laws of Radiation" (Jeans, 1905f), and compare the reference to this paper on p. 133, above.
17. On this subject see Brush, "Foundations of Statistical Mechanics" (Brush, 1967), pp. 162–168.

18. *Ibid.*, pp. 168–177.
19. The gap is, of course, filled by recourse to "natural radiation."
20. Max Planck, *Vorlesungen über die Theorie der Wärmestrahlung*, 1st ed. (Leipzig, 1906), p. 172, italics added.
21. The phrase was first introduced by Ehrenfest in his "Welche Züge der Lichtquantenhypothese spielen in der Theorie der Wärmestrahlung eine wesentliche Rolle?," *Ann. d. Phys.*, *36* (1911), 91–118; reprinted in his *Collected Scientific Papers* (Amsterdam, 1959), 185–212, edited by M. J. Klein. At the start of that paper Ehrenfest enumerates the characteristics of thermal radiation basic to his investigation, and he includes as item IV, "Avoidance of the Rayleigh–Jeans-Catastrophe in the Ultraviolet."
22. M. J. Klein discusses Ehrenfest's visit to Leyden in his, *Paul Ehrenfest: Volume 1: The Making of a Theoretical Physicist* (Amsterdam, New York, London, 1970), pp. 45f.
23. The relevant entry is number 361 (and perhaps the two or three more cryptic ones immediately following) in the second of Ehrenfest's research notebooks. These notebooks, of which there are thirty-six, constitute the series ENB:1 on deposit at the Museum Boerhaave, Leyden, and, on microfilm, at the various Archives for the History of Quantum Physics. Each notebook records the date at which it was opened (sometimes given according to both the Western and Russian calendars) and contains roughly three-hundred numbered entries of various length, a few of them also dated. Experiment suggests that the dates of other entries can be established to within a week or so by linear interpolation. Item numbers are consecutive from one notebook to the next, and they run from 1 to 999, after which a new series begins. Since all citations of the notebooks, below, are to the series ENB:1, the series designation is hereafter omitted, only volume and item number being supplied. The entry mentioned above would, for example, be cited as II-361.
24. The Ehrenfest notebooks containing materials relevant to these two published papers are numbers V, VI, and VII, opened respectively on 10 March 1905, 12 July 1905, and 27 January 1906. Doubtless the clues they provide would, together with much other material, permit a detailed reconstruction of the development of Ehrenfest's thought, a task which his stature and status would sufficiently justify. But it would be difficult, especially working from microfilms which are not always clear. Written for his own use, Ehrenfest's notes are generally exceedingly condensed, make use of drastic abbreviation, and include occasional words and phrases in shorthand (Gabelsberger). In addition, the subject often changes entirely from entry to entry. In working with the notebooks, I have therefore confined attention to items that seemed directly relevant to the two papers of present interest. For help in deciphering shorthand in a few entries of likely importance, I am much indebted to my colleague Victor Lange.
25. For examples see Ehrenfest, V–354, 369; VI–532; VII–668.
26. Ehrenfest, V–266. There are also some hints of this general approach in II–361, written after Ehrenfest's return from Leyden in 1903.
27. V–285.
28. For examples see, Ehrenfest, V–292, 368; VI–*453–456*, *466*. (The italics

eliminate an ambiguity in references to some of the numbered items in notebook VI where item 546 is inadvertently followed by 447, from which the numbers again increase. The second series of numbers, which duplicates the first until 547 is reached, is here italicized.) Note also the interesting list of analogies between radiation theory and mechanics in V–400.

29. The reference to Jeans occurs in Ehrenfest, V–407. For the later centrality of the Rayleigh–Jeans problem see below.
30. "Zur schw-Strahl. / 1. Merkwürdig dassnur *ein* Maximum," Ehrenfest, V–333.
31. "Planck fand viele Functionen die stets zunehmen—und danach verschiedene stationäre Zustande—Klarstellen!!", Ehrenfest, V–337. "Boltzmanns H fällt in jenen Fallen wo Entropie überhaupt definiert ist mit ihr zusammen —aber wie bei Planck? // Plancks Stationarität-Formel ist für $dS/dt = 0$ hinreichend aber nicht nothwendig!", V–340. The sign, //, indicates a break between paragraphs within a numbered entry and is used consistently. The sign, /, is used to separate items on different lines when the distinction seems relevant, as in the preceding note.
32. "*Abstracte Entropietheorie*," Ehrenfest, V–342.
33. "Kann man ernstl. glauben dass die Grösse der Electr. Ladung allein schon dafür sorgt, dass ein vorgegeb. Quantum Totalenergie auf eine bestimmte Art zerzaust werde—Versuch das Gegentheil zu beweisen," Ehrenfest, VI–444.
34. In late June (V–364) Ehrenfest calls for the use of Lorentz's dimensional considerations, the one element of his November paper not already specified above. For an example of its use see V–371.
35. Ehrenfest, "Über die physikalischen Voraussetzungen der Planck'schen Theorie der irreversiblen Strahlungsvorgänge," *Wiener Ber.*, 114 (1905), 1301–1314; *Collected Scientific Papers*, pp. 88–101. The passages quoted are from pp. 1303f.; 90f.
36. *Ibid.*, p. 1313; 100. Much of the relevant text is quoted above, p. 138.
†37. Immediately after completing the preceding argument, Ehrenfest notes that Planck has, in fact, required a satisfactory entropy function to reach an absolute, not simply a relative, maximum. But, he goes on, that criterion is empty, for in the absence of some means for identifying one of them with *the* entropy, there is still no way to choose between Σ_1 and Σ_2. (*Ibid.*, pp. 1309f.; 96f.)
†38. Ehrenfest claims that Planck's theory is compatible with an infinite number of different entropy functions, but to produce them he requires two independent functions, Σ_1 and Σ_2, the others being generated presumably by linear combination. For the existence of the first pair he simply cites Planck's earlier work, ignoring Planck's hope, expressed as late as 1901 (above, p. 115), that consideration of the most general possible case of cavity radiation would eliminate all but one. That hope was vain, of course, but Ehrenfest's argument gave Planck no reason to abandon it.
39. Ehrenfest, VI–*464* reads in full: "Allgemeine Theorie der Quasi-Entropien / 1. Stets wachs Funct. / 2. Einzigartigkeit des Endzustandes / 3. Complexionentheorie."
40. "Voraussetzungen der Planck'schen Theorie" (Ehrenfest, 1905), p. 1305; 92.

VI-*452* opens: "Es ist nicht so weither mit der Unabh des Endzustandes eines thermischen Systems von den speziellen Anfangsbed."
41. Ehrenfest, VI-*456*. The model is discussed again ten items later, a likely index of how seriously Ehrenfest was considering the problems the model raised. Note that what intervenes between the two descriptions is the call for a general theory of quasi-entropies.
42. Ehrenfest, VI-*457*: "H nimmt nur dann zu wenn die Mol do $d\omega$ verschiedene Erlebnisse haben − / a) Stösse unterein / b) [Stösse] auf eine sehr borstige Wand / ??!! c) Enorm lange Zeit −" The last item, c), is emphasized by being enclosed in a box. Like the preceding entry, this one is soon taken up again, in this case at VI-*469*.
43. Ehrenfest, VI-*462*: "Heuristisch für Thermod-kinet Theorie der Hohlraumstrahlung: // Partialschwing laufen ohne einander zubei / Resonatoren [in left hand column] // Molekule versch Arten laufen ohne jeden Zusammenstoss / katalyt-Substanzen [right hand column]".
44. Ehrenfest, VI-*515, 516*: In these entries Ehrenfest begins to speak of the model as a "Stossfrei" gas, and the term thereafter recurs.
45. Ehrenfest, VII-619, 631, 632; the title of the second item is, "Vereinfachtes Modell für Quasi H-Theorem." "Zur Planckschen Strahlungstheorie," *Phys. ZS*, **7** (1906), 528–532; *Collected Scientific Papers*, pp. 120–124. The discussion of the model occurs on pp. 529f.; 121f.
†46. That Ehrenfest did see in November the point he made explicit in his June paper is suggested by the juxtaposition of entries in his notebooks during the former month. In VI-*457*, quoted above, he points out that collisions between molecules and rough walls will cause entropy to increase even in the absence of collisions between molecules. VI-*459* then opens, "Nachweiss dass es *auch* in einem Resonatorfreien Spiegelraum Function Σ gibt die fortwähr zunehmen" (italics added). Finally, in VI-*462*, he introduces the parallel outlined above between collision-free molecules and individual vibration modes.
47. S. H. Burbury, "On Irreversible Processes and Planck's Theory in Relation Thereto," *Phil. Mag.*, **3** (1902), 225–240, quotation from p. 237. Note the previous mention of this paper on p. 136, above, as well as Larmor's 1902 reference to the impotence of resonators, quoted in the same place. Ehrenfest's notebooks contain many references to Burbury, but none relates unambiguously to this passage. Three postcards and a letter written by Burbury during 1905 are included in Ehrenfest's correspondence, but they mostly report only that reprints are on the way. The letter of 26 October does indicate that by then Ehrenfest had read Burbury's 1902 paper, but the parts that interested him are not identified.
48. "Planckschen Strahlungstheorie" (Ehrenfest, 1906), pp. 528f.; *Collected Scientific Papers*, pp. 120f.
49. Ehrenfest, V-352, 353, 373.
†50. That Ehrenfest at this time believed a resonator reradiated both at the frequency of the driving field and also at its natural frequency is suggested by the following fragment from V-401, written perhaps two weeks after the third entry just cited: "In einem bestimmten Augenblick im ganzen Raum kleine Strahlstöpselchen durcheinandergesteckt – und Resonatoren schwingen

und senden Stöpsel und empfangen ebensolche / Trifft ein Stöpsel auf Resonator so: 1. er lauft weiter / 2. Resonator sendet Stöpsel neuer Art / 3. Resonator sendet alter Art aus."

51. Ehrenfest, V-*354, 369*. "Planckschen Strahlungstheorie" (Ehrenfest, 1906), p. 529; *Collected Scientific Papers*, p. 121.
52. See above, p. 157, and note that the same phrase is elsewhere repeated.
53. Ehrenfest, VI-*461*: "Solche Resonatoren anzugeben die jede einfallende Welle in ein kleines Spectrum verwandeln (von Temp abh-) (Diffgl nicht linear)." VI-*465*: "Schwarzen Körper: jedes vorgeg. Energiequant in ganz best. Spectrum (von T abhängig) verwandelt."
54. Ehrenfest, VI-*497, 510, 523*. In these entries Ehrenfest isolates the difficulty of identifying Fourier components of the field from measurements of radiation *intensity*, asks whether it is legitimate in theory to call on monochromatic filters, proposes an analysis of the intensity spectrum produced by the apparatus of Lummer and Pringsheim, and suggests replacing Planck's definition of intensity with the "grating definition."
55. "Planckschen Strahlungstheorie" (Ehrenfest, 1906), p. 529nn.; *Collected Scientific Papers*, p. 121nn. As will appear below, Ehrenfest's identification of the source of the confusion is not quite right, a fact which increases the likelihood that these notebook entries are relevant.
56. *Ibid.* Ehrenfest's early awareness of the possible need for a non-linear resonator equation is indicated above. For the possible intervention of molecules, see note 63, below, and remember that Ehrenfest had dealt with molecules interacting with springs from an early date.
†57. Ehrenfest's reference to Planck's "Conclusion" occurs on the first page of his paper, which was submitted for publication on 28 June 1906. By that date he must have spent some time with the *Lectures*, for his article includes numerous other references to its text. In view of the dating of Planck's preface, one wonders whether the book can possibly have appeared in time for Ehrenfest to acquire it through normal channels. (Easter day in 1906 was 15 April, only six weeks before Ehrenfest's article was submitted.) Very likely Planck had sent him proof sheets in response to their correspondence. But that cannot be taken for granted. Einstein's review of the book for the *Annalen der Physik, Beiblätter, 30* (1906), 764–766, appears in the fifteenth (mid-August?) of the twenty-four annual installments of that journal. Planck's book must therefore have been issued with surprising speed.
58. *Wärmestrahlung* (Planck, 1906a), p. 220.
59. *Ibid.*, p. 28.
†60. Ehrenfest might well have raised the issue of energy redistribution in a letter to Planck accompanying a reprint of his November paper. That would help to explain the particular nature of the latter's slip. One cannot, of course, entirely bar the possibility that Planck, stimulated by Ehrenfest's 1905 paper, discovered the impotence of resonators for himself. But it seems extremely unlikely. The letter discussed in Chapter V (p. 132) indicates that Planck did not take Ehrenfest's argument seriously, and he had, in any case, known for some years that the electromagnetic entropy function was not unique. Ehrenfest's paper does not at all suggest the impotence of

resonators, so that a reader who followed Planck's footnote would inevitably have been puzzled, a difficulty Planck could have eliminated easily if he had recognized it.

61. For a helpful treatment of the problem of small oscillations and the use of normal coordinates, see Herbert Goldstein, *Classical Mechanics* (Reading, Mass., 1950), Chapter 10.
62. Ehrenfest, VII-691: "Energievertheil. über [?] Hauptschwingungen.... Ein H-Theorem auf dieser Basis." The question mark in square brackets takes the position of a scrawled sign that probably represents a spring. That sign appears to be repeated within the part of the entry replaced by ellipsis above. There Ehrenfest speaks of determining the effect of impacts, presumably molecular, on a spring coordinate of which the amplitude and phase had previously been given.
63. Ehrenfest, VII-782: "Durch zurückgehen auf Hauptschwing des Systems Aether + Resonatoren zeigen dass so nie eine Aender der 'Farbenvertheil' herauskommen kann— / Wohl aber durch Molekulstösse." Note the last phrase, which is not quoted in the text above.
64. *Wärmestrahlung* (Planck, 1906a), p. 175.
65. "Planckschen Strahlungstheorie" (Ehrenfest, 1906), p. 529; *Collected Scientific Papers*, p. 121.
66. *Ibid.*, p. 530; 122.
67. See above, pp. 51f. The general formula for entropy given there differs from the one embodied in equation (8) by the use of velocity, rather than momentum, coordinates and in the exclusion of the effects of spatial distribution. Though the more general formulation has older roots, a convenient contemporary source was Chapter 3 of the second volume of Boltzmann's *Gastheorie*.
68. The coincidence of results is not, of course, surprising. Ehrenfest has taken for granted that equal areas of phase space are equally probable, a conclusion that cannot be justified without resort to some form of ergodic hypothesis. The same hypothesis will, however, justify appeals to the equipartition theorem. Ehrenfest surely understood this equivalence by the time he published his famous encyclopedia article in 1912, but, as indicated in Chapter II, above, it was not well understood before that date. In 1906 it may still have been obscure to Ehrenfest.
69. "Planckschen Strahlungstheorie" (Ehrenfest, 1906), p. 531; *Collected Scientific Papers*, p. 123.
70. *Ibid.*, p. 532; 124.
71. *Ibid.*

Notes to Chapter VII

1. Albert Einstein, "Zur Theorie der Lichterzeugung und Lichtabsorption," *Ann. d. Phys.*, 20 (1906), 199–206, quotation on p. 202.
2. M. J. Klein, "Thermodynamics in Einstein's Thought," *Science*, 157 (1967), 509–516.
3. Albert Einstein, "Folgerungen aus den Capillaritätserscheinungen," *Ann. d.*

Phys., *4* (1901), 513–523, and "Über die thermodynamische Theorie der Potentialdifferenz zwischen Metallen und vollständig dissozierten Lösungen ihrer Salze und über eine elektrische Methode zur Erforschung der Molekularkräfte," *Ann. d. Phys.*, *8* (1902), 798–814.

4. J. Willard Gibbs, *Elementary Principles in Statistical Mechanics* (New York and London, 1902).
5. Albert Einstein, "Kinetische Theorie des Wärmegleichgewichtes und des zweiten Hauptsatzes der Thermodynamik," *Ann. d. Phys.*, *9* (1902), 417–433. Einstein permits the potential to be a slowly varying function of time. The remark about the generalizability of his treatment occurs on p. 427.
6. Albert Einstein, "Eine Theorie der Grundlagen der Thermodynamik," *Ann. d. Phys.*, *11* (1903), 170–187.
7. *Ibid.*, p. 171.
†8. The main difficulty with Einstein's argument is that his physicality condition is much too strong. The time required for a physical system to return close to its original microscopic configuration is vastly longer than the time required for typical observations. The reason that the averages which correspond to observables are nevertheless stable is that a system spends the overwhelming majority of its time in regions which yield the same average values over short intervals. Einstein's conception of a stable trajectory also presents difficulties. Apparently he had in mind some generalization of Boltzmann's H-theorem, stable trajectories emerging only after H has reached its maximum value. But, as indicated in Chapter II, the H-theorem is only statistical; the individual systems governed by equations of motion never do become "stationary" but continue forever to retrace configurations corresponding to all possible values of H.

Though they were not widely known, there are many precedents for the use of space (and thus ensemble) averages as a means of resolving the problems presented by time averages. The conceptual difficulties in relating the two and in justifying the substitution of one for the other play an essential role in the history of gas theory, one which badly needs historical study. They are traceable from Boltzmann's first papers on a mechanical derivation of the second law, and they lead, among other places, to an almost total misunderstanding in Britain of (Maxwell, 1879), the first paper to introduce ensembles. (I am indebted to Bruce R. Wheaton and Kathryn Olesko for discussion of these points.) There is also much precedent for Einstein's difficulties with this part of his argument as well as for his resort to ensembles, though their use was neither common nor at all well understood before the assimilation of Gibbs's book. Boltzmann had followed (Maxwell, 1879) in, for example, (Boltzmann, 1885, 1887) as well as in *Gas Theory*, *II* (Boltzmann, 1898). Gibbs knew most or all of these works; Einstein probably only the last.

9. *Statistical Mechanics* (Gibbs, 1902), Chapter 10.
10. For the following see, "Grundlagen der Thermodynamik" (Einstein, 1903), pp. 182–185.
11. *Ibid.*, 174–182. These are the elements which make the subject of Einstein's papers and of Gibbs's book "statistical thermodynamics" rather than just "statistical mechanics." There is very little earlier precedent for them.

12. Einstein, in fact, writes $1/(4h)$ for the mean energy of a molecule, a slip corrected in his next paper.
†13. These connections are suggested by Klein's "Thermodynamics in Einstein's Thought" (Klein, 1967), p. 511. Note also how directly, given the question about χ and its role in the definition of temperature, equation (4), above, suggests the treatment of energy fluctuations in a small body in equilibrium contact with a much larger one. But see also note 15, below.
†14. In one way or another virtually all earlier treatments of statistical mechanics had implicitly or explicitly assumed the number of particles or of systems to be so great that the law of large numbers applied without question. See Chapter I, note 60, above, for what currently appears to be the first reference to anything like fluctuation phenomena.
†15. Albert Einstein, "Zur allgemeinen molekularen Theorie der Wärme," *Ann. d. Phys.*, *14* (1904), 354–362, where the new definition of $\omega(E^*)$ is introduced at once, thus eliminating a significant imperfection of Einstein's previous paper. Temperature and entropy characterize individual systems, not just ensembles. But Einstein's definition of temperature depends, through equations (5) and (6), on the ensemble parameter δE^*. The same parameter is implicitly present in the quantities T and χ of equation (7), Einstein's definition of entropy. These difficulties disappear from the present paper where Einstein for the first time treats fluctuations. Formally, their disappearance is due to the new definition of $\omega(E^*)$. Conceptually, it is due to Einstein's inversion of his previous treatment of thermometers. In the fluctuation paper, an individual system is always considered to be in equilibrium with a relatively infinite environment of specified temperature T_0. The probability of a specified state of the individual system can then be computed from the equation $dW = C \exp(E/2\chi T_0) \, dp_1 \cdots dp_n$, where the probability is again conceived as the fraction of time the system spends in $dp_1 \cdots dp_n$. It is thus possible that the need to eliminate δE^* from the definitions of temperature and entropy also played a role in directing Einstein's attention to fluctuations. In this connection, note the discussion in the text of the limits of integration of equation (7).
16. *Ibid.*, p. 360.
17. *Ibid.*, pp. 360f. Einstein's paper on Brownian motion, a second physical phenomenon permitting the evaluation of χ, appeared in the next year.
18. See above, pp. 110f., 130–133.
19. Albert Einstein, "Über einen die Erzeugung und Verwandlung des Lichtes betreffenden heuristischen Gesichtspunkt," *Ann. d. Phys.*, *17* (1905), 132–148. For a more detailed discussion of this paper see M. J. Klein, "Einstein's First Paper on Quanta," *The Natural Philosopher*, *2* (1963), 59–86.
20. "Erzeugung und Verwandlung des Lichtes" (Einstein, 1905), p. 136.
21. *Ibid.*, p. 136f. Rayleigh had made the same point about the derivation of atomic constants from radiation theory in his (1905b).
22. See especially the derivation of equation (IV-7) on p. 100, above.
23. "Erzeugung und Verwandlung des Lichtes" (Einstein, 1905), p. 139.
24. Throughout his paper Einstein speaks of low intensity rather than of high frequency radiation. Since his derivations demand only that the value of

ν/T be appropriately high, the two are technically equivalent, but Einstein's choice of phrase is probably significant. For high intensity, with many light particles present, fluctuations disappear. Maxwell's equations, conceived as applicable to the mean behavior of the field, are then relevant. For a sketch of the later development of the idea of light particles, see M. J. Klein, "Einstein and the Wave-Particle Duality," *The Natural Philosopher, 3* (1964), 3–49.
25. Carl Seelig, *Albert Einstein: Eine dokumentarische Biographie* (Zurich, Stuttgart, Wien, 1953), pp. 173–176.
26. "Lichterzeugung und Lichtabsorption" (Einstein, 1906a), quotation on the first page.
27. Here and hereafter I substitute Planck's k for Einstein's R/N. Einstein's adoption of Planck's more compact notation takes place by stages in the two papers of 1909 considered below.
28. *Ibid.*, p. 202.
29. *Ibid.*, p. 203.
30. Indication of the nature of Einstein's efforts is provided in the closing paragraphs of the two 1909 papers cited immediately below. His correspondence provides additional detail about his changing attitude towards the problem. See also, Russell McCormmach, "Einstein, Lorentz, and the Electron Theory," *Historical Studies in the Physical Sciences, 2* (1970), 41–87, especially pp. 69–81.
31. Albert Einstein, "Zur gegenwärtigen Stand des Strahlungsproblems," *Phys. ZS, 10* (1909), 185–193, quotation from p. 186.
32. *Ibid.*, pp. 182f.
33. Albert Einstein, "Über die Entwicklung unserer Anschauungen über das Wesen und die Konstitution der Strahlung," *Phys. ZS, 10* (1909), 817–825, quotation from p. 822.
†34. Having in 1905 considered the entropy of radiation distributed in accordance with the Wien law, Einstein in his two papers of 1909 took up the corresponding problem for the Planck distribution. He then showed that the mean square fluctuation in a black-body cavity could be represented as the sum of two terms, one dominant at high frequency, the other at low. The first corresponded to particle-like behavior; the second to fluctuations produced by the interference of waves.

Notes to Chapter VIII

1. Joseph Larmor, "On the Statistical Theory of Radiation," *Phil. Mag., 20* (1910), 350–353; *Mathematical and Physical Papers by Sir Joseph Larmor*, Vol. 2 (Cambridge, Eng., 1929), pp. 413–415.
2. Armin Hermann, "Laue, Max von," in *Dictionary of Scientific Biography*, ed. C. C. Gillispie, Vol. 8 (New York, 1973), 50–53; Carl Seelig, *Albert Einstein: Eine dokumentarische Biographie* (Zurich, 1954), pp. 91f.
3. Laue to Einstein, 2 June 1906: "Zugleich danke auch ich Ihnen für den Korrekturbogen Ihrer inzwischen in den Annalen erschienenen Abhandlung, die ich mit sehr viel Interesse gelesen habe und der ich, wie ich jetzt ausführen

will, vollauf beistimme. / ...Wenn Sie am Anfang Ihrer letzten Antwort Ihren heuristischen Gesichtspunkt dahin formulieren, dass Strahlungsenergie nur in gewissen endlichen Quanten absorbiert und emittiert werden kann, so weiss ich nichts dagegen einzuwenden; auch alle Ihre Anwendungen stimmen mit dieser Fassung überein. Nun ist dies keine Eigentümlichkeit der elektromagnetischen Vorgänge im Vakuum, sondern der absorbierenden oder emittierenden Materie. Die Strahlung besteht daher nicht aus Lichtquanten, wie in §6 der ersten Arbeit steht, sondern verhält sich nur im Energieaustausch mit der Materie so, wie wenn sie daraus bestünde." Original in the Einstein archive at the Institute for Advanced Study, Princeton, New Jersey.
4. Ibid.
5. Einstein's 1906 paper is very seldom cited in the contemporary literature and Ehrenfest's not at all. They were, however, almost certainly known by both Planck and Lorentz before 1908.
6. H. A. Lorentz, "The Theory of Radiation and the Second Law of Thermodynamics," Proc. Amsterdam, 3 (1901), 436–450. This English version of a Dutch paper presented in the preceding year is reprinted in Lorentz's Collected Papers, Vol. 6 (The Hague, 1938), pp. 265–279. For a few further details see above p. 132f.
7. H. A. Lorentz, "Boltzmann's and Wien's Laws of Radiation," Proc. Amsterdam, 3 (1901), 607–620; Collected Papers, Vol. 6, pp. 280–292. "On the Emission and Absorption by Metals of Rays of Heat of Great Wavelengths," Proc. Amsterdam, 5 (1903), 666–685; Collected Papers, Vol. 3 (The Hague, 1936), pp. 155–176.
8. Ibid., p. 667; 156.
9. Ibid., p. 678; 168f.
10. Lorentz to Wien, 6 June 1908: "Ich habe mir nun in den letzten Jahren mit diesem Problem unaufhörlich den Kopf zerbrochen." From a typed transscript prepared by Wien or at his direction and now deposited at the Deutsches Museum, Munich. The location of the original is currently unknown. More about this important letter will be said below.
11. H. A. Lorentz, "La thermodynamique et les théories cinétiques," Bulletin des Séances de la Société Française de Physique, 1905, pp. 35–63; Collected Papers, Vol. 7 (The Hague, 1934), pp. 290–316. Quotation on p. 60; 313.
12. H. A. Lorentz, "Le partage de l'énergie entre la matière pondérable et l'éther," Atti del IV Congresso Internazionale dei Matematici (Roma, 6–11, Aprile 1908), 3 vols. (Rome, 1909), I, 145–165; Collected Papers, Vol. 7, pp. 317–346. Also reprinted in an improved form in Nuovo Cimento, 16 (1908), 5–34, and Revue Général des Sciences, 20 (1909), 14–26.
13. Ibid., p. 147; 320.
14. Ibid., pp. 160f.; 337.
15. Ibid., p. 163; 341. The two following pages are a note added before publication, for which see below.
16. Wien to Sommerfeld, 18 May 1908. The German text of this passage has been reproduced and discussed by Armin Hermann in his Frühgeschichte der Quantentheorie (1899–1913), (Mosbach in Baden, 1969), p. 50. Hermann also discusses some of the other material introduced below, but his evaluation

of Lorentz's position and its importance differs significantly from the one developed here. The full text of the letter is available on microfilm in the various Archives for the History of Quantum Physics. It was made available for filming by Dr.-Ing. Ernst Sommerfeld who intended that it be deposited with his father's other papers in the Deutsches Museum, Munich.

17. O. Lummer and E. Pringsheim, "Über die Jeans–Lorentzsche Strahlungsformel," *Phys. ZS.*, *9* (1908), 449–450. Note that, though the Rayleigh–Jeans law had been announced three years earlier, Lummer and Pringsheim had not bothered to point out its impossibility until Lorentz took it up.

18. H. A. Lorentz, "Zur Strahlungstheorie," *Phys. ZS.*, *9* (1908), 562–563.

19. Lorentz to Wien, 6 June 1908, the letter first introduced in note 10, above. "Eine solche neue Annahme ist nun eben die der Elementarquanten von Energie. An und für sich habe ich nichts dagegen; ich gebe Ihnen sofort zu, dass man gerade mit solchen neuen Auffassungen neue Fortschritte macht und dass vieles zu Gunsten der Hypothese spricht. Auch würde ich sofort bereit sein, die Hypothese rückhaltslos anzunehmen, wenn ich nicht auf eine Schwierigkeit stiesse. Dieselbe liegt darin, dass diejenigen Resonatoren, deren λ merklich kleiner als λ_m ist, nach der Formel von Planck nicht einmal ein einziges Energieelement erhalten. M.a.W., einige dieser Resonatoren (unter Umständen die Mehrzahl) müssten gar keine Energie besitzen, und doch sind sie, ebensogut wie die übrigen, der forwährenden Erregung durch die elektromagnetischen Wellen des Äthers ausgesetzt. Es is nämlich zu bemerken, dass nach der Theorie von Planck die Resonatoren in ganz stetiger Weise (ohne dass von einem endlichen Elementarquantum die Rede ist) Energie von dem Äther erhalten oder an ihn abgeben können. Indes möchte ich jetzt auf diese Frage nicht näher eingehen; ich hoffe bald zu vernehmen, wie sich Prof. Planck selbst zu demselben stellt."

20. H. A. Lorentz, "Die Hypothese der Lichtquanten," *Phys. ZS.*, *11* (1910), 349–354; *Collected Papers*, Vol. 7, pp. 374–384. Quotation from p. 350; 374f.

21. Max Planck, "Hendrik Antoon Lorentz," *Naturwissensch.*, *16* (1928), 549–555; III, 343–349. Quotation from pp. 552f.; 346f. Planck added a note to the printed version pointing out that the argument had originated with Lummer and Pringsheim.

22. H. A. Lorentz, "Über das Gleichgewicht der lebendigen Kraft unter Gasmolekülen," *Wiener Ber. II*, *95* (1887), 115–152; *Collected Papers*, Vol. 6, 74–111. Other relevant papers by Lorentz will be found in this and the preceding volume of his papers.

23. "Lorentz" (Planck, 1928), p. 553; 347.

24. Einstein to J. Laub, 19 May 1909: "Mit H. A. Lorentz habe ich gegenwärtig eine überaus interessante Korrespondenz über das Strahlungsproblem. Ich bewundere diesen Mann wie keinen anderen, ich möchte sagen, ich liebe ihn." The original was copied by Carl Seelig, and a copy of his transcription is in the Einstein archive at the Institute for Advanced Study, Princeton, New Jersey.

25. Ehrenfest had pointed out that Planck's model could alter the energy spectrum if molecules collided with resonators (pp. 162, 166, above). Planck, on the very last page of his *Lectures*, made the same point about a model in

which moving resonators collided among themselves. Lorentz's proof of the Rayleigh–Jeans law covered these cases, though I have no concrete evidence that that characteristic was especially noted at the time. Two years later, in the course of a more general investigation, Einstein and his student L. Hopf showed more explicitly that resonators attached to moving molecules would still yield the Rayleigh–Jeans law in their "Statistische Untersuchung der Bewegung eines Resonators in einem Strahlungsfeld," *Ann. d. Phys.*, *33* (1910), 1105–1115.

26. During this seven-year period Planck's single paper on the black-body problem is "Bemerkung über die Konstante des Wienschen Verschiebungs-gezetses," *Verh. d. D. Phys. Ges.*, *8* (1906), 695–696; II, 136–137. It is a brief comment on a recent determination of the constant $\lambda_m T$, and provides no evidence about Planck's attitudes. For the paper which does, see immediately below.

27. Max Planck, "Zur Dynamik bewegter Systeme," *Berl. Ber.*, 1907, pp. 542–570; *Ann. d. Phys.*, *26* (1908), 1–34; II, 176–209. The quotation is from the first sentence of the paper, which was presented on 13 June 1907. That Planck includes both laws of thermodynamics shows that the "ponderable matter" to be evacuated from the cavity did not include his resonators or, at least, the black speck required to ensure equilibrium. See p. 163, above, for Planck's insistence on this point in his *Lectures*. It recurs in a still more relevant form in a letter from Planck to Lorentz written on 10 July 1909: "Noch ein Wort über den Fall eines von jeglichen Materie u. jeglichen Electronen entblössten, allein von Strahlung erfüllten Hohlraumes. Hier gibt es nach meiner Anschauung überhaupt *keine Möglichkeit*, die Annäherung an einen stationären Endzustand, an eine normale Energieverteilung, auch nun zu verstehen. Dazu bedarf es ganz notwendig etwa eines Kohleteilchens." For the source of this and other letters from Planck to Lorentz, see note 30, below.

28. That Planck refers only to "the principles of mechanics," not to those of either electromagnetic theory or statistical mechanics, might justify a weaker reading of this passage. But Planck must have had electromagnetic theory as well as mechanics in mind, or the passage makes no sense. In addition, his use of "mechanics" seems often to include statistical mechanics as well, for example in his comment of 1909 (p. 199, below) where Planck describes as mechanically impossible the task of explaining the quantum of action. Besides, if Planck is not including his own theory, what can he mean by "recent...theoretical research on thermal radiation"? But see also note 44, below.

29. On this subject see, Chapter V, note 42, above.

30. Planck to Lorentz, 1 April 1908: "Natürlich wird es mich ungemein interessiren aus Ihrem Vortrag in Rom zu erfahren, wie Sie über die grosse Frage nach der Energieverteilung zwischen Aether und Materie denken. Dass die Elektronentheorie ohne Einführung neuer Hypothesen mit Notwendigket zu den Jeanschen Schlussfolgerungen führt, ist mir sehr plausibel, und ich glaube, dass es nur nützlich ist, wenn diese Punkt mit aller Schärfe zum Ausdruck gebracht wird." The original of this letter and of the others from

Planck to Lorentz mentioned here are deposited at the Algemeen Rijksarchief, The Hague, and are also available on microfilm at the depositories of the Archive for History of Quantum Physics.

31. Compare pp. 15f., above.

32. Planck to Lorentz, 7 October 1908: "Die erstere Annahmo [h characteristic for events in the free ether] wäre offenbar die radikalere; sie müsste zur Aufhebung der Maxwellschen Feldgleichungen führen.... / Ich sehe noch keinen zwingenden Grund ein, von der Annahme der absoluten Stetigkeit des freien Aether und aller Vorgänge darin abzugeben. Dann ist also das Wirkungselement h eine Eigenschaft der Resonatoren."

33. *Ibid.* "Ich denke mir nun den Vorgang so: Im freien Aether gehorchen die Lichtwellen genau der Maxwellschen Feldgleichungen. Deshalb ist aber die Amplitude einer Lichtschwingung, auch im stationären Strahlungsfelde, keineswegs mathematisch constant, sondern sie ist starken Schwankungen unterworfen, aus denen erst durch Mittelwertsbildung der constante messbare Werth der Strahlungsintensität K sich zusammensetzt. Befindet sich nun ein Resonator in einem solchen stationären Strahlungsfelde, so wird er von der dort befindlichen freien Strahlung zu Schwingungen angeregt. Aber— und nun kommt das Wesentliche—diese Erregung gehorcht nicht dem einfachen bekannten Pendelgesetz, sondern es existirt eine gewisse Reizschwelle: auf ganz kleine Erregung spricht die Resonator überhaupt nicht an, und wenn er auf grösseren anspricht, thut er dies nur so, dass seine Energie ein ganzes Vielfaches des Energieelement hν ist, so dass der Momentanwerth der Energie stets durch ein solcher ganzer Vielfacher dargestellt wird.

"Zusammenfassend möchte ich also sagen: ich mache zwei Voraussetzungen:

1) die Energie des Resonators in einem bestimmten Augenblick ist ghν (g ganze Zahl oder Null)

2) die in einem Zeitraum von vielen Billionen Schwingungen vom Resonator emittirte und absorbierte Energie und auch die mittlere Energie des Resonators ist ebensogross, als ob für die Resonatorschwingungen die gewöhnliche Pendelgleichung gelten würde.

"Diese beiden Voraussetzungen scheinen mir nicht unverträglich zu sein. Ich habe diesen Gedanken in einer Anmerkung, zu §109 meiner Vorlesungen über Wärmestrahlung, angedeutet."

34. *Vorlesungen über die Theorie der Wärmestrahlung* (Planck, 1906a), p. 108n.

35. *Ibid.*, pp. 154, 221.

†36. Planck's letters to Lorentz of 16 June and 10 July 1909 continue to urge that h be attributed to resonators, not the ether (compare the letter of 7 October 1908 cited in note 32, above, as well as the passage quoted immediately below). Lorentz appears to have conceded the point in a letter of [3]0 July 1909, towards which his preliminary notes are preserved with Planck's letters to him: "Thus, unless someone has a fortunate brainstorm, we should seek h in the radiating particles. I agree also that one should not speak of the entropy of ether alone and that its value depends on the condition of the matter." ("Dus zoolang niet iemand een gelukkigen inval heeft, zullen wij h in de stralende deeltjes zoeken. Dan geef ik ook toe dat van entropie van

aether op zichzelf geen sprake is, en dat de waarde daarvan met de eigenschappen der materie samenhangt.")
37. Planck to Lorentz, 16 June 1909: "Nun übt ein freies Electron nur dann einem Einfluss auf den Aether aus, wenn es seine Geschwindigkeit (an Grösse oder Richtung) ändert. Dies geschieht hauptsächlich beim Zusammenstoss, entweder mit ponderabeln Moleküln oder mit anderen Electronen. Ueber die Gesetze eines solchen Zusammenstosses wissen wir so gut wie nichts, und es ist nach meine Meinung eine ungegründete Annahme, dass dieselben nach den Hamiltonschen Gleichungen vor sich gehen. Vielmehr ist hier nur eine solche Annahme zulässig, die zu Consequenzen führt, die mit den Tatsachen übereinstimmen. Und eine solche Annahme scheint mir die folgende zu sein. *Der Energieaustausch zwischen Electronen und freiem Aether vollzieht sich stets nur nach ganzzahligen Quanten hv.* Das gilt sowohl für freie Electronen als auch für solche, die um Gleichgewichtslagen schwingen, wie in meinen Resonatoren." Compare Planck's letters to Lorentz on 10 July 1909, 7 January 1910, and 25 December 1913. All of them object to Lorentz's continuing attempts to preserve Maxwell's equations for free electrons and to restrict h to a special sort of particle.
38. *Phys. ZS.*, *10* (1909), 825, from the published version of the discussion which followed Einstein's Salzburg lecture, "Über die Entwicklung unserer Anschauungen über das Wesen und die Konstitution der Strahlung" (Einstein, 1909b). Italics added.
39. Max Planck, "Zur Theorie der Wärmestrahlung," *Ann. d. Phys.*, *31* (1910), 758–768; II, 237–247. Quotation on p. 764; 243.
40. *Ibid.*, p. 760; 239.
41. The title and subtitles of Planck's Chapter 3 illustrate the way in which he interchanges terms in the *Wärmestrahlung*. "Oscillator" appears in the overall title and the first subtitle; "resonator" replaces it in the next two subtitles.
42. Planck to Lorentz, 7 January 1910: "Freilich sagen Sie mit vollem Recht, dass ein solcher Resonator sich seines Names nicht mehr würdig zeigt, und dies hat mich bewogen, dem Resonator seinen Ehrennamen abzuerkennen und ihn allgemeiner 'Oscillator' zu nennen (wie auch in dem beigefügten Entwurf geschehen ist)."
43. Note particularly the title of Planck's (1901b), "Über die Elementarquanta der Materie und Elektricität," a usage quite standard in the German literature. The phrase "quantum of action" first appears in the *Wärmestrahlung* (Planck, 1906a), p. 154, and is thereafter used repeatedly in both his letters and his published writings. Unlike the term "resonator," it does not disappear after Planck concedes discontinuity.
†44. The first of these letters is Planck to Ehrenfest, 6 July 1905, reproduced on p. 132, above. The second is Planck to Wien, 2 March 1907, and the relevant paragraph should be quoted in full. "Ihre Messungen der von den einzelne Molekülen ausgesandten Strahlung interessieren mich natürlich sehr. Sollte sich aber der Umstand, dass diese Strahlung einen erheblich kleineren Betrag als das Elementarquantum der Energie angibt, nicht dadurch erklären lassen, dass nicht alle Moleküle gleichmässig emittieren, sondern dass

eine ganze Menge Moleküle überhaupt garnicht emittieren, während andere wieder ein volles Elementarquantum oder mehrere Elementarquanta emittieren? Dann auf der ungleichmässigen u. unregelmässigen Vertheilung der Energie auf die einzelne Moleküle beruht ja die ganze Berechnung der Strahlungsentropie und der Elementarquantums." The original of this letter is at the Staatsbibliothek Preussischer Kulturbesitz in Berlin, and a microfilm copy is available at some of the depositories of the Archive for History of Quantum Physics.

Here Planck not only uses the term "energy quantum" but also speaks of energy emission's occurring in whole quanta. Perhaps he had therefore already accepted the reinterpretation of his theory provided in 1906 by Einstein and Ehrenfest. The only evidence for a later date is the otherwise strange but admittedly inconclusive opening sentence in the paper on relativity he submitted three months later. Planck need, however, only be accepting the vocabulary and mode of analysis to be used by Wien in his interesting paper, "Über die absolute, von positiven Ionen ausgestrahlte Energie und die Entropie der Spektrallinien," *Ann. d. Phys.*, **23** (1907), 415–438. (Since that paper was not received by the journal until 9 May 1907, Planck was presumably commenting on a draft or on a description in a letter.)

45. The relevant part of Planck's Salzburg comment is quoted on p. 199, above. His Columbia lectures are *Acht Vorlesungen über theoretische Physik* (Leipzig, 1910), translated by A. P. Wills as *Eight Lectures on Theoretical Physics* (New York, 1915). On p. 95 (trans. pp. 95f.), at the very end of an extended discussion of black-body theory, Planck points out that electron theory will have to be supplemented to permit a full derivation of his results. He then mentions several authors who think it necessary to abandon Maxwell's equations for radiation in a pure vacuum and concludes: "I think, on the other hand, that it is not yet necessary to proceed in so revolutionary a manner, but that one may seek the meaning of the energy quantum $h\nu$ entirely in the interactions by which the resonators influence each other." Elsewhere these lectures, delivered at a time when Planck's attitude towards his theory was in flux, closely follow the pattern of his 1906 *Lectures* and are correspondingly uninformative concerning points of present interest.

46. For Ehrenfest see pp. 138, 168, above; for Einstein his (1905), p. 143; for Laue p. 189, above; and for Wien his (1907a), p. 417, the paper that, in a preliminary form, evoked Planck's second known use of the phrase.

47. See above, p. 99.

48. Wilhelm Wien, "Über eine Berechnung der Wellenlänge der Röntgenstrahlen aus dem Planckschen Energie-Element," *Göttinger Nachr.*, 1907, pp. 598–601, where the quoted phrase occurs on p. 599.

49. *Ibid.*

50. "Entropie der Spektrallinien" (Wien, 1907a), p. 433.

51. "Wellenlänge der Röntgenstrahlen" (Wien, 1907b), p. 599.

52. Wien to Sommerfeld, 15 June 1908: "Lorentz hat seinen Irrthum in bezug auf die Strahlungstheorie eingesehen und dass die Annahme von Jeans unhaltbar ist. Nun liegt allerdings der Fall insofern nicht ganz einfach, als

in der That es so scheint als ob die Maxwellsche Theorie für die Atome verlassen werden müsste. Ich habe Ihnen daher noch ein Problem zu stellen. Nämlich zu prüfen wei weit die statistische Mechanik und der Beweis von Lorentz fest begründet ist, dass ein den Maxwellsche Gleichungen (beziehentlich denen der Elektronentheorie) gehorchenden System auch der Satz der 'equipartition of energy' gehorchen muss, woraus eben das Jeanssche Gesetz zu folgen wäre. Nämlich eine Beschränkung der Freiheitsgrade, wie sie das Plancksche Energieelement verlangt, müsste doch auch eine elektromagnetische Deutung erlangen. Nun sieht es mir fast so aus als ob eine solche unmöglich wäre, als ob eben diese Beschränkung Zusatzkräfte erfordern (feste Verbindung und dergleichen) die nicht ins Maxwellsche Systeme passen. Wenn dass wirklich so liegt, so brauchte man sich nicht weiter den Kopf über eine Deutung des Energieelements und eine Darstellung der Spektralserien auf electromagnetischer Grundlage zu zerbrechen sondern müsste eine Ergänzung der Maxwellschen Gleichungen innerhalb der Atome zu finden suchen. Mir ist die ganze statistische Mechanik nicht so geläufig, dass ich ein sicheres Urtheil über den Grade ihre Zuverlässigkeit bilden könnte." For the source of this letter see note 16, above.

53. Wilhelm Wien, "Theorie der Strahlung," *Encyklopädie der mathematischen Wissenschaften*, (Leipzig, 1909-26), Vol. 5, Pt. 3, pp. 282-357; quotation on pp. 356f. The sketch of the consequential Rome Congress is on p. 333.
54. *Ibid.*, especially p. 320 which contains the prescient remark about difficulties in accounting for dispersion.
55. *Ibid.*, p. 357.
56. James Jeans, "Zur Strahlungstheorie," *Phys. ZS.*, *9* (1908), 853-855.
57. James Jeans, "On Non-Newtonian Mechanical Systems, and Planck's Theory of Radiation," *Phil. Mag.*, *20* (1910), 943-954.
58. Jeans's paper was, "Rapport sur la théorie cinétique de la chaleur spécifique d'après Maxwell et Boltzmann," *La Théorie du rayonnement et les quanta*, eds. P. Langevin and M. de Broglie (Paris, 1912), pp. 53-73. The discussion follows at once, and the quotation is on p. 76. Other remarks in both his text and the discussion indicate that by 1911 Jeans thought the obstacles to his non-equilibrium account of black-body phenomena nearly insurmountable. But he did not publicly renounce his earlier efforts until late in 1913 (*British Association Reports*, 1913, pp. 376-381), and he was widely believed to be opposed to the developing quantum theory until that date. After it he became Britain's leading proponent of the quantum.
59. H. A. Lorentz, "Alte und neue Fragen der Physik," *Phys. ZS.*, *11* (1910), 1234-1257; *Collected Papers*, Vol. 7, pp. 205-257.
60. *Ibid.*, p. 1248; 238.
61. Above, pp. 103f.

Notes to Chapter IX

1. M. J. Klein, "Einstein and the Wave-Particle Duality," *The Natural Philosopher*, *3* (1964), 3-49, discusses the two most significant subsequent contributions of the black-body problem to the evolution of the quantum theory.

Both began with conceptually novel derivations of Planck's law, the first by Einstein in 1916, the second by S. N. Bose in 1925.

†2. As many citations in this volume have already indicated, individual physicists in the early twentieth century often published identical or closely related articles in several journals at very nearly the same time. Where the numbers involved are small, a count of authors therefore provides a more reliable and stable index of activity than a count of articles. Note, however, that, though an individual author may appear only once each year in the count of total quantum authors, he may, in that same year, appear in more than one of the three categories ("black-body," "specific heat," and "other quantum") into which the quantum literature is here divided. The count of total authors is therefore sometimes smaller than that gained by summing the number of authors in the three sub-categories.

†3. The *Fortschritte* appeared in three sizable volumes per year, each elaborately divided and subdivided. Preliminary experimentation suggested, however, that all or virtually all quantum papers are abstracted in Sections: III. 1, 4, 5; IV. 1, 9–11, 14; V. 1 (with addendum); or VI. 1, 2, 6, 7. These were systematically inspected for the years 1905 to 1914, inclusive. Excepting a few items about which information was available from other sources, papers were counted as contributions to the quantum literature only if their title or abstract included a reference to Planck's theory, the quantum of action, or something else of the sort. Experiments on the black-body spectrum were not, for example, counted as quantum papers simply by virtue of their subject.

4. For normal growth patterns of science and its specialities see D. J. de S. Price, *Little Science, Big Science* (New York and London, 1963), especially Chapter 1.

5. Peter Debye "Der Wahrscheinlichkeitsbegriff in der Theorie der Strahlung," *Ann. d. Phys.*, *33* (1910), 1427–1434. This paper is sometimes described as the first to derive Planck's distribution law by applying combinatorials directly to the vibration modes of the electromagnetic field. But Ehrenfest (1906) had previously indicated how that result was to be achieved (above, pp. 167–169), and J. Weiss's, "Über das Plancksche Strahlungsgesetz (vorläufige Mitteilung)," *Phys. ZS.*, *10* (1909), 193–195, is a clumsy attempt to achieve the same end.

6. For the genesis of Debye's paper and reactions to it see pp. 9–12 of the interview with Debye conducted by T. S. Kuhn and George Uhlenbeck on 3 May 1962. A transcript is deposited in the Archive for History of Quantum Physics.

7. Paul Ehrenfest, "Welche Züge der Lichtquantenhypothese spielen in der Theorie der Wärmestrahlung eine wesentliche Rolle?," *Ann. d. Phys.*, *36* (1911), 91–118; *Collected Scientific Papers* (Amsterdam, 1959), 185–212.

8. Poincaré's earliest publication on this subject is "Sur la théorie des quanta," *C.R.*, *153* (1911), 1103–1108; reprinted in *Oeuvres de Henri Poincaré*, Vol. 9 (Paris, 1954), pp. 620–625. The analysis there summarized is developed at length in "Sur la théorie des quanta," *Journal de Physique Théorique et Appliquée*, *2* (1912), 5–34; *Oeuvres*, Vol. 9, pp. 626–653. For a full account of

Poincaré's view and of his role in the acceptance of the quantum theory see Russell McCormmach, "Henri Poincaré and the Quantum Theory," *Isis*, *58* (1967), 37–55. Note, however, that McCormmach's account of the impact of Poincaré's work on James Jeans may require some qualification in the light of the discussion on pp. 204f., above.

9. Albert Einstein, "Die Plancksche Theorie der Strahlung und die Theorie der spezifischen Wärme," *Ann. d. Phys*, *22* (1907), 180–190.

10. Albert Einstein, "Berichtigung zu meiner Arbeit: 'Die Plancksche Theorie der Strahlung, etc.'," *Ann. d. Phys.*, *22* (1907), 800.

11. The sketch which follows is based primarily on the following four works: H. E. Roscoe and Alexander Classen, *Lehrbuch der anorganischen Chemie*, Vol. I, 3rd ed. (Braunschweig, 1895), esp. pp. 114–121; Richard Börnstein and Wilhelm Meyerhoffer, *Landolt–Börnstein, physikalisch-chemische Tabellen*, 3rd ed. (Berlin, 1905), pp. 383–386; Walther Nernst, *Theoretische Chemie vom Standpunkte der Avogadroschen Regel und der Thermodynamik*, 3rd–6th eds. (Stuttgart, 1900, 1903, 1906 [also 1907], 1909); and B. Weinstein, *Thermodynamik und Kinetik der Körper*, Vol. II (Braunschweig, 1903), esp. pp. 210–226. The first two books (though not necessarily the editions here cited) are mentioned in Einstein's paper as the sources of the data there used. Nernst's text was examined both for its special authority and for its author's role, after 1906, in changing the long-accepted interpretation of existing data on specific heats. The pages cited from Weinstein are a section on "Die specifischen Wärmen der festen Körper." The author concludes that the specific heats of solids quite generally increase with temperature, but does not suppose that the variation is ordinarily large or that, large or not, it conflicts with the kinetic theory of heat.

12. The relevant table is reproduced in all of the works cited in the preceding footnote excepting Nernst's *Theoretische Chemie*. It displays a variation of over 700 percent in the specific heat of diamond. The 1905 edition of *Landolt–Börnstein* records some very recent (1903) measurements that show a variation of almost 400 percent in the specific heat of chromium, but other known variations are by at most a factor of two, usually far less.

13. For the role of chemical thermodynamics in calling attention to Einstein's work on specific heats see also, M. J. Klein, "Einstein, Specific Heats, and the Early Quantum Theory," *Science*, *148* (1965), 173–180.

14. A number of these previous attempts are discussed in Walther Nernst's *Experimental and Theoretical Applications of Thermodynamics to Chemistry* (New York, 1907), pp. 54–57, and his *The New Heat Theorem*, trans. Guy Barr from the 2nd German ed. (London, 1926), Chapter 1.

15. Walther Nernst, "Über die Berechnung chemischer Gleichgewichte aus thermischen Messungen," *Göttinger Nachr.*, 1906, pp. 1–40.

†16. In this form, Nernst's theorem does *not* require that specific heat approach zero with absolute temperature. But extrapolation from existing measurements in combination with the kinetic theory led Nernst to conclude that his theorem could hold only if, near absolute zero, the specific heats of solids were far lower than had previously been thought. He estimated their value as 1.5 calories per mole degree. In 1911, as the quantum and new

measurements persuaded him that specific heats went to zero with temperature, he restated his heat theorem in the form: $\lim(\mathrm{d}A/\mathrm{d}T) = 0$, from which $\lim(\mathrm{d}U/\mathrm{d}T) = 0$ follows. See, for example, his "Der Energieinhalt fester Stoffe," *Ann. d. Phys.*, *36* (1911), 395–439.

17. *Theoretische Chemie*, 5th ed. (Nernst, 1906b), pp. 177–179. Note that Dulong and Petit's generalization, which in the preceding section had been a law (*Gesetz*), has been demoted to a rule (*Regel*). For Nernst, of course, the demotion had occurred during 1906 or perhaps late 1905. In the sixth of his Silliman Lectures, delivered at Yale University in November 1906, he said: "Numerous measurements by different experimenters have shown, in full agreement with each other, that the atomic heats in the solid state decrease greatly at low temperatures" (*Applications of Thermodynamics to Chemistry* (Nernst, 1907), p. 63). The evaluation of experiment in his earlier writings is, of course, quite different.

†18. Walther Nernst, "Über neuere Probleme der Wärmetheorie," *Berl. Ber.*, 1911/I, pp. 65–90, esp. p. 80. Just how accessible the shortcomings of the Dulong–Petit law were is suggested by U. Behn's "Über die specifische Wärme einiger Metalle bei tiefen Temperaturen," *Ann. d. Phys.*, *66* (1898), 237–244, a paper that Nernst first cited in the new section added to his *Theoretische Chemie* in 1906. Though the specific heats reported by Behn remained in the range 4.0 to 6.3 calories per mole degree as temperature varied, the closing paragraphs of his article include the following sentence: "If one represents graphically the decrease of specific heat with temperature, it seems possible that all the curves cross at 0° absolute, so that the specific heats all take the same very small value (0 ?) there." See also Behn's next installment, "Über die specifische Wärme der Metalle, des Graphits und einiger Legirungen bei tiefen Temperaturen," *Ann. d. Phys.*, *1* (1900), 257–269.

19. *Theoretische Chemie*, 6th ed. (Nernst, 1909), p. 700.

20. Walther Nernst, "Untersuchungen über die spezifische Wärme bei tiefen Temperaturen. II.," *Berl. Ber.*, 1910/I, pp. 262–282; quotation on last page.

21. Walther Nernst, "Revue sur la détermination de l'affinité chimique à partir des données thermiques," *Journal chim. phys.*, *8* (1910), 228–267, esp. pp. 234–237.

22. "Probleme der Wärmetheorie" (Nernst, 1911a), p. 86.

23. Einstein to Laub, 16 March 1910: "Die Quantentheorie steht mir fest. Meine Voraussagungen inbetreff der spezifischen Wärmen scheinen sich glänzend zu bestätigen. Nernst, der eben bei mir war und Rubens sind eifrig mit der experimentellen Prüfung beschäftigt, sodass man bald darüber orientiert sein wird." The original was copied by Carl Seelig, and a copy of his transcription is in the Einstein archive at the Institute for Advanced Study, Princeton, N.J.

24. From p. 7 of the interview with Hevesey conducted on 25 May 1962 by T. S. Kuhn and E. Segrè. A transcript is deposited in the Archive for History of Quantum Physics.

25. Nernst to Solvay, 26 July 1910: "Wie es scheint, stehen wir zur Zeit mitten in einer umwälzenden Neugestaltung der Grundlagen, auf welchen die bisherige kinetische Theorie der Materie beruht.

"Einerseits führt diese Theorie in ihrer konsequenten Durchbildung, wie bisher von keiner Seite bestritten wurde, zu einer Strahlungsformel, deren Gültigkeit allen Erfahrungen widerspricht; andererseits folgen aus der gleichen Theorie gewisse Sätze über die specifische Wärme (Konstanz der specifischen Wärme der Gase gegenüber Aenderungen der Temperature, Gültigkeit der Regel von Dulong und Petit bis zu den tiefsten Temperaturen), die ebenfalls durch viele Messungen völlig widerlegt werden.

"Wie insbesondere Planck und Einstein gezeigt haben, verschwinden diese Widersprüche, wenn man der Bewegung der Elektronen und Atome bei Schwingungen um eine Ruhelage gewisse Schranken auferlegt (Lehre der Energiequanten); aber diese Auffassung entfernt sich wiederum so sehr von den bisher benutzten Bewegungsgleichungen materieller Punkte, dass mit ihrer Annahme zweifellos eine weitgehende Reformation unserer bisherigen Fundamentalanschauungen verbunden sein muss." Quoted from the original letter in the Solvay archives on pp. 7f. of an unpublished manuscript by Jean Pelseneer, the latter available on microfilm in the Archive for History of Quantum Physics. An accompanying letter proposes a special conference on the subject, for which see below.

26. "Probleme der Wärmetheorie" (Nernst, 1911a), where the quoted phrase appears on p. 81. The earlier parts of this interesting public lecture include an account of the difficulties of converting a laboratory from high pressure to low temperature research.
27. Max Planck, "Eine neue Strahlungshypothese," *Verh. d. D. Phys. Ges.*, *13* (1911), 138–148; II, 249–259. Planck's first reference to Einstein's theory of specific heats occurs on p. 146; 257.
28. H. A. Lorentz, "Sur la théorie des éléments d'énergie," *Arch. Néerland.*, *2* (1912), 176–191; *Collected Papers*, Vol. 6 (The Hague, 1938), pp. 152–167.
29. A Sommerfeld, "Das Plancksche Wirkungsquantum und seine allgemeine Bedeutung für die Molekularphysik," *Phys. ZS.*, *12* (1911), 1057–1068. W. Nernst, "Über ein allgemeines Gesetz, das Verhalten fester Stoffe bei sehr tiefen Temperaturen betreffend," *ibid.*, pp. 976–978. F. Hasenöhrl, "Über die Grundlagen der mechanischen Theorie der Wärme," *ibid.*, pp. 931–935.
30. Alfred Landé relates that, despite Sommerfeld's insistence that they were wasting their time, he and other advanced students of physics at Munich in the years 1912–14 devoted much time to the invention of atomic models or other devices designed to solve "the quantum riddle." (See p. 4 of the interview conducted by T. S. Kuhn and J. L. Heilbron on 5 March 1962 and deposited in the Archive for History of Quantum Physics.) Such efforts must have continued far longer at many other centers, particularly those outside of Germany, but they seldom left a published record.
31. "Theorie der spezifischen Wärme" (Einstein, 1907a), p. 184.
32. "Grundlagen der mechanischen Theorie der Wärme" (Hasenöhrl, 1911), p. 933.
33. *La Théorie du rayonnement et les quanta: Rapports et discussions de la réunion tenue à Bruxelles, du 30 octobre au 3 novembre 1911*, ed. P. Langevin and M. de Broglie (Paris, 1912), p. 293. See also Poincaré's remark in the discussion of Planck's paper (p. 120). He asked how one would unambiguously decompose

the energy of a three-dimensional oscillator, and Planck replied that no relevant quantum hypothesis had yet been formulated. For the radiation case, to which Planck had previously restricted himself, none was relevant.

34. Walther Nernst, "Zur Theorie der spezifischen Wärme und über die Anwendung der Lehre von den Energiequanten auf physikalisch-chemische Fragen überhaupt," *ZS. f. Elektrochem.*, *17* (1911), 265–275.
35. See above, pp. 147f.
36. Niels Bjerrum, "Über die spezifische Wärme der Gase," *ZS. f. Elektrochem.*, *17* (1911), 731–734.
37. Niels Bjerrum, "Über die ultraroten Absorptionsspektren der Gase," *Festschrift W. Nernst zu seinem fünfundzwanzigjährigen Doktorjubiläum gewidmet von seinen Schülern* (Halle, 1912), pp. 90–98.
38. Albert Einstein, "L'état actuel du problème des chaleurs spécifiques," *Théorie du rayonnement et les quanta* (Compendia, 1912), pp. 407–435. Einstein's remark about his unsuccessful attempt to treat the rotating dipole is on pp. 418f.
39. *Ibid.*, p. 447.
40. P. Weiss, "Über die rationalen Verhältnisse der magnetischen Momente der Moleküle und das Magneton," *Phys. ZS.*, *12* (1911), 935–952, including discussion printed on the last page. When Gans, in the discussion, showed how to explain the existence of a unit of magnetism by quantizing rotations, Weiss responded that Einstein had recently made the same suggestion to him. "Über die ultraroten Absorptionsspekren" (Bjerrum, 1912). J. W. Nicholson, "The Constitution of the Solar Corona. II.," *Month. Not.*, *72* (1912), 677–692.
41. Albert Einstein, ["Discussion"], *Théorie du rayonnement et les quanta* (Compendia, 1912), p. 450.
42. For Ehrenfest's route to and development of adiabatic invariance, see M. J. Klein, *Paul Ehrenfest* (Amsterdam, 1970), pp. 245–251, 257–292. On Ehrenfest's ignorance of the Lorentz–Einstein exchange at the first Solvay Congress, see especially p. 269n.
43. Albert Einstein, "Über einen die Erzeugung und Verwandlung des Lichtes betreffenden heuristischen Gesichtspunkt," *Ann. d. Phys.*, *17* (1905), 132–148. For a sketch of the argument, see pp. 180–182, above.
44. Philipp Lenard, "Über die lichtelektrische Wirkung," *Ann. d. Phys.*, *8* (1902), 149–198. A useful discussion of this paper and of the state of experimentation on the photoelectric effect is included in Bruce R. Wheaton, "The Photoelectric Effect and the Origin of the Quantum Theory of Free Radiation," unpublished M. A. dissertation (University of California, Berkeley, 1971), especially Chapter 11, and "Philipp Lennard and the Photoelectric Effect, 1889–1911," *Historical Studies in the Physical Sciences*, *9* (1978), 299–322.
45. A. F. Joffé, "Eine Bemerkung zu der Arbeit von E. Ladenburg: 'Über Anfangsgeswindigkeit und Menge der photoelektrischen Elektronen u.s.w.,'" *Ann. d. Phys.*, *24* (1907), 939–940. R. Ladenburg, "Die neueren Forschungen über die durch Licht- und Röntgenstrahlen hervorgerufene Emission negativer Elektronen," *Jahrb. d. Radioakt.*, *6* (1909), 425–484. R. A. Millikan, "A Direct Determination of 'h'," *Phys. Rev.*, *4* (1914), 73–75; "New Tests of

Einstein's Photo-Electric Equation," *Phys. Rev.*, *6* (1915), 55; and "A Direct Photoelectric Determination of Planck's 'h'," *Phys. Rev.*, *7* (1916), 355–388. For these papers and for attitudes towards Einstein's photoelectric equation before they were written, see (Wheaton, 1971), Chapter 13.

46. J. Franck and G. Hertz, "Über einen Zusammenhang zwischen Quantenhypothese und Ionisierungsspannung," *Verh. d. D. Phys. Ges.*, *13* (1911), 967–971, is a programmatic introduction to a continuing series that can easily be traced through the index to the journal. It reaches a first important climax in "Über Zusammenstösse zwischen Gasmolekülen und langsamen Elektronen," *ibid.*, *15* (1913), 373–390, where the authors show that, below a certain critical energy, electrons collide elastically with gas molecules. The critical level at which absorption of energy begins is, they suppose, governed by a quantum mechanism. A second important climax was reached the following year in "Über die Erregung der Quecksilberresonanzlinie 253.6 $\mu\mu$ durch Elektronenstösse," *Verh. d. D. Phys. Ges.*, *16* (1914), 512–517. In that paper Frank and Hertz reported that they had divided h into the energy at which inelastic collisions begin in mercury vapor, and that they then looked for and found a strong spectral line at the corresponding frequency. At first they believed that the energy at which inelastic collisions began corresponded to ionization, but that interpretation was soon set aside. For some early reactions to these experiments, see pp. 6–8 of the interview by T. S. Kuhn with Leonard Loeb on 7 August 1962, now deposited in the Archive for History of Quantum Physics.

47. Wilhelm Wien, "Über die absolute, von positiven Ionen ausgestrahlte Energie und die Entropie der Spektrallinien," *Ann. d. Phys.*, *23* (1907), 415–438. The reference to a "quantum of energy" occurs on p. 432, and it is in correspondence about this paper that Planck makes his second known use of that term.

48. Wilhelm Wien, "Über eine Berechnung der Wellenlänge der Röntgenstrahlen aus dem Planckschen Energie-Element," *Göttinger Nachr.*, 1907, pp. 598–601.

49. Johannes Stark, "Elementarquantum der Energie, Modell der negativen und der positiven Elektrizität," *Phys. ZS.*, *8* (1907), 881–884. Much additional information about Stark's involvement in the early quantum theory will be found in Armin Hermann's *Frühgeschichte der Quantentheorie (1899–1913)*, (Mosbach in Baden, 1969), Chapter 4, and in the articles there cited.

50. Johannes Stark, "Beziehung des Doppler-Effektes bei Kanalstrahlen zur Planckschen Strahlungstheorie," *Phys. ZS.*, *8* (1907), 913–919, and "Neue Beobachtungen an Kanalstrahlen in Beziehung zur Lichtquantenhypothese," *Phys. ZS.*, *9* (1908), 767–773.

51. F. Reiche, *Die Quantentheorie, ihr Ursprung und ihre Entwicklung* (Berlin, 1921), pp. 30, 182.

52. Johannes Stark, "Zur Energetik und Chemie der Bandenspektra," *Phys. ZS.*, *9* (1908), 85–94.

53. Johannes Stark, "Weitere Bemerkungen über die thermische und chemische Absorption im Bandenspektrum," *Phys. ZS.*, *9* (1908), 889–894. The last three pages of this paper are: "II. Anwendung der Lichtquantenhypothese

auf die Photochemie." For another step towards use of the quantum in chemistry see, Fritz Haber, "Elektronenemission bei chemischen Reaktionen," *Phys. ZS.*, *12* (1911), 1035–1044.
54. Johannes Stark, "Über Röntgenstrahlen und die atomistische Konstitution der Strahlung," *Phys. ZS.*, *10* (1909), 579–586. Planck in 1910 associated Larmor and J. J. Thomson with Einstein and Stark ("Zur Theorie der Wärmestrahlung," *Ann. d. Phys.*, *31* (1910), 758–768; *Physikalische Abhandlungen*, II, pp. 237–247). But he had apparently misunderstood Larmor (see above, pp. 136f.), and Thomson's notion of the structure of light was both older than and very different from Einstein's. For Thomson's theory see, Russell McCormmach, "J. J. Thomson and the Structure of Light," *British Journal for the History of Science*, *3* (1967), 362–387. Much additional information is included in Bruce R. Wheaton, "On the Nature of X and Gamma Rays. Attitudes towards Localization of Energy in the 'New Radiation,' 1896–1922," unpublished Ph.D. dissertation (Princeton University, 1978).
55. Johannes Stark, "Zur experimentellen Entscheidung zwischen Ätherwellen- und Lichtquantenhypothese. I. Röntgenstrahlen," *Phys. ZS.*, *10* (1909), 902–913.
56. Arnold Sommerfeld, "Über die Verteilung der Intensität bei der Emission von Röntgenstrahlen," *Phys. ZS.*, *10* (1909), 969–976. Note that Stark's article had been received by the journal on 16 November and was printed in the issue dated 22 November. Sommerfeld's elaborate answer was received on 6 December and appeared in the issue dated the 15th of that month. The controversy which followed is discussed in detail by Armin Hermann in "Die frühe Diskussion zwischen Stark und Sommerfeld über die Quantenhypothese (1)," *Centaurus*, *12* (1967), 38–59.
57. J. Stark, "Zur experimentellen Entscheidung zwischen der Lichtquantenhypothese und der Ätherimpulstheorie der Röntgenstrahlen," *Phys. ZS.*, *11* (1910), 24–31. A. Sommerfeld, "Über die Verteilung der Intensität bei der Emission von Röntgenstrahlen," *Phys. ZS.*, *11* (1910), 99–101.
58. "Emission von Röntgenstrahlen" (Sommerfeld, 1909), p. 970.
59. Wien to Sommerfeld, 27 December 1909. "Mit Ihrem Artikel, über die Röntgenstrahlen bin ich sehr einverstanden. Nur meine ich, dass auch das Energie-element auf elektromagnetische Vorgänge zurückgeführt werden muss, wenn wir *einen* durchsichtigen Mechanismus der Emission auf elektron[ische] Grundlage kennen. Dass es ein elektromagnetische und ein nicht elektromagnetische Röntgenemission gibt werden Sie wohl auch nicht annehmen." The full text is available on microfilm in the Archive for History of Quantum Physics; the original should be in the collection of the Deutsches Museum.
60. Arnold Sommerfeld, "Über die Struktur der γ-Strahlen," *Münchener Ber.*, 1911, pp. 1–60; quotation on p. 4.
61. *Ibid.*, pp. 24f. Discussion of the relevance of Planck's theory continues to p. 34 and is taken up again on pp. 39–42 where Sommerfeld, in a footnote, explicitly retracts his earlier statement that the quantum plays no role in x-rays produced by electron deceleration.

62. *Ibid.*, pp. 29–33.
†63. Planck discusses Sommerfeld's γ-ray paper at length in a letter to its author dated 6 April 1911 (on microfilm in the Archive for History of Quantum Physics; original presumably now in the Deutsches Museum). The following passage is representative: "Der bedeutsamste Fortschritt scheint mir zu liegen in der Erweiterung der Bedeutung des h für unperiodische Vorgänge. In meinem seitherigen Untersuchungen habe ich immer nur Oszillatoren von bestimmter Schwingungszahl ν betrachtet, und daraus ergibt sich ein bestimmtes Energieelement hν. Nimmt man aber einen Oszillator, der kein ausgeprägte Periode besitzt, so existiert für ihn auch kein bestimmtes Energieelement, und man muss auf die primäre Bedeutung von h zurückgehen.... Jedenfalls aber ist mir Ihr Vorgehen ausserordentlich sympathisch und scheint mir auch für die Zukunft viel zu versprechen."
64. "Das Plancksche Wirkungsquantum" (Sommerfeld, 1911b), followed by a page of discussion; the quotation is from the first page. Sommerfeld's theory and its development are also discussed in (Hermann, 1969), Chapt. 6, and in papers there cited. See also, Sigeko Nisio, "Sommerfeld's Theory of the Photoelectric Effect," *Proceedings of the XIVth International Congress of the History of Science*, 4 vols. (Tokyo, 1975). II, 302–304, and "Comment: Sommerfeld's Quantum Theory of 1911," *ibid.*, IV, 232–235.
65. "Das Plancksche Wirkungsquantum" (Sommerfeld, 1911b), p. 1063.
66. Lorentz, at the first Solvay Congress, commented on the existence of these two approaches (h a fundamental vs. h a derived constant), concluding that he saw little difference between them. (*Théorie du rayonnement et les quanta*, (Compendia, 1912), pp. 124f.)
67. A. E. Hass, "Über die elektrodynamische Bedeutung des Planck'schen Strahlungsgesetzes und über eine neue Bestimmung des elektrischen Elementarquantums und der Dimensionen des Wasserstoffatoms," *Wiener Ber. II*, 119 (1910), 119–144; much abridged reports on the theory appeared also in *Phys. ZS.*, 11 (1910), 537–538, and *Jahrb. d. Radioakt.*, 7 (1910), 261–268. For additional details about Haas's work see (Hermann, 1969), Chapter 5, and articles there cited.
†68. That Haas should, before the introduction of Rutherford's scattering theory and the subsequent invention of the concept of atomic number, have identified a one-electron model with atomic hydrogen is strange. The last paragraph of his introduction states that his paper employs "the Thomson model of a hydrogen atom" as "a special case of an optical resonator." But Thomson was consistently reluctant to make any precise claims about the number of electrons in particular sorts of atoms (see, for example, J. J. Thomson, *The Corpuscular Theory of Matter* (London, 1907), Chapter 6). More important, comparisons of the structural properties of Thomson models with those of the periodic table suggested that the number of electrons in oxygen, for example, was likely to be around sixty-five. Thomson's work on scattering led to somewhat lower values, varying within the range $n = A$ to $n = 3A$, with n the number of electrons and A atomic weight. These values were widely thought to be somewhat low. Only around Rutherford's group at Manchester, and then only from about 1911, was n regularly taken to be

approximately $A/2$ and hydrogen identified as a one-electron atom. (On this whole subject see J. L. Heilbron, "The Scattering of α and β Particles and Rutherford's Atom," *Archive for History of Exact Sciences*, *4* (1968), 247–307). Haas, whose aim was to produce a simple atomic model that would behave like a Planck resonator, may not have meant literally his reference to hydrogen. The one-electron case was the simplest model: hydrogen could be presumed the simplest atom.

69. (Hermann, 1969), p. 108, suggests that Haas identified $h\nu$ with potential rather than total energy, But, its textual inaccuracy aside, that identification would make the total energy emitted during ionization equal to $2h\nu$, not the behavior of a Planck resonator. That Hermann nevertheless emerges with Haas's value for h is due to a compensating error in his formula for potential energy.

70. The kinetic energy of Haas's model does have a maximum on the atom's surface, a fact which suggests that his demonstration could readily have been salvaged by those who took his results seriously.

71. "Alte und neue Fragen" (Lorentz, 1910b), pp. 1251ff.; *Théorie du rayonnement et les quanta*, (Compendia, 1912), pp. 121–124.

72. A. Schidlof "Zur Aufklärung der universellen elektrodynamischen Bedeutung der Planckschen Strahlungskonstanten h," *Ann. d. Phys.*, *35* (1911), 90–100. E. Wertheimer, "Die Plancksche Konstant h und der Ausdruck $h\nu$," *Phys. ZS.*, *12* (1911), 408–412.

73. J. J. Thomson, "On the Theory of Radiation," *Phil. Mag.*, *20* (1910), 238–247. This model is described and its influence discussed in J. L. Heilbron and T. S. Kuhn, "The Genesis of the Bohr Atom," *Historical Studies in the Physical Sciences*, *1* (1969), 211–290.

74. J. W. Nicholson, "The Constitution of the Solar Corona. II," and "The Constitution of the Solar Corona, III," *Month. Not.*, *72* (1912), 677–692, 729–739. These are the parts of a longer series in which Nicholson introduced and began to exploit Planck's constant. Russell McCormmach, "The Atomic Theory of John William Nicholson," *Archive for History of Exact Sciences*, *3* (1966), 160–184, describes Nicholson's work in detail.

75. "Verhältnisse der...Moleküle und das Magneton" (P. Weiss, 1911); P. Langevin, "La théorie cinétique du magnétisme et les magnétons," *Théorie du rayonnement et les quanta* (Compendia, 1912), pp. 393–404, where the model is introduced on pp. 402–404.

76. "Grundlagen der mechanischen Theorie der Wärme" (Hasenöhrl, 1911) sketches steps towards the introduction of an atomic model, and these were promptly exploited to explain the Balmer formula by K. F. Herzfeld in "Über ein Atommodell, das die Balmer'sche Wasserstoffserie aussendet," *Wiener Ber. II*, *121* (1912), 593–601. F. A. Lindemann developed special models to account, first, for the behavior of specific heats at low temperature and, then, for the selective photoeffect: "Über die Berechnung molekularer Eigenfrequenzen," *Phys. ZS.*, *11* (1910), 609–612, and "Über die Berechnung der Eigenfrequenzen der Elektronen im selektiven Photoeffekt," *Verh. d. D. Phys. Ges.*, *13* (1911), 482–488. Doubtless there are a few other examples of this, in any case not well defined, category of work.

77. Planck to Nernst, 11 June 1910: "Ich bin nun nach meinen Erfahrungen der Ansicht dass dies Bewusstsein der dringenden Notwendigkeit einer Reform kaum bei der Hälfte der von Ihnen in Aussicht genommenen Teilnehmer lebhaft genug ist um sie zu einem Besuche der Conferenz zu veranlassen.... Von der ganzen Reihe der von Ihnen genannten glaube ich nur dass ausser uns Einstein, Lorentz, W. Wien und Larmor sich ernstlich für die Sache interessieren." Quoted from the original in the Solvay archives on p. 6 of an unpublished manuscript, by Jean Pelseneer, further described in note 25 above. The estimated number of names of invitees is from the same source.
78. These numbers derive from an examination of the titles recorded in the register book of the Münchener Physikalisches Mittwochs-Colloquium, December 1908 to May 1939, on film in the Archive for the History of Quantum Physics. Note that there may have been discussions of Planck's work in the period before December 1908. Also, in November 1909, Debye talked on "Lichtelektrische Untersuchungen," a topic which may, but need not, have involved the quantum. The minute book of the $\nabla^2 V$ Club at Cambridge University (also on film in the Archive for History of Quantum Physics) records discussions of Planck's work at the 7th and 13th meetings (academic year 1901/02) after which the topic disappears until the 54th and 58th meetings (1910 and 1911). Even after that, however, explicit references to quantum topics are quite rare, in marked contrast to the situation at Munich.
79. See above, p. 215.
80. "Neue Beobachtungen an Kanalstrahlen" (Stark, 1908b), and A. Einstein, "Über die Entwicklung unserer Anschauungen über des Wesen und die Konstitution der Strahlung," *Phys. ZS.*, 10 (1909), 817–825.
81. In the survey from the *Fortschritte der Physik*, French names first appear in 1911. There are two in that year and four in each of the next two. For Poincaré's work, see above, p. 210.
82. *Report of the British Association*, 1913 (Compendia, 1913), pp. 376–386.
83. Jeans's *Report* (London, 1914) was published as a ninety-page pamphlet by the Physical Society. A second, entirely revised, edition appeared from the same source in 1924.

Notes to Chapter X

1. Planck to Lorentz, 7 October 1908. Quoted in context on p. 198, above.
2. Planck to Lorentz, 7 January 1910: "Die Unstetigkeit muss irgendwie einmal hineinkommen; sonst ist man rettungslos den Hamiltonschen Gleichungen und der Jeanschen Theorie ausgeliefert. Also habe ich die Unstetigkeit verlegt an den Punkt, wo sie am wenigsten schaden kann, auf die Erregung der Oszillatoren. Das Abklingen kann dann stetig erfolgen mit constanter Dämpfung." This is the letter, also discussed on pp. 200f., above, in which Planck announces that he is giving up the term "resonator" in favor of the "more general name 'oscillator'."
3. Max Planck, "Zur Theorie der Wärmestrahlung," *Ann. d. Phys.*, 31 (1910), 758–768; *Physikalische Abhandlungen und Vorträge*, II, 237–247. The quoted phrases appear on pp. 766, 768; II, 245, 247. The italics have been added.

4. Max Planck, "Eine neue Strahlungshypothese," *Verh. d. D. Phys. Ges.*, *13* (1911), 138–148; II, 249–259. Quotation from p. 142; II, 253.
5. *Ibid.*, p. 143; II, 254.
6. Max Planck, "Über die Begründung des Gesetzes der schwarzen Strahlung," *Ann. d. Phys.*, *37* (1912), 642–656; II, 287–301.
†7. Max Planck, "Zur Hypothese der Quantenemission," *Berl. Ber.*, 1911, pp. 723–731; II, 260–268. "La loi du rayonnement noir et l'hypothèse des quantités élémentaires d'action," *La Théorie du rayonnement et les quanta*, ed. P. Langevin and M. de Broglie (Paris, 1912), pp. 93–114; II, 269–286; in German under title "Die Gesetze der Wärmestrahlung und die Hypothese der elementaren Wirkungsquanten." The second paragraph of the first of these papers indicates why Planck switched from single- to multi-quantum emission. In the initial version of his second theory, equiprobable regions were not, he points out, of the anticipated size $h\nu$. Precisely what he has in mind is not clear, but it presumably relates to the following circumstance. In the later version of Planck's second theory the probability that an oscillator in ring n of the phrase plane passes to ring $n + 1$ is always $1 - \eta$, independent of n. In the initial version the corresponding probability is $1 - n\eta$, a quantity which decreases with increasing n and which could, in principle, even become negative.
8. P_n is a value of the distribution function, not a probability. Compare formula (II-3). I have here corrected a misprint in the original.
9. "Begründung des Gesetzes" (Planck, 1912a), p. 645; II, 290. The original is in italics.
10. *Ibid.*
11. Max Planck, *Vorlesungen über die Theorie der Wärmestrahlung*, 2nd ed., revised (Leipzig, 1913), pp. 99–109. Prior to this point the texts of the two editions have been virtually identical.
12. *Ibid.*, p. 105.
13. *Ibid.*, pp. 114f., to be compared with p. 134 in the first edition (Planck, 1906).
14. See above, Chapt. III, and pp. 116f.
15. *Wärmestrahlung*, 1st ed. (Planck, 1906a), p. 197. These formulas differ slightly from equations (III-12), introduced by Planck in 1899, because of a minor redefinition of the quantity δ_ν.
16. *Wärmestrahlung*, 2nd ed. (Planck, 1913), p. 186.
17. Max Planck, "Zur Geschichte der Auffindung des physikalischen Wirkungsquantums," *Naturwissensch.*, *31* (1943), 153–159; III, 255–267. Planck there speaks of "the hypothesis of 'natural radiation,' the content of which depends on the total incoherence of the individual harmonic partial vibrations from which a wave of thermal radiation is composed" (p. 155; II, 259). In his other autobiographical writings Planck simply equates natural radiation with molecular disorder.
18. *Wärmestrahlung*, 2nd ed. (Planck, 1913), p. 131.
19. Max Planck, *Vorlesungen über die Theorie der Wärmestrahlung*, 4th ed., revised (Leipzig, 1921). The third edition (1919) had been a verbatim reprint of the second.

NOTES TO PP. 244-247

†20. As p. 199, above, may suggest, however, comparisons of the relative conservatism of individuals are often equivocal. Lorentz believed that the energy of Planck's oscillators were necessarily restricted to integral multiples of $h\nu$, and he was in that respect more radical than Planck. But Lorentz also believed that Planck's oscillators were some special sort of particle and that the restriction on their energy had no application to electron theory; in that respect he was the conservative.

21. See, for example, J. H. Jeans, *Report on Radiation and the Quantum-Theory* (London, 1914), p. 83; E. P. Adams, "The Quantum Theory," *Bulletin of the National Research Council*, Vol. 1, Pt. 5 (October 1920), pp. 301–381, esp. pp. 311f.; F. Reiche, *Die Quantentheorie, ihr Ursprung und ihre Entwicklung* (Berlin, 1921), pp. 30f.

22. "Eine neue Strahlungshypothese" (Planck, 1911a), p. 148; II, 259. See also the reference to radioactivity in "Begründung des Gesetzes" (Planck, 1912a), p. 653; II, 298.

23. Above, p. 199. For Planck's view of the quantization of free-electron motions see "Eine neue Strahlungshypothese" (Planck, 1911a), pp. 146f.; II, 257f.

24. *Ibid.*

25. "La loi du rayonnement noir" (Planck, 1912b), pp. 283f.

26. Wien to Stark, 5 and 7 November 1911. Though unknown when the original printed catalogue was prepared, these letters are available on microfilm 81 of the various Archives for the History of Quantum Physics. The originals are deposited at the Deutsches Museum, Munich.

27. Information on this subject can be retrieved through the index entry, "Planck, Max," in R. H. Stuewer, *The Compton Effect: Turning Point in Physics* (New York, 1975).

28. "Eine neue Strahlungshypothese" (Planck, 1911a), p. 146; II, 257.

29. A. Einstein and O. Stern, "Einige Argumente für die Annahme einer molekularen Agitation beim absoluten Nullpunkt," *Ann. d. Phys.*, 40 (1913), 551–560, quotation from the last page. Note that this treatment of specific heats does not call for the quantization of rotational energy; E_r, above, is simply equated with the energy specified by the old or new form of Planck's distribution law and the result solved for ν as a function of T. Using some earlier results due to Einstein and Hopf, the authors, in the second part of their paper, show how to derive the Planck law without resort to discontinuity, on the assumption that the zero-point energy is precisely $h\nu$. Their closing sentence is, however, "It seems doubtful that the other difficulties will be overcome without the hypothesis of quanta."

30. H. Kamerlingh-Onnes and W. H. Keesom, "Über die Translationsenergie in einatomigen Gasen beim absoluten Nullpunkt," *Vorträge über die kinetische Theorie der Materie und der Elektrizität*, ed. M. Planck, P. Debye, *et al.* (Leipzig, 1914), pp. 193f. This short piece was a contribution to the discussion at the well-attended series of Wolfskehl lectures held in May 1913. Much useful information on the early history of the zero-point energy is included in A. Eucken, "Die Entwicklung der Quantentheorie vom Herbst 1911 bis Sommer 1913," an "Anhang" to the German edition of the Proceedings of

the First Solvay Congress: *Die Theorie der Strahlung und der Quanten, Verhandlungen auf einer von E. Solvay einberufenen Zusammenkunft...*, ed. A. Eucken (Halle, 1914), pp. 371–405.

31. Arnold Eucken, "Über den Quanteneffekt bei einatomigen Gasen und Flüssigkeiten," *Berl. Ber.*, 1914, pp. 682–693, quotation from p. 683. Compare also, O. Sackur, "Die spezifische Wärme der Gase und die Nullpunktsenergie," *Verh. d. D. Phys. Ges.*, **16** (1914), 728–734.

†32. The divorce between the zero-point energy and Planck's second theory was not, however, by any means complete before the advent of wave mechanics and Fermi statistics. Look, for example, at A. Byk, "Quantentheorie der molaren thermodynamischen Zustandsgrössen," in the authoritative *Handbuch der Physik*, ed. H. Geiger and K. Scheel, Vol. 9, *Theorien der Wärme* (Berlin, 1926), pp. 301–340. It includes a section, "Nullpunktsenergie" (pp. 324–326), most of which is devoted to a comparative discussion of the two standard versions of Planck's theory. The section concludes with the sentence: "For purposes of comparison the first and second Planck quantum theories of material structures have been specially highly developed in application to the specific heat of hydrogen, but the results have not decisively favored either theory."

†33. For details and documentation relevant to this discussion of the Bohr atom, see J. L. Heilbron and T. S. Kuhn, "The Genesis of the Bohr Atom," *Historical Studies in the Physical Sciences*, **1** (1969), pp. 211–290. Note, however, that footnote 145 of that paper explicitly denies the suggestion of T. Hirosige and S. Nisio ("Formation of Bohr's Theory of Atomic Constitution," *Japanese Studies in History of Science*, **3** (1964), 6–28) that the emission mechanism of Planck's second theory provided the model that Bohr employed in his first explanation of the generation of a spectral series. Because we considered only the first (the single-quantum emission) version of Planck's second theory, my co-author and I entirely missed the likely relevance of this aspect of the revised second theory to an important puzzle posed by Bohr's early work.

34. "Bohr Atom" (Heilbron and Kuhn, 1969), p. 268, from p. 4 of N. Bohr, "On the Constitution of Atoms and Molecules [Part I]," *Phil. Mag.*, **26** (1913), 1–25.

35. (Heilbron and Kuhn, 1969), pp. 266–274.

36. Planck had, in fact, already made use of the high-energy correspondence between classical and quantum formulas in "Eine neue Strahlungshypothese" (Planck, 1911a), p. 144; II, 255, a paper which Bohr also cites. That a technique resembling the Correspondence Principle had played a role in the development of Planck's second theory was first pointed out by Max Jammer, *The Conceptual Development of Quantum Mechanics* (New York, St. Louis, etc., 1966), p. 50.

37. See above, p. 129.

38. *Wärmestrahlung*, 2nd ed. (Planck, 1913), pp. 124f.

39. *Ibid.*, p. 131, a passage previously quoted on p. 243, above. The term "interaction" is puzzling in this context. Perhaps Planck is thinking only of interactions mediated by the radiation field.

40. *Ibid.*, p. 136.

41. Max Planck, "Die Quantenhypothese für Molekeln mit mehreren Freiheitsgraden (Erste Mitteilung)," *Verh. d. D. Phys. Ges.*, *17* (1915), 407–418; II, 349–360. The relevant discussion occurs on pp. 409f.; 351f.
42. Max Planck, "Die physikalische Struktur des Phasenraumes," *Ann. d. Phys.*, *50* (1916), 385–418; II, 386–419.
43. A. Sommerfeld, "Zur Theorie der Balmerschen Serie," *Münchener Ber.*, 1915, pp. 425–458, explicitly calls on Planck's phase-plane treatment as model. J. Ishiwara, "Die universelle Bedeutung des Wirkungsquantums," *Tôkyô Sûgaku-Buturigakkawi Kizi* [*Proceedings of the Tokyo Mathematical Physical Society*], *8* (1915), 106–116, takes its point of departure from the statement: "People currently tend...to explain the occurrence of h in terms of the existence of fixed finite elementary cells in phase space, the magnitude of which is the same for all elementary cells of equal probability." William Wilson, "The Quantum-Theory of Radiation and Line Spectra," *Phil. Mag.*, *29* (1915), 795–802, cites the second edition of Planck's *Lectures* but otherwise shows no direct evidence of conceptual dependence on the second theory.
44. Marcel Brillouin, ["Discussion"], *Théorie du rayonnement et les quanta* (Compendia, 1912), p. 451, italics in original.
†45. See above, p. 246. Though I am aware of no relevant direct evidence, there is a more important respect in which Planck's second theory may have influenced Einstein. Planck's constant η is a transition probability, the first to be used in the quantum theory. Its introduction, furthermore, both transformed and simplified the derivation of Planck's distribution law in ways that Einstein is likely to have valued. After the invention of the Bohr atom, it would have been natural to extend the use of transition probabilities to absorption and perhaps to spontaneous emission as well. There is, in short, a natural route from Planck's second theory to Einstein's 1916 derivation of the distribution law from simple postulates concerning transition probabilities.
46. Max Planck, "Eine veränderte Formulierung der Quantenhypothese," *Berl. Ber.*, 1914, pp. 918–923; II, 330–335. Quotation from opening page. Sometimes known as Planck's "third theory," this treatment was rapidly shown to be untenable by A. D. Fokker, "Die mittlere Energie rotierender elektrischer Dipole im Strahlungsfeld," *Ann. d. Phys.*, *43* (1914), 810–820. Planck never attempted to revise it, but returned instead to his second theory.
47. N. Bohr, "On the Quantum Theory of Line-Spectra," *Kongelige Danske Videnskabenes Selskabs Skrifter, Naturvidenskabelig og Mathematisk Afdeling*, Ser. 8, Vol. 4, (Copenhagen, 1918), pp. 1–100. (This is Parts I and II of a projected four-part work. A summary of Part III was printed in 1922 as pp. 101–118 of the preceding.) A. Sommerfeld, *Atombau und Spektrallinien* (Braunschweig, 1919) and many later editions.
48. See note 33, above.
49. Planck to Ehrenfest, 23 May 1915: "Natürlich werde Ich Ihnen sehr gerne einen Correcturabzug meiner Publication über die rotierenden Dipole schicken.... Selbstverständlich geht es in meiner Arbeit nicht ohne Hypothese ab, und ich halte es wohl für möglich, dass Ihre Hass gegen die

Nullpunksenergie sich auf die von mir eingeführte elektrodynamische Emissions-hypothese überträgt, die doch zur Nullpunktsenergie führt. Aber was tun? Ich meinerseits hass die Unstetigkeit der Energie noch mehr als die Unstetigkeit der Emission. Ihren und Ihrer werten Gattin besten Gruss!" Available on the microfilm of the Scientific Correspondence of Paul Ehrenfest at the various Archives for the History of Quantum Physics. The original is deposited at the Museum Boerhaave, Leyden. For Ehrenfest's attitude towards the zero-point energy see, M. J. Klein, *Paul Ehrenfest* (Amsterdam and London, 1970), Chapter 11, esp. pp. 267f.

50. Siegfried Valentiner, *Die Grunglagen der Quantentheorie in elementarer Darstellung*, 2nd ed., enlarged (Braunschweig, 1919).

51. "The Quantum Theory" (Adams, 1920). The revised and enlarged second edition was published in the *Bulletin of the National Research Council*, Vol. 7, Pt. 3 (November 1923), pp. 1–109. J. H. Jeans, *Report on Radiation and the Quantum Theory*, 2nd ed. (London, 1924). For the reference in the first edition, see note 21, above.

52. I have examined only the recent reprint: *Theorie der Wärmestrahlung, Vorlesungen von Max Planck*, 6th ed. (Leipzig, 1966). The quotation is from the "Vorwort zur fünften Auflage" on p. x.

BIBLIOGRAPHY

The bibliography that follows is restricted to works actually used in the preparation of this volume. Most of them have already been cited in the notes, but often in an abbreviated form to which this bibliography serves as key. The bibliography also amplifies the notes with respect to such matters as publication (including multiple appearances), date of submission, and the availability of English translations. With respect to none of these matters, however, has an attempt been made to ensure completeness. Almost certainly there are additional English translations of some of the published primary sources listed below, and some were probably also reprinted in journals not listed here. The portion of the bibliography devoted to published primary sources—arranged alphabetically by author and chronologically for each author—may also be helpful while reading the book. It provides easy access to the order in which closely spaced papers were composed (or, in any case, submitted for publication), a sort of detail on which some parts of the book's argument depend.

The list of secondary sources includes most of the recent scholarly literature that has provided points of departure for this volume. The exceptions are articles by Hermann and Kangro of which the contents have since been largely incorporated in their books. Some of these articles include additional detail, but the books (Hermann, 1969; Kangro, 1970) are the places to start, and the articles which they largely supplant are cited in them. A full bibliography of the older secondary literature will be found in (Kangro, 1970). This one should be regarded as a supplement to it.

The citations of published primary sources employ the abbreviated journal titles used in the notes and discussed on p. 255, above. In other respects the method of citation is straightforward, and only one of its aspects requires mention here. Many of the books included in the list of primary sources appeared in several editions, and the differences between editions have sometimes proved important to this volume's argument, most notably in the cases of Planck's *Lectures* and Nernst's *Theoretical Chemistry*. In such cases, relevant individual editions have been listed as new publications, so that the same title may appear several times in the list of an author's publications. References to editions that have not merited treatment as new works are given in abbreviated form: older editions with the first of those cited in full, more recent ones with the last of the cited editions.

Two more remarks, both directed to the list of published primary sources, complete this introduction to the bibliography. It includes a few collective works, which it would have been misleading to place under the name of an individual author or editor. These have been grouped together under the heading "Compendia," located alphabetically between Clausius and Culverwell. Finally, where one exists, a standard edition of collected works by an individual author is the first entry under that author's name. Except where otherwise noted, all listed articles, but no listed books, by that author are reprinted in the corresponding collection of his works.

Secondary sources

Agassi, Joseph
- 1967   "The Kirchhoff–Planck Radiation Law," *Science, 156* (1967), 30–37.

D'Agostino, Salvo
- 1975   "Hertz's Researches on Electromagnetic Waves," *Historical Studies in the Physical Sciences, 6* (1975), 261–323.

Blackmore, J. T.
- 1972   *Ernst Mach* (Berkeley, Los Angeles: University of California Press, 1972).

Bork, A. M.
- 1966   "Physics just before Einstein," *Science, 152* (1966), 597–603.

Brout, R.
- 1956   "Statistical Mechanics of Irreversible Processes. Part VIII: Boltzmann Equation," *Physica, 22* (1956), 509–524.

Brush, S. G.
- 1957a   "The Development of the Kinetic Theory of Gases. I. Herapath," *Annals of Science, 13* (1957), 188–198.
- 1957b   "The Development of the Kinetic Theory of Gases. II. Waterston," *Annals of Science, 13* (1957), 273–282.
- 1965   *Kinetic Theory, Selected Readings in Physics*, 2 vols. (Oxford, New York, etc.: Pergamon, 1965–66).
- 1967   "Foundations of Statistical Mechanics, 1845–1915," *Archive for History of Exact Sciences, 4* (1967), 145–183.
- 1969   "Maxwell, Osborne Reynolds, and the Radiometer," *Historical Studies in the Physical Sciences, 1* (1969), 105–125. With C. W. F. Everitt.
- 1970   "The Wave Theory of Heat," *British Journal for the History of Science, 5* (1970), 145–167.

1974 "The Development of the Kinetic Theory of Gases, VIII. Randomness and Irreversibility," *Archive for History of Exact Sciences, 12* (1974), 1–88.

1976 *The Kind of Motion We Call Heat: A History of the Kinetic Theory of Gases in the 19th Century* (Amsterdam: North Holland; New York: Elsevier, 1976), a collection which includes several of the above.

Cardwell, D. S. L.

1971 *From Watt to Clausius* (Ithaca: Cornell University Press, 1971; London: Heinemann Educational Books, 1971).

Daub, E. E.

1970a "Waterston, Rankine, and Clausius on the Kinetic Theory of Gases," *Isis, 61* (1970), 105–106.

1970b "Maxwell's Demon," *Studies in History and Philosophy of Science, 1* (1970), 213–227.

Doran, B. G.

1975 "Origin and Consolidation of Field Theory in Nineteenth-Century Britain: From the Mechanical to the Electromagnetic View of Nature," *Historical Studies in the Physical Sciences, 6* (1975), 133–260.

Dugas, René

1959 *La Théorie physique au sens de Boltzmann* (Neuchatel: Griffon, 1959).

Everitt, C. W. F.

1969 See (Brush, 1969).

1974 "Maxwell, James Clerk" in *Dictionary of Scientific Biography*, ed. C. C. Gillespie, Vol. 9 (New York: Scribner's, 1974), pp. 198–230; also published as *James Clerk Maxwell* (New York: Scribner's, 1975).

Forman, P. L.

1969 See (Kuhn, 1967).

1975 "Physics *circa* 1900," *Historical Studies in the Physical Sciences, 5* (1975), 5–185. With J. L. Heilbron and Spencer Weart. Included here as an essential resource, this pioneering monograph appeared too late to influence the form of this book.

Fox, Robert

1971 *The Caloric Theory of Gases from Lavoisier to Regnault* (Oxford: Clarendon Press, 1971).

Garber, Elizabeth

1976 "Some Reactions to Planck's Law, 1900–1914," *Studies in History and Philosophy of Science, 7* (1976), 89–126.

Goldberg, Stanley

1976 "Max Planck's Philosophy of Nature and His Elaboration of the Special Theory of Relativity," *Historical Studies in the Physical Sciences, 7* (1976), 125–160.

ter Haar, D.

1967 *The Old Quantum Theory* (Oxford: Pergamon, 1967).

Heilbron, J. L.

1968 "The Scattering of α and β Particles and Rutherford's Atom," *Archive for History of Exact Sciences, 4* (1968), 247–307.

1969 "The Genesis of the Bohr Atom," *Historical Studies in the Physical Sciences*, *1* (1969), 211–290. With T. S. Kuhn. See also (Kuhn, 1967; Forman, 1975).

Hermann, Armin

1967 "Die frühe Diskussion zwischen Stark und Sommerfeld über die Quantenhypothese (1)," *Centaurus*, *12* (1967), 38–59.

1969 *Frühgeschichte der Quantentheorie (1899–1913)* (Mosbach in Baden: Physik-Verlag, 1969). Translated as *The Genesis of the Quantum Theory*, trans. C. W. Nash (Cambridge, Massachusetts: M.I.T. Press 1971).

1973a "Laue, Max von," in *Dictionary of Scientific Biography*, ed. C. C. Gillispie, Vol. 8 (New York: Scribner's, 1973), pp. 50–53.

1973b *Max Planck in Selbstzeugnissen und Bilddokumenten* (Hamburg: Rowohlt, 1973).

Hiebert, E. N.

1968 *The Conception of Thermodynamics in the Scientific Thought of Mach and Planck*, Wissenschaftlicher Bericht Nr. 5/68, Ernst Mach Institut (Freiburg i. Br., [1968]).

1971 "The Energetics Controversy and the New Thermodynamics," in *Perspectives in the History of Science and Technology*, ed. D. H. D. Roller (Norman, Oklahoma: University of Oklahoma Press, 1971), pp. 67–86.

Hirosige, Tetu

1964 "Formation of Bohr's Theory of Atomic Constitution," *Japanese Studies in History of Science*, *3* (1964), 6–28. With Sigeko Nisio.

1969 "Origins of Lorentz' Theory of Electrons and the Concept of the Electromagnetic Field," *Historical Studies in the Physical Sciences*, *1* (1969), 151–209.

1970 "The Genesis of the Bohr Atom Model and Planck's Theory of Radiation," *Japanese Studies in History of Science*, *9* (1970), 35–47. With Sigeko Nisio.

Jammer, Max

1966 *The Conceptual Development of Quantum Mechanics* (New York, St. Louis, etc.,: McGraw-Hill, 1966).

Kac, Mark

1959 *Probability and Related Topics in Physical Science* (London and New York: Interscience Publishers, 1959).

Kangro, Hans

1970 *Vorgeschichte des Planckschen Strahlungsgesetzes* (Wiesbaden: Steiner, 1970). Translated as *History of Planck's Radiation Law* (London: Taylor and Francis, 1976).

1972 *Original Papers in Quantum Physics*, annot. Kangro, trans. D. ter Haar and S. G. Brush (London: Taylor and Francis, 1972).

Klein, M. J.

1962 "Max Planck and the Beginnings of Quantum Theory," *Archive for History of Exact Sciences*, *1* (1962), 459–479.

1963a "Planck, Entropy, and Quanta, 1901–1906," *The Natural Philosopher*, *1* (1963), 83–108.

1963b "Einstein's First Paper on Quanta," *The Natural Philosopher, 2* (1963), 59–86.
1964 "Einstein and the Wave-Particle Duality," *The Natural Philosopher, 3* (1964), 3–49.
1965 "Einstein, Specific Heats, and the Early Quantum Theory," *Science, 148* (1965), 173–180.
1966 "Thermodynamics and Quanta in Planck's Work," *Physics Today, 19,* No. 11 (1966), 23–32.
1967 "Thermodynamics in Einstein's Thought," *Science, 157* (1967), 509–516.
1970a "Maxwell, His Demon, and the Second Law of Thermodynamics," *American Scientist, 58* (1970), 84–97.
1970b *Paul Ehrenfest: Volume 1: The Making of a Theoretical Physicist* (Amsterdam and London: North-Holland; New York: Elsevier, 1970).
1972 "Mechanical Explanation at the End of the Nineteenth Century," *Centaurus, 17* (1972), 58–82.
1973 "The Development of Boltzmann's Statistical Ideas," *The Boltzmann Equation: Theory and Applications,* ed. E. G. D. Cohen and W. Thirring, *Acta Physica Austraica,* Suppl. X (Vienna and New York: Springer, 1973), pp. 53–106.

Kuhn, T. S.
1959 "Conservation of Energy as an Example of Simultaneous Discovery," in *Critical Problems in the History of Science,* ed. Marshall Clagett, (Madison: University of Wisconsin Press, 1959), pp. 321–356.
1967 *Sources for History of Quantum Physics: An Inventory and Report* (Philadelphia: American Philosophical Society, 1967). With J. L. Heilbron, P. L. Forman, and Lini Allen.
1969 See (Heilbron, 1969).

Mach, Ernst
1896 *Die Principien der Wärmelehre* (Leipzig: Barth, 1896). Second ed. (1900). Third ed. (1923).

McCormmach, Russell
1966 "The Atomic Theory of John William Nicholson," *Archive for History of Exact Sciences, 3* (1966), 160–184.
1967a "J. J. Thomson and the Structure of Light," *British Journal for the History of Science, 3* (1967), 362–387.
1967b "Henri Poincaré and the Quantum Theory," *Isis, 58* (1967), 37–55.
1970a "H. A. Lorentz and the Electromagnetic View of Nature," *Isis, 61* (1970), 459–497.
1970b "Einstein, Lorentz, and the Electron Theory," *Historical Studies in the Physical Sciences, 2* (1970), 41–87.
1972 "Hertz, Heinrich Rudolf," in *Dictionary of Scientific Biography,* ed. C. C. Gillispie, Vol. 6 (New York: Scribner's, 1972), pp. 340–350.

Nisio, Sigeko
1975a "Sommerfeld's Theory of the Photoelectric Effect," *Proceedings of the XIVth International Congress of the History of Science,* Vol. 2 (Tokyo: Science Council of Japan, 1975), pp. 302–304.

1975b "Comment: Sommerfeld's Quantum Theory of 1911," *Proceedings of the XIVth International Congress of the History of Science*, Vol. 4 (Tokyo: Science Council of Japan, 1975), pp. 232–235. See also (Hirosige, 1964, 1970).

Price, D. J. de S.
1963 *Little Science, Big Science* (New York and London: Columbia University Press, 1963).

Reiche, Fritz
1921 *Die Quantentheorie, ihr Ursprung und ihre Entwicklung* (Berlin: Springer, 1921). Translated as *The Quantum Theory*, trans. H. Hatfield and H. Brose (London: Methuen, 1922).

Rosenfeld, Léon
1936 "La première phase de l'évolution de la théorie des quanta," *Osiris*, *2* (1936), 149–196.

Scheel, Karl
1926 "Physikalische Literatur," *Handbuch der Physik*, ed. H. Geiger and K. Scheel, Vol. 1 (Berlin: Springer, 1926), pp. 180–186.

Seelig, Carl
1954 *Albert Einstein: Eine dokumentarische Biographie* (Zurich, Stuttgart, Vienna: Europa, 1954).

Seigel, Daniel
1976 "Balfour Stewart and Gustav Robert Kirchhoff: Two Independent Approaches to 'Kirchhoff's Radiation Law'," *Isis*, *67* (1976), 565–600

Stuewer, R. H.
1975 *The Compton Effect: Turning Point in Physics* (New York: Science History, 1975).

Weart, Spencer
1975 See (Forman, 1975).

Wheaton, B. R.
1971 "The Photoelectric Effect and the Origin of the Quantum Theory of Free Radiation," unpublished M. A. dissertation (University of California, Berkeley, 1971).

1978a "On the Nature of X and Gamma Rays. Attitudes toward Localization of Energy in the 'New Radiation,' 1896–1922," unpublished Ph.D. dissertation (Princeton University, 1978).

1978b "Philipp Lenard and the Photoelectric Effect, 1889–1911," *Historical Studies in the Physical Sciehces*, *9* (1978), 299–322.

Whittaker, E. T.
1951 *History of the Theories of Aether and Electricity: The Classical Phase*, revised and enlarged ed. (Edinburgh, London, etc.: Nelson, 1951), reprinted (New York, Harper Torch books, 1960).

Woodruff, A. E.
1966 "William Crookes and the Radiometer," *Isis*, *57* (1966), 188–198.

Published primary sources

Adams, E. P. (1878–1956)
1920   "The Quantum Theory," *Bulletin of the National Research Council*, 1, Pt. 5 (1920), 301–381. Second ed., *ibid.*, 7, Pt. 3 (1923), 1–109.

Behn, U. (1868–1908)
1898   "Über die specifische Wärme einiger Metalle bei tiefen Temperaturen," *Ann. d. Phys.*, *66* (1898), 237–244, received 21 July 1898.
1900   "Über die specifische Wärme der Metalle, des Graphits und einiger Legirungen bei tiefen Temperaturen," *Ann. d. Phys.*, *1* (1900), 257–269, received 11 January 1900.

Bjerrum, Niels (1879–1958)
1911   "Über die spezifische Wärme der Gase," *ZS. f. Eleckrochem.*, *17* (1911), 731–734, presented 27 May 1911.
1912   "Über die ultraroten Absorptionsspektren der Gase," *Festschrift W. Nernst zu seinem fünfundzwanzigjährigen Doktorjubiläum gewidmet von seinen Schülern* (Halle: Knapp, 1912), pp. 90–98, dated March 1912.

Bohr, Niels (1885–1962)
1913   "On the Constitution of Atoms and Molecules [Part I]," *Phil. Mag.*, *26* (1913), 1–25, dated 5 April 1913.
1918   "On the Quantum Theory of Line-Spectra," *Kongelige Danske Videnskabenes Selskabs Skrifter, Naturvidenskabelig og Mathematisk Afdeling*, Series 8, Vol. 4, number 1, parts 1–3 (Copenhagen: Høst & Søn, 1918–22).

Boltzmann, Ludwig (1844–1906)
Wissenschaftliche Abhandlungen von Ludwig Boltzmann, ed. Fritz Hasenöhrl, 3 vols. (Leipzig: Barth, 1909). Reprinted (New York: Chelsea, 1968).
1866   "Über die mechanische Bedeutung des zweiten Hauptsatzes der Wärmetheorie," *Wiener Ber. II*, *53* (1866), 195–220, read 8 February 1866.
1868   "Studien über das Gleichgewicht der lebendigen Kraft zwischen bewegten materiellen Punkten," *Wiener Ber. II*, *58* (1868), 517–560, received 8 October 1868.
1872   "Weitere Studien über die Wärmegleichgewicht unter Gasmolekülen," *Wiener Ber., II*, *66* (1872), 275–370, presented 10 October 1872. Trans. in (Brush, 1965), II, 88–175.
1876   "Über den Zustand des Wärmegleichgewichtes eines Systems von Körpern mit Rücksicht auf die Schwerkraft. I.," *Wiener Ber. II*, *73* (1876), 128–142, received 27 January 1876.
1877a   "Bemerkungen über einige Probleme der mechanischen Wärmetheorie," *Wiener Ber. II*, *75* (1877), 62–100, presented 11 January 1877. Partial trans. in (Brush, 1965), II, 188–193.
1877b   "Über die Beziehung zwischen dem zweiten Hauptsatze der mechanischen Wärmetheorie und der Wahrscheinlichkeitsrechnung respektive den Sätzen über das Wärmegleichgewicht," *Wiener Ber. II*, *76* (1877), 373–435, presented 11 October 1877.

BIBLIOGRAPHY

1878 "Weitere Bemerkungen über einige Probleme der mechanischen Wärmetheorie," *Wiener Ber. II*, *78* (1878), 7–46, presented 6 June 1878.

1879 "Erwiderung auf die Bemerkung des Hrn. Oskar Emil Meyer," *Ann. d. Phys.*, *8* (1879), 653–655.

1880 "Erwiderung auf die notiz des Hrn. O. E. Meyer: 'Über eine Veränderte Form' usw.," *Ann. d. Phys.*, *11* (1880), 529–534.

1881 "Referat über die Abhandlung von J. C. Maxwell 'Über Boltzmann's Theorem betreffend die mittlere Verteilung der lebendigen Kraft in einem System materieller Punkte," *Annalen der Physik, Beiblätter*, *5* (1881), 403–417; *Phil. Mag.*, *14* (1882), 299–312.

1883 "Über das Arbeitsquantum, welches bei chemischen Verbindungen gewonnen werden kann," *Wiener Ber. II*, *88* (1883), 861–896, presented 18 October 1883. Reprinted in *Ann. d. Phys.*, *22* (1884), 39–72.

1884a "Über eine von Hrn. Bartoli entdeckte Beziehung der Wärmestrahlung zum zweiten Hauptsatze," *Ann. d. Phys.*, *22* (1884), 31–39, dated March 1884.

1884b "Ableitung des Stefan'schen Gesetzes betreffend die Abhängigkeit der Wärmestrahlung von der Temperatur aus der elektromagnetischen Lichttheorie," *Ann. d. Phys.*, *22* (1884), 291–294.

1885 "Über die Eigenschaften monozyklischer und anderer damit verwandter Systeme," *Journal für reine und angewandte Mathematik*, *98* (1885), 68–94, dated 9 October 1884.

1887 "Über die mechanischen Analogien des zweiten Hauptsatzes der Thermodynamik," *Journal für reine und angewandte Mathematik*, *100* (1887), 201–212, dated September 1885.

1894 "Über den Beweis des Maxwellschen Geschwindigkeitsverteilungsgesetzes unter Gasmolekülen," *Münchener Ber.*, *24* (1894), 207–210, presented 5 May 1894; *Ann. d. Phys.*, *53* (1894), 955–958.

1895a "Nochmals das Maxwellsche Verteilungsgesetz der Geschwindigkeiten," *Münchener Ber.*, *25* (1895), 25–26, presented 5 January 1895.

1895b "On Certain Questions of the Theory of Gases," *Nature*, *51* (1894–95), 413–415, issue of 28 February 1895.

1895c "Nochmals das Maxwellsche Verteilungsgesetz der Geschwindigkeiten," *Ann. d. Phys.*, *55* (1895), 223–224, issue of 1 May 1895. Significantly revised version of (Boltzmann, 1895a).

1895d "On the Minimum Theorem in the Theory of Gases," *Nature*, *52* (1895), 221, dated 20 June 1895.

1896a "Entgegnung auf die wärmetheoretischen Betrachtungen des Hrn. E. Zermelo," *Ann. d. Phys.*, *57* (1896), 773–784, dated 20 March 1896. Trans. in (Brush, 1965), II, 218–228.

1896b *Vorlesungen über Gastheorie. I. Theil: Theorie der Gase mit einatomigen Molekülen, deren Dimensionen gegen die mittlere Weglänge verschwinden* (Leipzig: Barth, 1896). Trans. in *Lectures on Gas Theory*, trans. S. G. Brush (Berkeley and Los Angeles: University of California Press, 1964).

1897a "Zu Hrn. Zermelo's Abhandlung 'Über die mechanische Erklärung irreversibler Vorgänge,'" *Ann. d. Phys.*, *60* (1897), 392–398, dated 16 December 1896. Trans. in (Brush, 1965), II, 238–245.

1897b "Über irreversible Strahlungsvorgänge," *Berl. Ber.*, 1897, pp. 660–662, presented 17 June 1897.

1898 *Vorlesungen über Gastheorie. II. Theil: Theorie van der Waals'; Gase mit zusammengesetzten Molekülen; Gasdissociation; Schlussbemerkungen* (Leipzig: Barth, 1898). Trans. with Boltzmann, 1896b.

Born, Max (1882–1970)

1910 See (Lorentz, 1910b).

Bryan, G. H. (1864–1928)

1894 "Report on the Present State of our Knowledge of Thermodynamics. Part II.—The Laws of Distribution of Energy and their Limitations," *Report of the British Association*, 1894, pp. 64–102, presented at the meeting of August 1894.

1906 ["Review of Planck's *Wärmestrahlung*"], *Nature*, 74 (1906), supplement to the issue of October 11, pp. iii–iv.

Burbury, S. H. (1831–1911)

1894a "On the Law of Distribution of Energy," *Phil. Mag.*, 37 (1894), 143–158, issue of January 1894.

1894b "Boltzmann's Minimum Function," *Nature*, 51 (1894–95), 78, dated 12 November 1894.

1894c "The Kinetic Theory of Gases," *Nature*, 51 (1894–95), 175f., dated 5 December 1894.

1899 *A Treatise on the Kinetic Theory of Gases* (Cambridge, England: Cambridge University Press, 1899).

1902 "On Irreversible Processes and Planck's Theory in Relation Thereto," *Phil. Mag.*, 3 (1902), 225–240, issue of February 1902.

1903 "On the Conditions necessary for Equipartition of Energy. (Note on Mr. Jean's Paper, *Phil. Mag.* November 1902.)," *Phil. Mag.*, 5 (1903), 134f., issue of January 1903.

Byk, Alfred (b. 1878)

1926 "Quantentheorie der molaren thermodynamischen Zustandsgrössen," in *Handbuch der Physik*, ed. H. Geiger and K. Scheel, Vol. 9 (Berlin: Springer, 1926), pp. 301–340.

Classen, Alexander (1843–1934)

1895 See (Roscoe, 1895).

Clausius, R. J. E. (1822–1888)

1854 "Über eine veränderte Form des zweiten Hauptsatzes der mechanischen Wärmetheorie," *Ann. d. Phys.*, 93 (1854), 481–506.

1857 "Über die Art der Bewegung welche wir Wärme nennen," *Ann. d. Phys.*, 100 (1857), 353–380, dated 5 January 1857. Trans. in (Brush, 1965), I, 111–134.

1858 "Über die mittlere Länge der Wege, welche bei Molecularbewegung gasförmigen Körper von den einzelnen Molecülen zurückgelegt werden, nebst einigen anderen Bemerkungen über die mechanischen Wärmetheorie," *Ann. d. Phys.*, 105 (1858), 239–258, dated 14 August 1858. Trans. in (Brush, 1965), I, 135–147.

1864 *Abhandlungen über die mechanische Wärmetheorie*, 2 vols. (Braunschweig: Vieweg, 1864–67). English trans. of Vol. 1: *The Mechanical Theory of Heat* ed. T. Archer Hirst (London: J. Van Voorst, 1867). French trans. of Vols. 1 and 2: *Théorie méchanique de la chaleur*, trans. F. Folie, 2 vols. (Paris: Lacroix, 1868–69).

1865 "Über verschiedenen für die Anwendung bequeme Formen der Hauptgleichungen der mechanischen Wärmetheorie," *Ann. d. Phys.*, *125* (1865), 353–400.

1876 *Die mechanische Wärmetheorie*, 2nd rev. and enlarged ed., 3 vols. (Braunschweig: Vieweg, 1876–89). English trans. of Vol 1, *The Mechanical Theory of Heat*, trans. W. R. Brown (London: Macmillan, 1879).

1879 *Die mechanische. Wärmetheorie* 3rd rev. ed. 3 vols. (Braunschweig: Vieweg, 1879–91). French trans. of Vols. 1 and 2, *Théorie méchanique de chaleur*, trans. F. Folie and E. Ronker (Paris: Lacroix, 1888–93). For details see Chapter I, n. 18, above.

Compendia

1900 *Rapports présentés au Congrès international de physique réuni à Paris en 1900* (Paris: Gauthier-Villars, 1900), Congress held August 1900.

1905 *Landolt-Börnstein, physikalisch-chemische Tabellen*, ed. Richard Börnstein and Wilhelm Meyerhoffer, 3rd ed. (Berlin: Springer, 1905). First ed. (1883). Second ed. (1894). Many subsequent revisions.

1912 *La Théorie du rayonnement et les quanta: Rapports et discussions de la réunion tenue à Bruxelles, du 30 octobre au 3 novembre 1911*, ed. P. Langevin and M. de Broglie (Paris: Gauthier-Villars, 1912). Papers presented at the Instituts Solvay, Institut international de physique, Conseil de physique. Also in German, *Die Theorie der Strahlung und der Quanten, Verhandlungen auf einer von E. Solvay einberufenen Zusammenkunft...*, ed. A. Eucken (Halle: Knapp, 1914).

1913 "Discussion on Radiation," *Report of the British Association*, 1913, pp. 376–386, presented 12 September 1913.

1914 *Vorträge über die kinetische Theorie der Materie und der Elektrizität*, ed. M. Planck, P. Debye, *et al.* (Leipzig: Teubner, 1914).

Culverwell, E. P. (1855–1931)

1894 "Dr. Watson's Proof of Boltzmann's Theorem on Permanence of Distributions," *Nature*, *50* (1894), 617, dated 12 October 1894.

Day, A. L. (1869–1960)

1902 "Measurement of High Temperature," *Science*, *15* (1902), 429–433, report by C. K. Wead, Secretary of Philosophical Society of Washington, of meeting held 15 February 1902.

Debye, Peter (1884–1966)

1910 "Der Wahrscheinlichkeitsbegriff in der Theorie der Strahlung," *Ann. d. Phys.*, *33* (1910), 1427–1434, received 12 October 1910.

Drude, Paul (1863–1906)

1906 *Lehrbuch der Optik*, 2nd ed. (Leipzig: Hirzel, 1906). First ed. (1900). Third ed. (1912).

Ehrenfest, Paul (1880–1933).

Collected Scientific Papers, ed. M. J. Klein (Amsterdam: North-Holland; New York: Interscience, 1959).

1905	"Über die physikalischen Voraussetzungen der Planck'schen Theorie der irreversiblen Strahlungsvorgänge," *Wiener Ber.* II, 114 (1905), 1301–1314, presented 9 November 1905.
1906	"Zur Planckschen Strahlungstheorie," *Phys. ZS.*, 7 (1906), 528–532, dated 28 June 1906.
1911	"Welche Züge der Lichtquantenhypothese spielen in der Theorie der Wärmestrahlung eine wesentliche Rolle?," *Ann. d. Phys.*, 36 (1911), 91–118, received 8 July 1911.
1912	"Begriffliche Grundlagen der statistischen Auffassung in der Mechanik," *Encyclopädie d. mathematischen Wissenschaften*, Vol. IV, Mechanik, ed. F. Klein and C. Müller, Pt. 4, Heft 6 (Leipzig: Teubner, 1912). With Tatiana Ehrenfest. Trans. as *The Conceptual Foundations of the Statistical Approach in Mechanics*, trans. M. J. Moravcsik (Ithaca: Cornell University Press, 1959).

Einstein, Albert (1879–1955)

1901	"Folgerungen aus den Capillaritätserscheinungen," *Ann. d. Phys.*, 4 (1901), 513–523, dated 13 December 1900.
1902a	"Über die thermodynamische Theorie der Potentialdifferenz zwischen Metallen und vollständig dissozierten Lösungen ihrer Salze und über eine elektrische Methode zur Erforschung der Molekularkräfte," *Ann. d. Phys.*, 8 (1902), 798–814, received 30 April 1902.
1902b	"Kinetische Theorie des Wärmegleichgewichtes und des zweiten Hauptsatzes der Thermodynamik," *Ann. d. Phys.*, 9 (1902), 417–433, received 26 June 1902.
1903	"Eine Theorie der Grundlagen der Thermodynamik," *Ann. d. Phys.*, 11 (1903), 170–187, received 26 January 1903.
1904	"Zur allgemeinen molekularen Theorie der Wärme," *Ann. d. Phys.*, 14 (1904), 354–362, dated 27 March 1904.
1905	"Über einen die Erzeugung und Verwandlung des Lichtes betreffenden heuristischen Gesichtspunkt," *Ann. d. Phys.*, 17 (1905), 132–148, received 18 March 1905. Trans. in *The World of the Atom*, Vol. 1, ed. H. A. Boorse and L. Motz (New York: Basic Books, 1966), pp. 544–557. Also trans. in D. ter Haar, *The Old Quantum Theory* (Oxford: Pergamon, 1967), pp. 91–107.
1906a	"Zur Theorie der Lichterzeugung und Lichtabsorption," *Ann. d. Phys.*, 20 (1906), 199–206, received 13 March 1906.
1906b	["Review of Planck's *Wärmestrahlung*,"] *Ann. d. Phys., Beiblätter*, 30 (1906), 764–766.
1907a	"Die Plancksche Theorie der Strahlung und die Theorie der spezifischen Wärme," *Ann. d. Phys.*, 22 (1907), 180–190, received 9 September 1906.
1907b	"Berichtigung zu meiner Arbeit: 'Die Plancksche Theorie der Strahlung, etc.'," *Ann. d. Phys.*, 22 (1907), 800, received 3 March 1907.
1909a	"Zur gegenwärtigen Stand des Strahlungsproblems," *Phys. ZS.*, 10 (1909), 185–193, dated 12 February 1909.
1909b	"Über die Entwicklung unserer Anschauungen über das Wesen und die Konstitution der Strahlung," *Phys. ZS*, 10 (1909), 817–825, received 14 October 1909.

1910 "Statistische Untersuchung der Bewegung eines Resonators in einem Strahlungsfeld," *Ann. d. Phys.*, *33* (1910), 1105–1115, received 29 August 1910. With L. Hopf.

1912 "L'état actuel du problème des chaleurs spécifiques," *La Théorie du rayonnement et les quanta* (See Compendia above, 1912), pp. 407–435.

1913 "Einige Argumente für die Annahme einer molekularen Agitation beim absoluten Nullpunkt," *Ann d. Phys.*, *40* (1913), 551–560, received 5 January 1913. With Otto Stern.

Eucken, Arnold (1884–1950)

1914a "Über den Quanteneffekt bei einatomigen Gasen und Flüssigkeiten," *Berl. Ber.*, 1914, pp. 682–693, presented 28 May 1914.

1914b "Die Entwicklung der Quantentheorie vom Herbst 1911 bis Sommer 1913," in *Die Theorie der Strahlung und der Quanten, Verhandlungen auf einer von E. Solvay einberufenen Zusammenkunft...*, ed. A. Eucken (Halle: Knapp, 1914), pp. 371–405.

Fokker, A. D. (b. 1887)

1914 "Die mittlere Energie rotierender elektrischer Dipole im Strahlungsfeld," *Ann. d. Phys.*, *43* (1914), 810–820, received 23 December 1913.

Franck, James (1882–1964)

1911 "Über einen Zusammenhang zwischen Quantenhypothese und Ionisierungsspannung," *Verh. d. D. Phys. Ges.*, *13* (1911), 967–971, received 31 October 1911. With Gustav Hertz.

1913 "Über Zusammenstösse zwischen Gasmolekülen und langsamen Elektronen," *Verh. d. D. Phys. Ges.*, *15* (1913), 373–390, received 25 April 1913. With Gustav Hertz.

1914 "Über die Erregung der Quecksilberresonanzlinie 253.6 $\mu\mu$ durch Elektronenstösse," *Verh. d. D. Phys. Ges.*, *16* (1914), 512–517, received 21 May 1914. With Gustav Hertz.

Gibbs, J. W. (1839–1903)

1902 *Elementary Principles in Statistical Mechanics, Developed with Special Reference to the Rational Foundation of Thermodynamics* (New York: Scribner's; London: Arnold, 1902). Reprinted (New York: Dover, 1960).

Haas, A. E. (1884–1941)

1910a "Über die elektrodynamische Bedeutung des Planck'schen Strahlungsgesetzes und über eine neue Bestimmung des elektrischen Elementarquantums und der Dimensionen des Wasserstoffatoms," *Wiener Ber. II*, *119* (1910), 119–144, presented 10 March 1910.

1910b "Über eine neue theoretische Methode zur Bestimmung des elektrischen Elementarquantums und des Halbmessers des Wasserstoffatoms," *Phys. ZS.*, *11* (1910), 537–538, received 24 March 1910.

1910c "Der Zusammenhang des Planckschen elementaren Wirkungsquantums mit dem Grundgrössen der Elektronentheorie," *Jahrb. d. Radioakt.*, *7* (1910), 261–268, received 27 March 1910.

Haber, Fritz (1868–1934)

1911 "Elektronenemission bei chemischen Reaktionen," *Phys. ZS.*, *12* (1911) 1035–1044, presented 27 September 1911; *Verh. d. Ges. Deutscher Naturforscher und Ärzte*, 1911, Pt. 1, pp. 215–229.

Hasenöhrl, F. (1874–1915)
1911 "Über die Grundlagen der mechanischen Theorie der Wärme," *Phys. ZS.*, *12* (1911), 931–935, presented 25 September 1911.
Helmholtz, H. von (1821–1894)
1902 *Vorlesungen über theoretische Physik*, Vol. 5: *Dynamik continuirlich verbreiteter Massen*, ed. O. Krigar-Menzel (Leipzig: Barth, 1902).
Hertz, Gustav (1887–1975)
See (Franck, 1911, 1913, 1914).
Hertz, H. R. (1857–1894)
1889 "Die Kräfte electrischer Schwingungen, behandelt nach der Maxwell'schen Theorie," *Ann. d. Phys.*, *36* (1889), 1–22 dated November 1888. Reprinted as Chapter 9 of (H. R. Hertz, 1892).
1892 *Untersuchungen über die Ausbreitung der elektrischen Kraft* (Leipzig: Barth, 1892). Second ed. (1894). Third ed. (1914). Trans. as *Electric Waves*, trans. D. E. Jones (London and New York: Macmillan, 1893). Reprinted (New York: Dover, 1962).
Herzfeld, K. F. (b. 1892)
1912 "Über ein Atommodell, das die Balmer'sche Wasserstoffserie aussendet," *Wiener Ber. II*, *121* (1912), 593–601, received 7 March 1912.
Hopf, Ludwig (1884–1939)
1910 See (Einstein, 1910).
Ishiwara, Jun (1881–1947)
1915 "Die universelle Bedeutung des Wirkungsquantums," *Tôkyô Sûgaku-Buturigakkawi Kizi* [*Proceedings of the Tokyo Mathematical Physical Society*], *8* (1915), 106–116, received 4 April 1915.
Jahnke, P. R. E. (1863–1921)
1900 See (Lummer, 1900b).
Jeans, J. H. (1877–1946)
1901 "The Distribution of Molecular Energy," *Phil. Trans.*, *196* (1901), 397–430, received 14 June 1900.
1902 "On the Conditions necessary for equipartition of Energy," *Phil. Mag.*, *4* (1902), 585–596, issue of November 1902.
1903 "The Kinetic Theory of Gases Developed from a New Standpoint," *Phil. Mag.*, *5* (1903), 597–620, issue of June 1903.
1904 *Dynamical Theory of Gases* (Cambridge, England: Cambridge University Press, 1904). Second ed. (1916). Third ed. (1921). Fourth ed. (1925), reprinted (New York: Dover, 1954).
1905a "On the Partition of Energy between Matter and Aether," *Phil. Mag.*, *10* (1905), 91–98, originally dated March, postscript 7 June, issue of July 1905.
1905b "The Dynamical Theory of Gases," *Nature*, *71* (1904–05), 601, 607, issue of 27 April 1905.
1905c "The Dynamical Theory of Gases and of Radiation," *Nature*, *72* (1905), 101–102, dated 20 May 1905.
1905d "On the Application of Statistical Mechanics to the General Dynamics of Matter and Aether," *Proc. Roy. Soc. London*, *76* (1905), 296–311, received 19 May 1905.

1905e "A Comparison between Two Theories of Radiation," *Nature*, *72* (1905), 293–294, issue of 27 July 1905.
1905f "On the Laws of Radiation," *Proc. Roy. Soc. London*, *76* (1905), 545–526, received 11 October 1905.
1908 "Zur Strahlungstheorie," *Phys. ZS.*, *9* (1908), 853–855, dated 3 October 1908.
1910 "On Non-Newtonian Mechanical Systems, and Planck's Theory of Radiation," *Phil. Mag.*, *20* (1910), 943–954, dated 17 August 1910.
1912 "Rapport sur la théorie cinétique de la chaleur spécifique d'après Maxwell et Boltzmann," *La Théorie du rayonnement et les quanta* (See Compendia above, 1912), pp. 53–73.
1913 ["Discussion,"] *Report of the British Association*, 1913, pp. 376–381.
1914 *Report on Radiation and the Quantum-Theory* (London: The Electrician, 1914). Second ed. (London: Fleetway, 1924).

Joffé, A. F. (1880–1960)
1907 "Eine Bermerkung zu der Arbeit von E. Ladenburg: 'Über Anfangsgeschwindigkeit und Menge der photoelektrischen Elektronen u.s.w.'," *Münchener Ber.*, *37* (1907), 279–280, presented 2 November 1907; *Ann. d. Phys.*, *24* (1907), 939–940, received 17 November 1907.

Kamerlingh-Onnes, Heike (1853–1926)
1914 "Über die Translationsenergie in einatomigen Gasen beim absoluten Nullpunkt," *Vorträge über die kinetische Theorie* (See Compendia above, 1914), pp. 193–194, presented April 1913. With W. H. Keesom.

Kayser, H. G. J. (1853–1949)
1902 *Handbuch der Spectroscopie*, Vol. 2 (Leipzig: Hirzel, 1902). Volume 1 appeared in 1900 and Vol. 3 in 1905. Many others appeared later.

Keesom, W. H. (1876–1956)
1914 See (Kamerlingh-Onnes, 1914).

Kirchhoff, G. R. (1824–1887)
Gesammelte Abhandlungen (Leipzig: Barth, 1882).
1859 "Über den Zusammenhang zwischen Emission and Absorption von Licht und Wärme," *Monatsberichte der Akademie der Wissenschaften zu Berlin*, 1859, pp. 783–787, dated 11 December 1859.
1860 "Über das Verhältnis zwischen dem Emissionsvermögen und dem Absorptionsvermögen der Körper für Wärme und Licht," *Ann. d. Phys.*, *109* (1860), 275–301, dated January 1860.
1894 *Vorlesungen über mathematische Physik*, Vol. 4: *Vorlesungen über die Theorie der Wärme*, ed. Max Planck (Leipzig: Teubner, 1894).

Kries, Johannes von (1853–1928)
1886 *Die Principien der Wahrscheinlichkeitsrechnung* (Freiburg: Akademische Verlagsbuchhandlung, 1886). Second edition (Tübingen: Mohr, 1927).

Kurlbaum, Ferdinand (1857–1927)
1901 See (Rubens, 1900, 1901).

Ladenburg, Rudolph (1882–1952)
1909 "Die neueren Forschungen über die durch Licht- und Röntgenstrahlen hervorgerufene Emission negativer Elektronen," *Jahrb. d. Radioakt.*, *6* (1909), 425–484, received 6 September 1909.

Langevin, Paul (1872–1946)
  1912 "La théorie cinétique du magnétisme et les magnétons," *La Théorie du rayonnement et les quanta* (See Compendia above, 1912), pp. 393–404.
Langley, S. P. (1834–1906)
  1886 "Observations on Invisible Heat-Spectra and the Recognition of Hitherto Unmeasured Wave-lengths, Made at the Allegheny Observatory," *Phil. Mag.*, 21 (1886), 394–409.
Larmor, Joseph, (1857–1942)
  Mathematical and Physical Papers by Sir Joseph Larmor, 2 vols. (Cambridge, England: Cambridge University Press, 1929).
  1902a "Radiation, Theory of," in *The [Eighth of the] New Volumes of the Encyclopaedia Britannica,...being Volume XXXII of the Complete Work* (London: Black, 1902), pp. 120–128. Not in the collected works.
  1902b "On the Application of the Method of Entropy to Radiant Energy," *Report of the British Association*, 1902, p. 546, presented 16 September 1902.
  1909 "The Statistical and Thermodynamical Relations of Radiant Energy," *Proc. Roy. Soc. London*, 83 (1909–10), 82–95, presented 18 November 1909.
  1910 "On the Statistical Theory of Radiation," *Phil. Mag.*, 20 (1910), 350–353, dated 4 July 1910.
Lenard, Philipp (1862–1947)
  1902 "Über die lichtelektrische Wirkung," *Ann. d. Phys.*, 8 (1902), 149–198. received 17 March 1902.
Lindemann, F. A. (1886–1957)
  1910 "Über die Berechnung molekularer Eigenfrequenzen," *Phys. ZS.*, 11 (1910), 609–612, received 25 June 1910.
  1911 "Über die Berechnung der Eigenfrequenzen der Elektronen im selektiven Photoeffekt," *Verh. d. D. Phys. Ges.*, 13 (1911), 482–488, received 16 June 1911.
Lorentz, H. A. (1853–1928)
  Collected Papers, 9 vols. (The Hague: Nijhoff, 1934–39).
  1887 "Über das Gleichgewicht der lebendigen Kraft unter Gasmolekülen," *Wiener Ber. II*, 95 (1887), 115–152, presented 20 January 1887.
  1901a "The Theory of Radiation and the Second Law of Thermodynamics," *Proc. Amsterdam*, 3 (1901), 436–450, presented 29 December 1900.
  1901b "Boltzmann's and Wien's Laws of Radiation," *Proc. Amsterdam*, 3 (1901), 607–620, presented 23 February 1901.
  1903 "On the Emission and Absorption by Metals of Rays of Heat of Great Wavelength," *Proc. Amsterdam*, 5 (1903), 666–685, presented 24 April 1903.
  1905 "La thermodynamique et les théories cinétique," *Bulletin des Séances de la Société Française de Physique*, 1905, pp. 35–63, conference held 27 April 1905.
  1908 "Zur Strahlungstheorie," *Phys, ZS.*, 9 (1908), 562–563, dated 19 July 1908.
  1909 "Le partage de l'énergie entre la matière pondérable et l'éther," *Atti del IV Congresso Internazionale dei Matematici (Roma, 6–11,*

Aprile 1908), 3 vols. (Rome: R. Accademia dei Lincei, 1909), I, pp. 145–165, presented April 8, 1908. Reprinted with revisions in *Nuovo Cimento*, *16* (1908), 5–34, and *Revue géneral des Sciences*, *20* (1909), 14–26.

1910a "Die Hypothese der Lichtquanten," *Phys. ZS.*, *11* (1910), 349–354, presented 17 April 1909.

1910b "Alte und neue Fragen der Physik," *Phys. ZS.*, *11* (1910), 1234–1257, received 2 November 1910 (condensation by Born of lectures delivered 24–29 October 1910).

1912a "Sur l'application au rayonnement du théorème de l'équipartition de l'énergie," *La Théorie du rayonnement et les quanta* (See Compendia above, 1912), pp. 12–39.

1912b "Sur la théorie des éléments d'énergie," *Arch. Néerland.*, *2* (1912), 176–191.

1913 ["Discussion,"] *Report of the British Association*, 1913, p. 385.

1916 *Les Théories statistiques en thermodynamique*, ed. L. Dunoyer (Leipzig and Berlin: Teubner, 1916). Lectures presented November 1912, revised 1913.

Loschmidt, Josef (1821–1895)

1876 "Über den Zustand des Wärmegleichgewichtes eines Systems von Körpern mit Rücksicht auf die Schwerkraft. I," *Wiener Ber. II*, *73* (1876), 128–142, presented 20 January 1876.

Lummer, Otto (1860–1925)

1899a "Die Vertheilung der Energie im Spectrum des schwarzen Körpers," *Verh. d. D. Phys. Ges.*, *1* (1899), 23–41, presented 3 February 1899. With Ernst Pringsheim.

1899b "Die Vertheilung der Energie im Spectrum des schwarzen Körpers und des blanken Platins," *Verh. d. D. Phys. Ges.*, *1* (1899), 215–235, presented 3 November 1899. With Ernst Pringsheim.

1900a "Über die Strahlung des schwarzen Körpers für lange Wellen," *Verh. d. D. Phys. Ges.*, *2* (1900), 163–180, concerning submission date see Chapter IV, n. 10. With E. Pringsheim.

1900b "Über die Spectralgleichung des schwarzen Körpers und des blanken Platins," *Ann. d. Phys.*, *3* (1900), 283–297, received 30 July 1900. With P. R. E. Jahnke.

1900c "Le rayonnement des corps noirs," *Rapports présentés au Congrès international de physique réuni à Paris en 1900* (See Compendia above, 1900), Vol. 2, pp. 41–99. Congress held in August 1900.

1908 "Über die Jeans–Lorentzsche Strahlungsformel," *Phys. ZS.*, *9* (1908), 449–450, dated May 1908. With Ernst Pringsheim.

Maxwell, J. C. (1831–1879)

Scientific Papers of James Clark Maxwell, ed. W. D. Niven, 2 vols. (Cambridge, England: Cambridge University Press, 1890). Reprinted (New York: Dover, 1952).

1860 "Illustrations of the Dynamical Theory of Gases," *Phil. Mag.*, *19* (1860), 19–32; *20* (1860), 21–37. Issues of January and July 1860. Partial reprint in (Brush, 1965), I, 148–171.

1866 "On the Dynamical Theory of Gases," *Phil. Mag.*, *32* (1866), 390–393, read 31 May 1866; *35* (1868), 129–145, 185–217; *Phil. Trans.*, *157* (1867), 49–88. Reprinted in (Brush, 1965), II, 23–87.

1871 *Theory of Heat* (London and New York: Longmans, Green, 1871). Second and 3rd eds. (1872). Fourth ed. (1875). Ninth ed. (1888). The 13th ed. (1899) was reprinted as late as 1916.

1873 *A Treatise on Electricity and Magnetism*, 2 vols. (Oxford: Clarendon Press, 1873). Second ed. (1881). Third ed. (1892). Reprint of third ed. (New York: Dover, 1954).

1878 "Tait's 'Thermodynamics'," *Nature*, *17* (1877–78), 257–259, 278–280. Issues of 31 January and 7 February 1878.

1879 "On Boltzmann's Theorem on the Average Distribution of Energy in a System of Material Points," *Trans. Cambridge Phil. Soc.*, *12* (1871–79), 547–570, presented 6 May 1878.

Meyer, O. E. (1834–1909)

1877 *Die kinetische Theorie der Gase* (Breslau: Maruschke, 1877); 2nd rev. ed. (1899); trans. R. E. Baynes (London, New York, and Bombay: Longmans, Green, 1899).

Michelson, W. A. (1860–1927)

1887 "Essai théorique sur la distribution de l'énergie dans les spectres des solides," *Journ. de Phys. et le Radium*, *6* (1887), 467–479.

Millikan, R. A. (1868–1953)

1914 "A Direct Determination of 'h'," *Phys. Rev.*, *4* (1914), 73–75, dated 24 April 1914.

1915 "New Tests of Einstein's Photo-Electric Equation," *Phys. Rev.*, *6* (1915), 55, presented 24 April 1915.

1916 "A Direct Photoelectric Determination of Planck's 'h'," *Phys. Rev.*, *7* (1916), 355–388, dated March 1916.

Nernst, Walther (1864–1914)

1900 *Theoretische Chemie vom Standpunkte der Avogadroschen Regel und der Thermodynamik*, 3rd ed. (Stuttgart: Encke, 1900). First ed. (1893). Second ed. (1898).

1903 *Theoretische Chemie vom Standpunkte der Avogadroschen Regel und der Thermodynamik*, 4th ed. (Stuttgart: Encke, 1903).

1906a "Über die Berechnung chemischer Gleichgewichte aus thermischen Messungen," *Göttinger Nachr.*, 1906, pp. 1–40, presented 23 December 1905.

1906b *Theoretische Chemie vom Standpunkte der Avogadroschen Regel und der Thermodynamik*, 5th ed. (Stuttgart: Encke, 1906). There is also a 1907 imprint of this edition.

1907 *Experimental and Theoretical Applications of Thermodynamics to Chemistry* (New York: Scribner's, 1907).

1909 *Theoretische Chemie vom Standpunkte der Avogadroschen Regel und der Thermodynamik*, 6th ed. (Stuttgart: Encke, 1909). Many later editions through 1926.

1910a "Untersuchungen über die spezifische Wärme bei tiefen Temperaturen II," *Berl. Ber.*, 1910/I, pp. 262–282, presented 17 February 1910.

1910b "Revue sur la détermination de l'affinité chimique à partir des données thermiques," *Journal chim. phys.*, *8* (1910), 228–267, dated March 1910.

1911a "Über neuere Probleme der Wärmetheorie," *Berl. Ber.*, 1911/I, pp. 65–90, presented 26 January 1911.

1911b "Zur Theorie der spezifischen Wärme and über die Anwendung der Lehre von den Energiequanten auf physikalisch-chemische Fragen überhaupt," *ZS. f. Elektrochem.*, *17* (1911), 265–275, received 21 February 1911.

1911c "Der Energieinhalt fester Stoffe," *Ann. d. Phys.*, *36* (1911), 395–439, received 17 August 1911.

1911d "Über ein allgemeines Gesetz, das Verhalten fester Stoffe bei sehr tiefen Temperaturen betreffend," *Phys. ZS.*, *12* (1911), 976–978, presented 27 September 1911.

1912 "Application de la théorie des quanta à divers problèmes physicochimiques," *La Théorie du rayonnement et les quanta* (See Compendia above, 1912), pp. 254–290.

1918 *Die theoretischen und experimentellen Grundlagen des neuen Wärmesatzes* (Halle: Knapp, 1918). Second ed. (1924). Trans. as, *The New Heat Theorem*, trans. Guy Barr from the 2nd German ed. (London: Methuen, 1926; New York: Dutton, 1926). Reprinted (New York: Dover, 1969).

Netto, E. (1846–1919)

1898 "Kombinatorik," *Encyklopädie der mathematischen Wissenschaften*, Vol. I, *Arithmetik und Algebra*, ed. W. F. Meyer, Pt. 1, Heft 1 (Leipzig: Teubner, 1898).

Nicholson, J. W. (1881–1955)

1912a "The Constitution of the Solar Corona. II," *Month. Not.*, *72* (1912), 677–692, dated 28 April 1912.

1912b "The Constitution of the Solar Corona, III," *Month. Not.*, *72* (1912), 729–739.

Paschen, Friedrich (1865–1947)

1885 "Über Gesetzmässigkeiten in Spectren fester Körper und über eine neue Bestimmung der Sonnentemperatur," *Göttinger Nachr.*, 1885, pp. 294–305.

1896 "Über Gesetzmässigkeiten in den Spectren fester Körper, erste Mittheilung," *Ann. d. Phys.*, *58* (1896), 455–492, dated May 1896.

Peierls, R. E. (b. 1907)

1929 "Zur kinetischen Theorie der Wärmeleitung in Kristallen," *Ann. d. Phys.*, *3* (1929), 1055–1101, received 24 October 1929.

Planck, Max (1858–1947)

Physikalische Abhandlungen und Vorträge, 3 vols. (Braunschweig: Vieweg, 1958).

1879 *Über den zweiten Hauptsatz der mechanischen Wärmetheorie* (Munich: Ackerman, 1879). Included in collected works.

1880 *Gleichgewichtszustände isotroper Körper in verschiedenen Temperaturen* (Munich: Ackerman, 1880). Included in collected works.

BIBLIOGRAPHY

1882 "Verdampfen, Schmelzen und Sublimieren," *Ann. d. Phys.*, *15* (1882), 446–475, dated December 1881.

1887a "Über das Princip der Vermehrung der Entropie. Erste Abhandlung," *Ann. d. Phys.*, *30* (1887), 562–582, dated December 1886; ".... Zweite Abhandlung," *ibid.*, *31* (1887), 189–203, dated February 1887; ".... Dritte Abhandlung," *ibid.*, *32* (1887), 462–503, dated July 1887.

1887b *Das Princip der Erhaltung der Energie* (Leipzig: Teubner, 1887).

1891 "Allgemeines zur neueren Entwicklung der Wärmetheorie," *Verhandlungen der Gesellschaft deutscher Naturforscher und Ärzte*, 1891, Pt. 2, pp. 56–61, presented at meeting in Halle, 21–25 September 1891; *ZS. f. phys. Chem.*, *8* (1891), 647–656.

1893 "Ein neues Harmonium in natürlicher Stimmung nach dem System von C. Eitz," *Verh. d. D. Phys. Ges.*, *12* (1893), 8–9.

1894 "Über den Beweis des Maxwellschen Geschwindigkeitsverteilungsgesetzes unter Gasmolekülen," *Münchener Ber.*, *24* (1894), 391–394, presented 3 November 1894; *Ann. d. Phys.*, *55* (1895), 220–222.

1895 "Absorption und Emission elektrischer Wellen durch Resonanz," *Berl. Ber.*, 1895, pp. 289–301, presented 21 March 1895; *Ann. d. Phys.*, *57* (1896), 1–14.

1896 "Über elektrische Schwingungen, welche durch Resonanz erregt und durch Strahlung gedämpft werden," *Berl. Ber.*, 1896, pp. 151–170, presented 20 February 1896; *Ann. d. Phys.*, *60* (1897), 577–599.

1897a "Über irreversible Strahlungsvorgänge. Erste Mitteilung," *Berl. Ber.*, 1897, pp. 57–68, presented 4 February 1897.

1897b "Über irreversible Strahlungsvorgänge. Zweite Mitteilung," *Berl. Ber.*, 1897, pp. 715–717, presented 8 July 1897.

1897c "Notiz zur Theorie der Dämpfung electrischer Schwingungen," *Ann. d. Phys.*, *63* (1897), 419–422, dated August 1897.

1897d "Über irreversible Strahlungsvorgänge. Dritte Mitteilung," *Berl. Ber.*, 1897, pp. 1122–1145, presented 16 December 1897.

1898 "Über irreversible Strahlungsvorgänge. Vierte Mitteilung," *Berl. Ber.*, 1898, pp. 449–476, presented 7 July 1898.

1899 "Über irreversible Strahlungsvorgänge. Fünfte Mitteilung (Schluss)," *Berl. Ber.*, 1899, pp. 440–480, presented 18 May 1899.

1900a "Über irreversible Strahlungsvorgänge," *Ann. d. Phys.*, *1* (1900), 69–122, received 7 November 1899.

1900b "Deduktion der Strahlungs-Entropie aus dem zweiten Haupsatz der Thermodynamik," *Verh. d. D. Phys. Ges.*, *2* (1900), 37, presented 2 February 1900.

1900c "Entropie und Temperatur strahlender Wärme," *Ann. d. Phys.*, *1* (1900), 719–737, received 22 March 1900.

1900d "Über eine Verbesserung der Wien'schen Spektralgleichung," *Verh. d. D. Phys. Ges.*, *2* (1900), 202–204, presented 19 October 1900. Trans. in (ter Haar, 1967; Kangro, 1972).

1900e "Zur Theorie des Gesetzes der Energieverteilung im Normalspectrum," *Verh. d. D. Phys. Ges.*, *2* (1900), 237–245, presented 14 December 1900. Trans. in (ter Haar, 1967; Kangro, 1972).

1901a	"Über das Gesetz der Energieverteilung im Normalspectrum," *Ann. d. Phys.*, *4* (1901), 553–563, received 7 January 1901.
1901b	"Über die Elementarquanta der Materie und der Elektricität," *Ann. d. Phys.*, *4* (1901), 564–566, received 9 January 1901.
1901c	"Über irreversible Strahlungsvorgänge (Nachtrag)," *Ann. d. Phys.*, *6* (1901), 818–831, received 16 October 1901.
1901d	"Über die Verteilung der Energie zwischen Aether und Materie," *Arch. Néerland.*, *6* (1901), 55–66, for Bosscha anniversary, 18 November 1901. Reprinted *Ann. d. Phys.*, *9* (1902), 629–641.
1906a	*Vorlesungen über die Theorie der Wärmestrahlung*, 1st ed. (Leipzig: Barth, 1906).
1906b	"Bemerkung über die Konstante des Wienschen Verschiebungsgesetzes," *Verh. d. D. Phys. Ges.*, *8* (1906), 695–696, presented 14 December 1906.
1907	"Zur Dynamik bewegter Systeme," *Berl. Ber.*, 1907, pp. 542–570, presented 13 June 1907; *Ann. d. Phys.*, *26* (1908), 1–34.
1910a	"Zur Theorie der Wärmestrahlung," *Ann. d. Phys.*, *31* (1910), 758–768, received 18 January 1910.
1910b	*Acht Vorlesungen über theoretische Physik* (Leipzig: Hirzel, 1910), originally delivered at Columbia University, 1909. Trans. as *Eight Lectures on Theoretical Physics*, trans. A. P. Wills (New York: Columbia University Press, 1915).
1911a	"Eine neue Strahlungshypothese," *Verh. d. D. Phys. Ges.*, *13* (1911), 138–148, presented 3 February 1911.
1911b	"Zur Hypothese der Quantenemission," *Berl. Ber.*, 1911, pp. 723–731, presented 13 July 1911.
1912a	"Über die Begründung das Gesetzes der schwarzen Strahlung," *Ann. d. Phys.*, *37* (1912), 642–656, received 14 January 1912.
1912b	"La loi du rayonnement noir et l'hypothèse des quantités élémentaires d'action," *La Théorie du rayonnement et les quanta* (See Compendia above 1912), pp. 93–114.
1913	*Vorlesungen über die Theorie der Wärmestrahlung*, 2nd ed. revised (Leipzig: Barth, 1913). Reprinted as 3rd ed. (1919). Fourth ed. revised (1921). Fifth ed. revised (1923). Reprinted as 6th ed. (1966). Second ed. trans. as *The Theory of Heat Radiation*, trans. Morton Masius (Philadelphia: Blakiston, 1914), reprinted (New York: Dover, 1959).
1914	"Eine veränderte Formulierung der Quantenhypothese," *Berl. Ber.*, 1914, pp. 918–923, presented 23 July 1914.
1915	"Die Quantenhypothese für Molekeln mit mehreren Freiheitsgraden (Erste Mitteilung)," *Verh. d. D. Phys. Ges.*, *17* (1915), 407–418, presented 5 November 1915.
1916	"Die physikalische Struktur des Phasenraumes," *Ann. d. Phys.*, *50* (1916), 385–418, received 13 April 1916.
1922	"Die Entstehung und bisherige Entwicklung der Quantentheorie," *Les Prix Nobel en 1919–1920* (Stockholm: Norstedt, 1922), pp. 1–14, read 2 June 1920. Trans. as "The Genesis and present state of develop-

ment of the quantum theory" in *Nobel Lectures... Physics, 1901–1921* (Amsterdam: Elsevier, 1967), pp. 407–420.

1928 "Hendrik Antoon Lorentz," *Naturwissensch.*, **16** (1928), 549–555.

1943 "Zur Geschichte der Auffindung des physikalischen Wirkungsquantums," *Naturwissensch.*, **31** (1943), 153–159, issue of 2 April 1943.

1948 *Wissenschaftliche Selbstbiographie* (Leipzig: Barth, 1948). Trans. F. Gaynor in *Scientific Autobiography and Other Papers* (New York: Philosophical Library, 1949), pp. 13–51.

Poincaré, Henri (1854–1912)

Oeuvres de Henri Poincaré, ed. under the auspices of the Académie des Sciences, 11 vols. (Paris: Gauthier-Villars, 1916–54).

1911 "Sur la théorie des quanta," *C. R.*, **153** (1911), 1103–1108, meeting of 4 December 1911.

1912 "Sur la théorie des quanta," *Journal de Physique Théorique et Appliquée*, **2** (1912), 5–34.

Pringsheim, Ernst (1859–1917)

1901 "Einfache Herleitung des Kirchhoff'schen Gesetzes," *Ver. d. D. Phys. Ges.*, **3** (1901), 81–84, presented 3 May 1901.

See also (Lummer, 1899a, 1899b, 1900a, 1908).

Lord Rayleigh [John William Strutt] (1842–1919)

Scientific Papers, 6 vols. (Cambridge, England: Cambridge University Press, 1899–1920). Reprinted (New York: Dover, 1964).

1900a "The Law of Partition of Kinetic Energy," *Phil. Mag.*, **49** (1900), 98–118, issue of January 1900.

1900b "Remarks upon the Law of Complete Radiation," *Phil. Mag.*, **49** (1900), 539–540, issue of June 1900.

1905a "The Dynamical Theory of Gases," *Nature*, **71** (1904–05), 559, issue of 13 April 1905.

1905b "The Dynamical Theory of Gases and Radiation," *Nature*, **72** (1905), 54–55, dated 6 May 1905.

1905c "The Constant of Radiation as Calculated from Molecular Data," *Nature*, **72** (1905), 243–244, dated 7 July 1905.

Roscoe, H. E. (1833–1915)

Lehrbuch der anorganischen Chemie, 3rd ed., 2 vols. (Braunschweig: Vieweg, 1895–97). With Alexander Classen. A reworking of materials previously published under other titles by Roscoe and Schorlemmer.

Rubens, Heinrich (1865–1922)

1900 "Über die Emission langwelliger Wärmestrahlen durch den schwarzen Körper bei verschiedenen Temperaturen," *Berl. Ber.*, 1900, pp. 929–941, presented 25 October 1900. With F. Kurlbaum.

1901 "Anwendung der Methode der Reststrahlen zur Prüfung des Strahlungsgesetzes," *Ann. d. Phys.*, **4** (1901), 649–666, received 10 February 1901. With F. Kurlbaum.

Rutherford, Ernest (1871–1937)

1929 "Note by Professor E. Rutherford," *Naturwissensch.*, **17** (1929), 483.

Sackur, Otto (1880–1914)
  1914 "Die spezifische Wärme der Gase und die Nullpunktsenergie," *Verh. d. D. Phys. Ges.*, *16* (1914), 728–734, presented 10 July 1914.

Schaefer, Clemens (1878–1968)
  1907 ["Review of Planck's *Wärmestrahlung*"], *Phys. ZS.*, *8* (1907), 224, received 27 January 1907.

Schidlof, Arthur (1877–1934)
  1911 "Zur Aufklärung der universellen elektrodynamischen Bedeutung der Planckschen Strahlungskonstanten h," *Ann. d. Phys.*, *35* (1911), 90–100, received 13 March 1911.

Sommerfeld, Arnold (1868–1952)
  Gesammelte Schriften, 4 vols. (Braunschweig: Vieweg, 1968).
  1909 "Über die Verteilung der Intensität bei der Emission von Röntgenstrahlen," *Phys. ZS.*, *10* (1909), 969–976, received 6 December 1909.
  1910 "Über die Verteilung der Intensität bei der Emission von Röntgenstrahlen," *Phys. ZS.*, *11* (1910), 99–101, received 21 January 1910.
  1911a "Über die Struktur der γ-Strahlen," *Münchener Ber.*, 1911, pp. 1–60, presented 7 January 1911.
  1911b "Das Plancksche Wirkungsquantum und seine allgemeine Bedeutung für die Molekularphysik," *Phys. ZS.*, *12* (1911), 1057–1068, presented 25 September 1911; *Verh. d. Ges. Deutscher Naturforscher und Ärzte*, Pt. 2 (1911), pp. 31–49.
  1915 "Zur Theorie der Balmerschen Serie," *Münchener Ber.*, 1915, pp. 425–458, presented 6 December 1915.
  1919 *Atombau und Spektrallinien* (Braunschweig: Vieweg, 1919). Many later editions.
  1948 "Gedächtsnisfeier der Physikalischen Gesellschaft in Württemberg-Baden zu Heidenheim am 15. November 1947," *Ann. d. Phys.*, *3* (1948), 3–6.

Stark, Johannes (1874–1957)
  1907a "Elementarquantum der Energie, Modell der negativen und der positiven Elektrizität," *Phys. ZS.*, *8* (1907), 881–884, received 30 October 1907.
  1907b "Über eine Berechnung der Wellenlänge der Röntgenstrahlen aus dem Planckschen Energie-Element," *Göttinger Nachr.*, 1907, pp. 598–601, presented 23 November 1907.
  1907c "Beziehung des Doppler-Effektes bei Kanalstrahlen zur Planckschen Strahlungstheorie," *Phys. ZS.*, *8* (1907), 913–919, received 2 December 1907.
  1908a "Zur Energetik und Chemie der Bandenspektra," *Phys. ZS.*, *9* (1908), 85–94, received 30 December 1907.
  1908b "Neue Beobachtungen an Kanalstrahlen in Beziehung zur Lichtquantenhypothese," *Phys. ZS.*, *9* (1908), 767–773, received 7 October 1908.
  1908c "Weitere Bemerkungen über die thermische und chemische Absorption im Bandenspektrum," *Phys. ZS.*, *9* (1908), 889–894, received 20 October 1908.

1909a "Über Röntgenstrahlen und die atomistische Konstitution der Strahlung, I. Röntgenstrahlen," *Phys. ZS.*, *10* (1909), 579–586, received 23 July 1909.
1909b "Zur experimentellen Entscheidung zwischen Ätherwellen- und Lichtquantenhypothese. I. Röntgenstrahlen," *Phys. ZS.*, *10* (1909), 902–913, received 16 November 1909.
1910 "Zur experimentellen Entscheidung zwischen der Lichtquantenhypothese und der Ätherimpulstheorie der Röntgenstrahlen," *Phys. ZS.*, *11* (1910), 24–31, received 28 December 1909.

Stefan, Josef (1835–1893)
1879 "Über die Beziehung zwischen der Wärmestrahlung und der Temperatur," *Wiener Ber. II*, *79* (1879), 391–428, presented 23 January 1879.

Stern, Otto (1888–1969)
1913 See (Einstein, 1913).

Tait, P. G. (1831–1901)
1877 *Sketch of Thermodynamics*, 2nd ed. (Edinburgh: Douglas, 1877), First ed. (1866).

Thiesen, M. F. (1849–1936)
1900 "Über das Gesetz der schwarzen Strahlung," *Verh. d. D. Phys. Ges.*, *2* (1900), 65–70, presented 2 February 1900.

Thomson, J. J. (1856–1940)
1907 *The Corpuscular Theory of Matter* (London: Constable; New York: Scribner's, 1907).
1910 "On the Theory of Radiation," *Phil. Mag.*, *20* (1910), 238–247, dated July 1910.

Thomson, William (1851–1923)
Mathematical and Physical Papers, 6 vols. (Cambridge, England: Cambridge University Press, 1882–1911).
1874 "The Kinetic Theory of the Dissipation of Energy," *Proc. Edinburgh*, *8* (1872–75), 325–334, presented 2 February 1874.

Valentiner, Siegfried (1876–1958)
1919 *Die Grundlagen der Quantentheorie in elementarer Darstellung*, 2nd ed. (Braunschweig: Vieweg, 1919). First ed. (1914).

Voigt, Woldemar (1850–1919)
1903 *Thermodynamik*, 2 vols. (Leipzig: Göschensche, 1903–04).

Watson, H. W. (1827–1903)
1876 *A Treatise on the Kinetic Theory of Gases* (Oxford: Clarendon, 1876). Second ed. (1893).

Weber, H. F. (1843–1912)
1888 "Untersuchungen über die Strahlung fester Körper," *Berl. Ber.*, 1888, pp. 933–957.

Weinstein, [Max] Bernhard (1852–1918)
1903 *Thermodynamik und Kinetik der Körper*, 3 vols. (Braunschweig: Vieweg, 1903).

Weiss, J. (n. d.)
1909 "Über das Plancksche Strahlungsgesetz (vorläufige Mitteilung)," *Phys. ZS.*, *10* (1909), 193–195, received 13 February 1909.

Weiss, Pierre (1865–1940)
- 1911 "Über die rationalen Verhältnisse der magnetischen Momente der Moleküle und das Magneton," *Phys. ZS.*, *12* (1911), 935–952, presented 25 September 1911; *Verh. d. Ges. Deutscher Naturforscher und Ärzte*, Pt. 2 (1911), pp. 50–77.

Wertheimer, Eduard (n.d.)
- 1911 "Die Plancksche Konstant h und der Ausdruck $h\nu$," *Phys. ZS.*, *12* (1911), 408–412, received 13 March 1911.

Wien, Wilhelm (1864–1928)
- 1893 "Eine neue Beziehung der Strahlung schwarzer Körper zum zweiten Hauptsatz der Wärmetheorie," *Berl. Ber.*, 1893, pp. 55–62, received 9 February 1893.
- 1896 "Über die Energievertheilung im Emissionsspectrum eines schwarzen Körpers," *Ann. d. Phys.*, *58* (1896), 662–669, dated June 1896.
- 1900 "Les lois théorique du rayonnement," *Rapports présentés au Congrès international de physique réuni à Paris en 1900* (See Compendia above, 1900), Vol. 2, pp. 23–40. Congress held in August 1900.
- 1907a "Über die absolute, von positiven Ionen ausgestrahlte Energie und die Entropie der Spektrallinien," *Ann. d. Phys.*, *23* (1907), 415–438, received 9 May 1907.
- 1907b "Über eine Berechnung der Wellenlänge der Röntgenstrahlen aus dem Planckschen Energie-Element," *Göttinger Nachr.*, 1907, pp. 598–601, presented 23 November 1907.
- 1909 "Theorie der Strahlung," *Encyclopädie der mathematischen Wissenschaften*, Vol. V, *Physik*, ed. A. Sommerfeld, Pt. 3, Heft 4 (Leipzig: Teubner, 1909).

Wilson, William (1875–1965)
- 1915 "The Quantum-Theory of Radiation and Line Spectra," *Phil. Mag.*, *29* (1915), 795–802, dated March 1915.

Zermelo, E. F. F. (1871–1923)
- 1896a "Über einen Satz der Dynamik und die mechanische Wärmetheorie," *Ann. d. Phys.*, *57* (1896), 485–494, dated December 1895. Trans. in (Brush, 1965), II, 208–217.
- 1896b "Über mechanische Erklärungen irreversibler Vorgänge," *Ann. d. Phys.*, *59* (1896), 793–801, dated 15 September 1896. Trans. in (Brush, 1965), II, 229–237.

Manuscript sources

Though always incomplete and by now also out of date, the inventory-report *Sources for History of Quantum Physics* (Kuhn, 1967) remains the indispensable published inventory of manuscripts relevant to historians of the quantum theory. It was the final report of an archival project which interviewed living participants in the development of quantum concepts, microfilmed their papers when available, and produced a preliminary survey of relevant manuscripts already on deposit. Some of the manuscripts to which the report leads are preserved on the project's microfilms at the various Archives for the History of Quantum Physics; others are identified, often most summarily, as available at one or another particular depository. All the depositories and many of the manuscripts that have proved relevant to this book can be located through that report, so that no function would be served by summarizing its contents here. This portion of the bibliography is therefore restricted to a few discursive remarks about presently relevant sources that have been discovered or made available in the ten years since the report was prepared. Together with a great deal else, this information will be included in the magistral survey of unpublished documents in twentieth-century physics currently being prepared by the Office for History of Science and Technology at the University of California, Berkeley, with the assistance of the Center for History of Physics of the American Institute of Physics in New York.

Shortly after the report went to press an additional set of significant manuscripts belonging to Arnold Sommerfeld was discovered in the home of his son. A selection was microfilmed and is available on microfilms #83 and #84 of the Archives for the History of Quantum Physics. More recently two major collections, both previously deposited at European centers have been filmed for the Archives. One is the Ehrenfest collection, of which the contents have very recently been reported in the *Catalogue of the Paul Ehrenfest Archive at the Museum Boerhaave, Leiden*, Communication 151 of the National Museum for the History of Science and Medicine (Leyden, 1977). The other is the papers of H. A. Lorentz, deposited at the Algemeen Rijksarchief at The Hague. Since that film was prepared, the Lorentz papers have been rearranged and catalogued, alterations which will make the collection far more accessible. It is to be hoped that a new set of films will shortly be prepared.

Finally, after this book was complete in draft, a previously unknown and very significant collection of letters from Planck to Wien was acquired by the Staatsbibliothek Preussischer Kulturbesitz (formerly the Preussische Staatsbibliothek) in Berlin. It contains 147 letters and two postcards written between 1900 and 1928. Most of them deal with editorial matters relating to the *Annalen der Physik*, of which the two men were co-editors from 1907 until Wien's death in 1928, and they are a rich source for the development of German physics. But little use has been made of them in this book. As emphasized in Chapter V, note 42†, they make no reference to Planck's theory before 1908, and the copious references thereafter have not seemed to alter views suggested by the sources from which the draft manuscript had been prepared before the collection became available.

AFTERWORD: REVISITING PLANCK

IT IS NOW over five years since the publication of *Black-body theory and the quantum discontinuity, 1894–1912*, a book which provides the most fully realized illustration of the concept of history of science basic to my historical publications. The same conception underlies my more philosophical writing—is, indeed, what ties these apparently disparate aspects of my work together. But what I see as the best and, technical difficulty aside, the most representative of my historical works has been widely received, even among those who praise it, as a misfit, a problem child, among my publications. Under those circumstances the invitation to prepare for *HSPS* a summary of the book's central thesis and of the main arguments relevant to its evaluation has been particularly welcome.[1] With the editors' encouragement, I shall try also to indicate what, beyond the historical facts of the matter, seems to me to be at stake. Need for a restatement of this sort results only partly from the technicality which has kept many commentators, friendly as well as hostile, from quite recognizing what the book is about. In some circles those difficulties have been compounded by misperceptions of the relationship between it and my earlier *Structure of scientific revolutions*.

Taken as a whole, the book is a narrative account of Planck's invention of the black-body theory known by his name and of that theory's development during the years when it and a closely-related theory of specific heats were the two exemplary applications of a still-to-be-developed quantum theory. Though the account is probably more thorough and detailed than previous treatments of its subject, its objective was of another sort: a fundamental reinterpretation of Planck's thought and of the stages in its gradual transformation. Most of the

---

Reprinted by permission from *HSPS* 14(2): 231–52. © 1984 by the Regents of the University of California.
 1. The final form of this paper has greatly benefited from suggestions directed to an earlier draft by Peter Galison, John Heilbron, Paul Horwich, Andrew Pickering, and Norton Wise.

detail in the book serves the reinterpretation. Only in the first chapter of Part I and the last chapter of Part II were extra topics included to make the narrative more nearly comprehensive and balanced.

In returning to the subject here, I have concentrated on the reinterpretation of Planck, abandoning both balance and the narrative mode. The paper that results consists of four parts. Section 1 outlines the principal elements of the new interpretation: the long-forgotten way in which Planck and many of his readers initially understood his blackbody theory. Section 2 summarizes the three main lines of argument favoring that interpretation. Section 3 concludes my return to Planck by analyzing those passages in his early black-body papers that have persistently been read as the introduction of quantization. In the final section, an addendum, I ruminate on the historiographic/philosophical position that these arguments both presuppose and reinforce. The treatment is skeletal throughout; fleshed out versions of all but Section 4 will be found in the book, together with documentation and acknowledgements. Parenthetical page references, below, indicate relevant passages in the book.

1. PLANCK'S FUNDAMENTAL INNOVATION

Planck, it has ordinarily been said, introduced at the end of 1900 the concept of a linear electrical oscillator with energy restricted to integral multiples of the energy quantum $h\nu$, ν being the oscillator frequency and h the universal constant later known by Planck's name. He had discovered that restricting energy levels to a discontinuous spectrum was essential to the derivation of the black-body radiation law he had introduced shortly before. Though Planck doubtless hoped that the discontinuity would prove eliminable and may even have conceived it as primarily a mathematical artifice, he as yet saw no route to his black-body law that did not require the hypothesis of discontinuity, of energy quantization. That view of Planck's development has been standard for years, and I am among those who have propagated it and based research upon it. But I am now quite certain it is wrong and that Planck's discovery should instead be described in the following manner.

In order to derive his black-body law Planck had to subdivide the energy continuum into cells or elements of size ϵ. Boltzmann had introduced such divisions in 1877 when presenting a probabilistic derivation of the entropy and velocity distribution of a gas, and Planck was following Boltzmann closely. But there was, Planck discovered, a crucial difference between the requirements he and Boltzmann had to place on these elements. For Boltzmann the precise size of ϵ made no difference. It did have to be large enough so that the cells would contain many molecules and also (though this requirement Boltzmann seems not to have recognized) small enough to permit the substitution of integrals for sums. But any value satisfying those constraints would do (pp. 59f.). Planck's derivations, in contrast, required that the cell size be proportional to oscillator frequency. With h as the constant of proportionality, Planck's cells were fixed at the small finite size $h\nu$.

Only by fixing the cell size in this way could Planck derive his distribution law. The restriction puzzled him. But it was for him a restriction on cell size, not on resonator energy, and it did not therefore bring to mind anything like quantization. With respect to energy, Planck's oscillators (he called them "resonators") were like Boltzmann's molecules. They moved freely through and between energy cells as required by Newton's and, in Planck's case, by Maxwell's laws (pp. 104f., 130-134). What differentiated Planck's derivation from Boltzmann's was not that the former required the violation of these or other laws of classical physics but rather the entry of the constant h, which Planck referred to as the "quantum of action" and which he would later describe as giving a physical structure to phase space (pp. 129, 250f.).

In fact, of course, Planck's law cannot be derived in this way. Planck failed to notice that his argument was valid only for frequencies such that $h\nu \ll kT$. But that failure was standard at the time. Boltzmann, whom Planck was following, had overlooked the equivalent approximation in his own derivation. Until Einstein pointed it out at Salzburg in 1909, neither he nor anyone else appears to have been aware that Planck's mathematics had slipped (pp. 185f.). And until that slip was discovered, Planck's view of what he had done was both coherent and independent of discontinuity.

2. THREE SORTS OF EVIDENCE

The preceding interpretation of Planck's early theory is clearly nonstandard. Also, it could be wrong. No single piece of available evidence demands it, and evidence incompatible with it could yet be discovered, a letter, for example, or an unpublished manuscript. As things now stand, however, evidence for the reinterpretation seems to me overwhelming. The relevant arguments take three main forms.

First, the reinterpretation makes the development of Planck's black-body research both more nearly continuous and also a deeper, more elegant piece of physics than it appears in the standard version. That development began in 1894 when Planck launched an attempt to apply to electromagnetic radiation a formulation of the second law of thermodynamics that he had originally introduced in his doctoral thesis. His main aim in doing so was to show that reversible equations (in this case those of wave theory) could be used to explain irreversible processes, and his approach was very close, though Planck may not at the start have known it, to one that Boltzmann had developed for gases in 1872.

Boltzmann's approach was not yet the probabilistic one mentioned in the previous section but an earlier one that required tracing molecular trajectories and averaging over the possible molecular collisions in a gas. More specifically, Boltzmann showed, or claimed to show, that if $f(q_i, p_i, t)\, dq_i\, dp_i$ was the fraction of gas molecules in the phase-space cell $dq_i\, dp_i$ at time t, then the integral over phase space of $f \log f$ was a function, $H(t)$, which could only decrease with time until it reached a minimum value that it thereafter retained. When the minimum was reached the gas was in equilibrium and the distribution function had acquired the Maxwellian form. That result constituted Boltzmann's H-theorem, which demonstrated, subject to a certain controversial restriction, the irreversible approach of a gas to equilibrium. Boltzmann carried it one crucial step further. Using the standard kinetic theory definitions of heat and temperature, he showed that S, the entropy of the gas in equilibrium, was just the negative of H. Decreases in H therefore corresponded precisely to increases in S (pp. 39–42). A second route to the distribution law was thus available. It could be obtained by applying the standard thermodynamic relation, $\partial S/\partial E = 1/T$.

Planck attempted to develop a similar theorem for radiation in a cavity that also contained damped resonators tuned to all radiation frequencies. He sought, that is, a function S of the field and resonator parameters that could only increase with time. By 1899 he had found one, though it, too, was subject to a restriction like Boltzmann's. Next, he attempted to complete the Boltzmann program, which he by then knew well, and at once encountered, as he would again in 1900, a key difference between the radiation and the gas-theoretical cases (pp. 73–86).

What Planck now sought was a function that specified the distribution of radiant energy over frequency at each temperature. But temperature does not appear in either Maxwell's equations or those governing a damped resonator, and Planck had no model, like that of kinetic theory, which permitted him to introduce it. He could not therefore directly derive a distribution function, nor could he, for the same reason, derive the entropy. But he could and did take an indirect route to those goals, one that he acknowledged from the start was open to question. He assumed that his function S, just because it increased monotonically to a maximum, was *the* thermodynamic entropy, and he then obtained a distribution function with the aid of the relationship $\partial S/\partial E = 1/T$. He had sought, he said, for other functions that, like entropy, could only increase monotonically with time, and he had found none. His function, he believed, was unique, and it must therefore coincide with the entropy of the radiation system (pp. 86-91).

The distribution law at which Planck arrived in this way was the so-called Wien law, which until late in 1899 corresponded well with available experiments. During the following twelve months, however, as measurements were extended further into the infra-red, the experimental accuracy of that law became more and more questionable, and other laws were proposed. The resulting uncertainties extended, of course, to Planck's derivation, and he immediately sought a more direct way of establishing an entropy function, his own or another, to insert in his treatment of irreversibility.

A first such attempt was announced in March and led again to the Wien law (pp. 92-97). Then, as evidence against that law continued to accumulate, Planck produced, first, a particularly promising candidate for the distribution law, and, then, in December 1900 and January 1901, a probabilistic derivation of that law. Those two derivation papers are the ones in which he is supposed to have introduced the concept of a discrete energy spectrum for resonators, thus of quantization and discontinuity. But the papers themselves make no explicit mention of such concepts, and his next relevant paper is not easily reconciled with the assumption that he nevertheless had them in mind.

That paper, which appeared late in 1901, was described in its title as a "Supplement" to the one in which, at the start of 1900, he had presented his proof of irreversibility and his demonstration that the Wien law would follow if his candidate for entropy function were unique. After a brief introduction, both the paragraphs and formulas of the "Supplement" were numbered to continue where those of the earlier paper had stopped. What he showed in those paragraphs was that his new entropy function, like the older one he had thought unique, could only increase monotonically to a maximum with time. The role of his new probabilistic argument was, as he saw it, simply to fill a gap in the theory he had completed in 1899. It demonstrated that the new function was *the* thermodynamic entropy, thus replacing the always questionable uniqueness assumption.

There followed a four-and-one-half year interval in which Planck published nothing on radiation theory. Then, in 1906, for the first edition of his *Lectures on the theory of thermal radiation,* he redeveloped precisely the treatment he had provided in 1900. The first three chapters of the *Lectures* derived the properties of cavity radiation and damped oscillators from the equations of "Electrodynamics and thermodynamics"; the fourth used probability theory to derive an entropy function; and the fifth, titled "Irreversible radiation processes," combined these results to show that the entropy function Planck had derived could only increase to a maximum with time. A bibliography of Planck's writings on thermal radiation concluded the volume (pp. 116–120). Neither in it nor in Planck's earlier papers is there any mention of discontinuity, any talk of a restriction on resonator energy, any formula like $U = nh\nu$, with U the energy of a single resonator and n an integer. Not until 1908 do formulations of that sort occur in Planck's known work, published or unpublished. What has nevertheless suggested that he conceived the energy spectrum as discrete is the manner in which he computes the probability of an energy distribution, the probability which he then converts to entropy by taking its logarithm. That evidence is striking, but it is also based on a persistent misunderstanding of Planck's derivation, a misunderstanding that will be the subject of the next section.

Glimpses of the possibility of the narrative just sketched are what initially persuaded me that Planck's first theory did not involve quantization or discontinuity. Told in this way, the story makes better historical sense than the long-standard version. First, as almost always happens in historical reinterpretation, the new narrative is more nearly continuous than its predecessor. The stages that prepare the way to fundamental innovation are seen to be more numerous than they seemed before, but also individually smaller, more fully prepared, more obviously within the reach of an exceptionally capable person. Consider the successive scholarly retellings of the story of Newton, Galileo, Darwin, or Freud.

Second, this reinterpretation eliminates a number of the apparent textual anomalies and inconsistencies that have led to talk of Planck's conservatism, of his confusion, and of his good luck in finding within Boltzmann's work the probabilistic formula he needed while failing entirely to see how properly to derive that formula from his model. Planck's so-called "second theory," a reformulation first announced in 1911, was not for him a retreat, as is often said, but rather the first theory from his pen to make use of discontinuity at all. His apparently unproblematic use of the equations of classical physics (in particular to derive the famous formula, $u_\nu = (8\pi\nu^2/c^3)\,U_\nu$, relating energy density in the field to resonator energy) within a theory that set these equations aside, was not an example of inconsistency, for such a theory did not then exist. And, as I shall argue in the next section, Planck understood both probability theory and its role in his derivation very well. In short, the Planck who appears in the reinterpretation is a better physicist—less a sleepwalker, deeper and more coherent—than the Planck of the standard story. Though the reinterpretation changes the nature of his contributions to the quantum theory, it does not at all diminish his role in that theory's creation.

A second sort of evidence for the reinterpretation is likely in the long run to prove decisive. Whatever Planck's theory may have been in 1900 and 1901, the theory presented in his 1906 *Lectures* is the continuum theory I have just attributed to him. My most learned and authoritative critic concedes this point. No other commentator of whom I am aware has challenged it. Either Planck held the same view when he introduced his theory at the end of 1900, or his view had changed in the six intervening years.

The last alternative is extremely unlikely. The evidence that Planck believed in a discrete energy spectrum in 1900 and 1901 is his method of calculating the probability of a particular distribution. He used the identical method in 1906, and described it in virtually identical words. The main presently relevant differences between the earlier and later treatments are, as will appear in Section 4, below, the inclusion in the latter of an occasional clarifying phrase. If Planck had changed his mind, found a continuum version of his previously discontinuous theory, would he not have altered his derivation, or at least his phraseology, accordingly?

There is another difficulty as well. By the turn of the century Planck's publication pattern was well established. Typically, he published reports on research programs in progress. Then, when he thought the program finished, he would sometimes present a systematic summary of the whole in an article or in a book. Between 1901 and 1906, however, he published nothing at all on black-body theory. What he did then publish in the way of new research dealt primarily with dispersion theory. An important footnoote in his *Lectures* suggests that he took up that topic in search of clues to an explanation of the universal constant h, but nothing in his papers suggests that he found any, and he makes no reference to this work in his lectures (p. 133). It is unlikely that Planck would have brought out a new version of his theory in book form without either publishing reports on the research that had led to it or mentioning that the version was new, not to be found in the papers listed at the end of his book. His behavior was very different when, in and after 1911, he did explicitly change his mind. Before describing his second theory in a much-revised second edition of his *Lectures,* Planck had, in the two preceding years, published four different papers on two different versions of his new viewpoint (pp. 236f.).

Turn, finally, to a third line of evidence, Planck's behavior at the time when, in my view, he was at last persuaded that his derivation demanded discontinuity. The point itself was not original with Planck: Einstein and Ehrenfest had insisted upon it in 1906 (pp. 166–169; 182–185). But Planck himself was not convinced for another two years. What then changed his mind appears to have been interaction with Lorentz who, in April of 1908, gave a major address on the black-body problem to an international assembly of mathematicians in Rome (pp. 189-196). It is in a letter to Lorentz about that lecture, a letter written in October 1908, that Planck's first known statement of the discreteness of the energy spectrum and the need for discontinuity occurs. The excitation of resonators, he writes,

> does not correspond to the simple known law of the pendulum; rather there exists a certain threshold: the resonator does not respond at all to very small excitations; if it responds to larger ones, it does so only in such a way that its energy is an integral multiple of the energy element $h\nu$, so that the instantaneous value of the energy is always represented by such an integral multiple.
>
> In sum, I might therefore say, I make two assumptions:
> 1) the energy of the resonator at a given instant is $gh\nu$ (g a whole number or 0);
> 2) the energy emitted and absorbed by a resonator during an interval containing many billion oscillations (and thus also the average energy of a resonator) is the same as it would be if the usual pendulum equation applied (p. 198).

These are the sorts of statements that one expects, but cannot find, anywhere in Planck's previous writings.

Statements of the same sort appear regularly in Planck's later correspondence and in his published writings after late 1909. Then, early in 1911, he presented a first version of his second theory, which restricted the discontinuity he had just acknowledged to the emission process (pp. 199f., 235f.). These conceptual changes were accompanied by two significant alterations in Planck's technical vocabulary. The phrase "energy element," which had referred to a mathematical subdivision of the energy continuum, was replaced by "energy quantum," the term for a physically separable and indivisible atom of energy. Only twice before had Planck used the latter phrase in his extant writings, both times in correspondence with people who used it themselves. From 1909, however, he used it regularly. Simultaneously, Planck banished the term "resonator," which had previously been his standard word for the hypothetical entities which, by absorption and reemission, redistributed energy in the field. Writing to Lorentz at the end of 1909, he said: "Of course you are right to say that such a resonator [one that responds discontinuously to stimulation] no longer deserves its name, and that has moved me to strip it of its title of honor and call it by the more general name 'oscillator'." Thereafter, Planck used the term oscillator exclusively (pp. 200f.).

Changes like these, which began only in late 1908, are a third sort of evidence for the proposed reinterpretation of Planck's early theory. There are others, for example his autobiographical narratives (pp. 131), but the three sketched here are primary.

3. THE COUNTER-EVIDENCE: PLANCK'S DERIVATION

What, then, is the evidence that favors the longstanding interpretation of Planck's first derivations of his black-body law? There is some, and it is strong enough so that two of Planck's most proficient early readers, Lorentz and Ehrenfest, described that derivation as positing discontinuity (p. 138). But I am nevertheless persuaded that that longstanding description rests on a misunderstanding of Planck's derivation, the first presentations of which were extremely obscure. And Lorentz, at least, corrected his description after the publication of Planck's *Lectures*. In 1908, for example, he wrote that "according to Planck's theory [elsewhere Lorentz calls it "Planck's first treatment of this subject"] resonators receive or give up energy to the ether in an entirely continuous manner (without there being any talk of a finite energy quantum)" (p. 139). Most other early readers, apparently including both Larmor and Einstein, had read Planck that way from the beginning (pp. 135–40, 182–186).

Look now at Planck's derivation, which is often said to be no derivation at all. Working backward from his distribution law Planck derived a corresponding entropy function. Looking for a probability W, of which the logarithm would be proportional to that entropy, he discovered an appropriate combinatorial formula, probably in an 1877 paper of Boltzmann's that he is known to have read. According to the standard account, however, he was unable to justify his use of that formula, and his remarks on the reasons for introducing it were no more than hand-waving. I have offered a new account of the derivation, one that makes it good physics, and that account seems to have been generally accepted. But what has not ordinarily been recognized about it—partly because I had not, at the time of writing, fully grasped the point myself—is the way in which the reinterpretation of Planck's derivation eliminates even the appearance of discontinuity from his early papers.

To understand either Planck's derivation or the standard way of misreading it, one must first understand Boltzmann's probabilistic derivation of the distribution law for gases. Boltzmann examines the ways of distributing total energy E over a gas consisting of N molecules. For this purpose, he divides the energy continuum into P cells of size ϵ, so that $P\epsilon = E$. Then he distributes the N molecules at random over these cells, asking at the end how many molecules have been placed in the first cell, how many in the second, how many in the k'th, and so on up to the P'th. The answer to that question is specified by a set of integers, $w_1, w_2, \cdots w_k, \cdots w_P$, and this set specifies a particular distribution of molecules along the energy continuum. Not just any set of w_k's will do, however. To meet the physical conditions of the problem, the w_k's must satisfy the so-called constraints on number of molecules and on total energy: they must, that is, satisfy the equations $\sum_k w_k = N$ and $\sum_k k w_k = P$. If, for example, $w_1 = N$, then all the molecules lie in the first cell, and all the other cells must be empty. Correspondingly, if a single molecule lies in the P'th cell, then it possesses all available energy, and all other molecules must lie in the first cell. There are, of course, a great many distributions that do satisfy these constraints and thus represent possible conditions of the gas. Each of them can, furthermore, since the molecules are indistinguishable among themselves, be achieved in numerous ways. Boltzmann asks in how many ways each possible distribution can be achieved, and records the now standard answer, $N!/w_1!w_2! \cdots w_P!$

That number is proportional to the probability of the corresponding distribution; the equilibrium distribution is the one with greatest probability; and the logarithm of its probability is proportional to the entropy of the gas. When the corresponding computations are carried through, they disclose a central feature of the formula for molecular entropy, one that separates Boltzmann's problem from Planck's. The cell size ϵ turns out to affect only the size of a disposable additive constant, so that ϵ may be eliminated from the entropy formula. (James Jeans and many more recent commentators are mistaken in saying that Boltzmann eliminates ϵ by letting it go to zero; indeed, for finite N, his derivation does not permit his taking the limit, since his cells must remain large enough to contain many molecules.) In an aside, Boltzmann notes also that the total number of ways of achieving all possible distributions of N molecules over P cells is just $(N+p-1)!/(N-1)!P!$, the very formula that Planck's derivation will require.

Planck's derivation has quite properly been taken to be modeled on Boltzmann's. But their problems are not the same, and there are two very different ways in which Planck's modeling could have been achieved. The traditional interpretation is, I believe, based on the wrong one: it takes Planck to have been asking for the most probable way of distributing some given energy E over N resonators all at the same frequency. His distribution law can be derived in that way. Lorentz did so in 1910; Planck adopted that derivation in 1913; and it has been standard since, a fact that has facilitated the traditional reading. But if that first interpretation is the one Planck had in mind between 1900 and 1906, then, though his method implies quantization, his derivation makes no sense. The second way of relating Planck's problem to Boltzmann's does make sense of the derivation, but it simultaneously eliminates the appearance of quantization.

Boltzmann had considered all the molecules of his gas. Planck considered all the resonators in his cavity, N at frequency ν, N' at frequency ν', N'' at frequency ν'', and so on. To the first set he attributed energy E, to the second E', to the third E'', and so on, again. The sum of all these energies was fixed at E_T; a distribution was a particular division of E_T among the resonators at each frequency; and it was the division of energy among particular frequencies that Planck varied to maximize the entropy of the system as a whole. This interpretation of Planck's derivation fits the texts closely, and it also models the problem Planck needed to solve: the distribution of energy, not over the resonators at a single frequency, but over sets of resonators at different frequencies. Unfortunately, the problem thus formulated is quite cumbersome, and Planck adopted a drastic shortcut to reach a solution. That shortcut, referred to in his first papers but only explained in his *Lectures*, is what made his first derivation papers so hard to understand.

Look now at what is at stake in the choice between these two ways of understanding Planck's derivation. According to the traditional view Planck should have begun by distributing P units of energy, ϵ or $h\nu$, at random over N resonators; next he should have computed the number of ways of achieving some particular distribution; and, finally, he should have determined the most probable distribution, computed the number of ways in which it could be obtained, and set the logarithm of that number proportional to entropy. Planck does take the first step, illustrating what a distribution means by supposing that the first resonator receives 7 elements, the second 38, the third 11, the fourth 0, the fifth 9, and so on, until all P are distributed. Those passages are the ones that provide evidence for quantization. But Planck's next step is to ask, not for the number of ways of obtaining some particular distribution, the number he ought then have maximized. Rather he asks for the total number of ways of obtaining all distributions, records the formula $(N+P-1)!/(N-1)!P!$ as the answer, and then, without further comment, sets its logarithm proportional to entropy. That is why so many of his later readers have felt that, having previously determined the combinatorial form his distribution law required, Planck substituted hand-waving for a derivation.

Now suppose that Planck, rather than distributing energy over resonators at a single frequency, were instead distributing it to the sets of resonators at different frequencies, E to those at frequency ν, E' to those at frequency ν', and so on. His problem would then be to discover in how many ways that could be done for a particular division, $E, E', E'', \cdots$, and then to discover which particular division could be achieved in the largest number of ways. The straightforward, but cumbersome, way to do that would be to begin as Boltzmann had begun. At each frequency, ν, ν', etc., divide the continuum of available energy, E, E', etc., into P, P', etc., units of size ϵ, ϵ', etc. Next, at each frequency, distribute the N, N', etc., resonators over the P, P', etc., cells and find out in how many ways the corresponding distributions could be achieved. It is only after this point that Planck's derivation begins to diverge from Boltzmann's, for Planck must next compute, not the most probable, but the total number of ways in which the N, N', etc., resonators can be distributed over each of the P, P', etc., cells. The product of all those numbers is the total number of ways of achieving the distribution that attributes energy E to the resonators at frequency ν, E' to those at ν', E'' to those at ν'', etc. And it is this product that must be maximized by varying E, E', E'', etc., subject to the constraint on total energy (pp. 103–110). The logarithm of the resulting number is, again, proportional to entropy, but in the radiation case, unlike the molecular, the dependence of entropy on cell size is not restricted to an additive constant. Cell size, thus, plays an essential role in determining the entropy of radiation, and ϵ cannot be dropped from the entropy formula.

If Planck had traveled that route he would have gotten the answer he did get, and no one would have suggested that he was restricting resonator energy to an integral number of elements or that he did not understand what he was doing. But Planck instead, after an extremely condensed sketch of the full method, pointed out that it would be "obviously very roundabout" (p. 106), and he announced that he was therefore resorting to a drastic shortcut. His problem did not require knowledge of the number of ways of achieving any particular distribution of N resonators over P energy cells, but only of the total number of ways in which all possible distributions of a given amount of energy could be achieved. That number could have been computed directly, by summing over the number of ways of achieving each possible Boltzmann distribution. But the computation required was lengthy, and its outcome was known to be the same as the number of ways in which P energy elements could be distributed over N resonators. Planck chose the latter, briefer technique of computation for the early accounts of his derivation.

Choice of that shortcut considerably obscured the conceptual structure of Planck's derivation, and the obscurity was compounded by the way he described his distribution procedure. He failed, that is, to point out that his distribution of energy over resonators was only a way of computing the *number* he needed, that it did not correspond to the distribution technique (resonators over energy) that produced the individual distributions over which his full derivation would have summed. But neither his choice of technique nor his imperfect description of it made his derivation unsound or restricted it to the case of a discrete energy spectrum. What Planck was computing was the total number of ways of distributing resonators of a given frequency to cells of size $h\nu$ along the energy continuum.

All of that can be seen in Planck's original papers of 1900 and 1901, but only with difficulty. The same interpretation is, however, inescapable if one turns to the first edition of Planck's *Lectures*. There, as in his papers, Planck writes of distributing P whole units, ϵ, of energy to N resonators, 4 to one of them, 3 to another, and so on. But he also provides a full description of the "roundabout" method, spells out the nature of his shortcut, and clarifies the phrase, "the number of resonators with energy of a given magnitude," by adding "(better: which lie in a given 'energy region')." Elsewhere, after dividing the phase plane of a one-dimensional harmonic oscillator into annular rings of area $\Delta U = h\nu$, Planck asks for "the probability that the energy of a resonator lies between U and $U + \Delta U$" (pp. 128–30). Here there can be no talk of quantization or discontinuity.

Two assorted remarks bring this explication of Planck's derivation to a close. First, there is one other putative difficulty with Planck's derivation, his apparent failure to maximize his count of the number of ways of realizing a particular distribution. That difficulty, too, proves to be only apparent. The function of maximization is to discover the equilibrium state, and there are other ways to do that, one of which Planck utilizes (pp. 107–108). Though he is straining the limits of turn-of-the-century physics, Planck is not groping in the dark. Second, as previously noted, Planck's result is incompatible with the continuity of resonator energy. He ought to have arrived at the Rayleigh-Jeans law. There is a mistake in his derivation, the one that Einstein pointed out in 1910. Boltzmann's method of counting ways of realizing a distribution is valid only if the distribution function is effectively constant within each cell. In Planck's case this requires that $h\nu \ll kT$, an approximation that does not generally hold. The difficulty remained invisible to all but Einstein, however, and even he did not recognize it until about 1910.

4. HISTORIOGRAPHIC/PHILOSOPHICAL ADDENDUM

I shall carry these argument sketches no further. Fuller versions are to be found in my book, together with some additional sorts of evidence. But *Black-body theory* has also raised issues of another sort. The concept of historical reconstruction that underlies it has from the start been fundamental to both my historical and my philosophical work. It is by no means original: I owe it primarily to Alexandre Koyré; its ultimate sources lie in neo-Kantian philosophy. But it is not everywhere accepted or even understood, and some significant responses to my reinterpretation of Planck seem less a reaction to its thesis or arguments than to the concepts of history and method that gave rise to them.

Twenty-five years ago, two pages from the end of *Structure of scientific revolutions*, I wrote: "If we can learn to substitute evolution-from-what-we-do-know for evolution-toward-what-we-wish-to-know, a number of vexing problems may vanish in the process." The intent of that remark was philosophic: to understand scientific progress we need not suppose that science moves closer and closer to the truth; the same phenomena follow from the assumption that science, at any time, simply evolves from its current position under the pressure of currently available argument and observation; if one adopts that more minimal explanation, then significant epistemological difficulties disappear. The same sentence has, however, an historiographic reading, which is for me the source of the philosophic one. Not accidentally, that reading is most fully expressed two pages from the start of the book, where I described the "historiographic revolution" whose implications I aimed to make explicit.

"Historians of science," I suggested in that place, "have begun to...ask, for example, not about the relation of Galileo's views to those of modern science, but rather about the relationship between his views and those of his group, i.e., his teachers, contemporaries, and immediate successors in the sciences. Furthermore, they insist upon studying the opinions of that group and other similar ones from the viewpoint—usually very different from that of modern science—that gives those opinions the maximum internal coherence and the closest possible fit to nature." The corresponding historiographic reading of the sentence from the end of the book is, then: if we can learn to refrain from attributing to past scientists a grasp of views that had not emerged when they wrote—from equating discoveries with their canonical formulations—vexing problems may vanish in the process. On this reading, the problems that vanish are mostly textual anomalies found in passages that do not quite make sense when read in ways that became standard only after they were written.

This point is often heard as a mere plea for scholarly accuracy: historians should be careful not to attribute to their subjects beliefs that conflict with things they said or wrote. But I have in mind both something more and something different. Creative scientists can be, and typically are, responsible for the emergence of beliefs that they did not hold themselves, at least not during the period when their discoveries were made. If one is to learn how those discoveries came about, how new knowledge emerged, then one must find out how the discoverers themselves thought about what they were doing. Often it turns out that not just their beliefs but their very modes of thought were different from the ones to which their discoveries gave rise, and the latter difference is central to *Structure of scientific revolutions*.

The wrenching experience of entering into an older mode of thought is the source of my references to gestalt switches and revolutions; difficulties in translating the discoverer's language into our own are what led me to write also of incommensurability; and paradigms were the concrete examples needed—since definition in words was impossible—to acquire the language of the older mode. I do my best, for urgent reasons, not to think in these terms when I do history, and I avoid the corresponding vocabulary when presenting my results. It is too easy to constrain historical evidence within a predetermined mold. If history (or ultimately philosophy) is to be learned from the texts that are my main sources, then I must minimize the role of prior conviction in my approach to them.[2] Often I do not know for some time after my historical work is completed the respects in which it does and does not fit *Structure*.

Nevertheless, when I do look back, I have generally been well satisfied by the extent to which my narrative fit the developmental schema that *Structure* provides. *Black-body theory* is no exception. The start of the revolution that produced the old quantum theory is moved from the end of 1900 to 1906; most of the book deals with the period before the revolution occurred. The preceding crisis, to the extent that there was one, resulted from the difficulties in reconciling Planck's derivation with the tenets of classical physics. Planck's change of vocabulary—from "resonator" to "oscillator" and from "element" to "quantum"—is the central symptom of incommensurability. It signals the changed meaning of the quantity $h\nu$ from a mental subdivision of the energy continuum to a physically separable atom of energy. That my critics continue to apply the term "energy quantum" to the pre-1906 papers and lectures in which Planck consistently used "energy element" reveals something of the difficulty of reversing the gestalt switch that took place during that year and those that followed. Among the numerous paradigms (in the sense of concrete examples) to be found in the book, Boltzmann's probabilistic derivation of the entropy of a gas is of particular interest, for it illustrates the problem to which the concept of paradigm was a response. The derivation was not reduced to rules but instead served as a model to be applied by means of analogy. As a result, when its application was transferred from gases to radiation, Planck and Lorentz could invent different analogies with which to effect the change.

These illustrations of the substantive applicability of *Structure* can be extended, but for this paper it is the book's historiographic applications

---

2. On this subject see also my essay review, "The halt and the blind: Philosophy and history of science," *British journal for the philosophy of science, 31* (1980), 181–192.

that are relevant. My way of using concepts like revolution and gestalt switch was drawn from and continues appropriately to represent what historians must often go through to recapture the thought of a past generation of scientists. Concerned to reconstruct past ideas, historians must approach the generation that held them as the anthropologist approaches an alien culture. They must, that is, be prepared at the start to find that the natives speak a different language and map experience into different categories from those that they themselves bring from home. And they must take as their objective the discovery of those categories and the assimilation of the corresponding language. "Whig history" has been the term reserved for failure in that enterprise, but its nature is better evoked by the term "ethnocentric."

What has preceded is prefatory, a quick overview of the close parallelism between an historiographic and an epistemological position. Perhaps the parallelism is obvious, but it has corollaries which are not. Entry into a discoverer's culture often proves acutely uncomfortable, especially for scientists, and sophisticated resistance to such entry ordinarily begins within the discoverer's own retrospects and continues in perpetuity. That resistance—with its manifestations, causes, and consequences—is the central concern of this closing section.

Systematic distortions of memory, both the discoverer's memory and the memory of many of his contemporaries, are a first manifestation of resistance. Another, regularly found among members of later generations, is the attribution of real or supposed anomalies in the discoverer's behavior to "confusion." I shall take up these techniques in turn, beginning with two examples of a discoverer's systematic misconstruction of the memory of his own discovery. Both are from my own experience, and both illustrate the suppression of concepts that played an essential role in the discovery and their replacement by the projection onto the past of concepts that emerged only in the discovery's aftermath.[3]

The first experience involves Otto Stern, the physicist who conceived, and in 1922 carried out with Walther Guerlac, the first direct demonstration of space-quantization. While the required apparatus was still under construction, Stern published a paper demonstrating that a beam of silver atoms passing through an inhomogeneous magnetic field would only be broadened if classical theory held but would be split in two if the atoms were spatially quantized. That splitting is what the experiment showed, and it followed, as Stern had shown, directly from current theory. Four years later, however, relevant theory changed

---

3. On this subject see also Section XI, "The invisibility of revolutions," in *The structure of scientific revolutions*.

drastically. To understand why the beam split in two then required a knowledge of electron spin, a conception not even dreamt of when the experiment was made. Furthermore, the later theory predicted that, in the absence of spin, the beam of silver atoms should be split, not into two parts, but into three. At the time of my meeting with Stern I did not yet know the theoretical paper, and I therefore asked why he had expected the beam to be split in two. He responded sharply that, before the discovery of spin, he and Gerlach obviously could not have known what to expect. Their experiment was done simply to see what would happen. In that respect, he insisted, the experiment was a fishing expedition.

My second example is more complex. The famous paper that announced the Bohr model of the hydrogen atom was submitted from Copenhagen on 6 March 1913 and published the following July, the first installment of a three-part series. I first read it during the fall of 1962 in preparation for interviews with its author. Not surprisingly, the paper includes a full description of the quantized Bohr model for the hydrogen atom as it would be taught in an elementary physics course today. But it also includes a number of phrases incompatible with that model. In particular, Bohr sometimes wrote as though the hydrogen spectrum were emitted by an electron falling into the ground state from outside the atom and strumming all the stationary states that it passed along the way.[4]

These anomalous remarks, together with Bohr's repeated assertion that he had not known the Balmer formula until February 1913, suggested an unexpected hypothesis, subsequently fully confirmed by the discovery of an unpublished manuscript. Many months before he attempted an explanation of spectra, Bohr had developed a quantized version of the Rutherford atom for chemical applications of the sort made familiar by J. J. Thomson. That model, which I was quite sure had had only a ground state, provided the basis for the second and third installments of the 1913 series. The first, which developed the Bohr model for hydrogen and derived the Balmer formula from it, was a last-minute insertion.

My first few interviews with Bohr dealt with the background for his atomic model, and I asked what sorts of connections he had made between the Rutherford atom and the quantum during the period before his attention was directed to the Balmer formula. He replied that he could not have had developed ideas on the subject before turning to spectra, and his assistant later reported to me that, after I had

---

4. These passages and much else are discussed in J. L. Heilbron and T. S. Kuhn, "The genesis of the Bohr atom," *HSPS*, *1* (1969), 211–290.

left the room, Bohr shook his head and said of our exchange, "Stupid question. Stupid question."

All that occurred at our first interview. For the next one, I included a similar question in a list submitted to Bohr in advance, and it was received in much the same way as the original. One last attempt to retrieve memories of an early quantized Rutherford atom occurred late in the third interview. This time, however, when Bohr said again that there could have been no concrete model without the Balmer formula in hand, I for the first time showed him the passages in his famous paper that had led me to enquire. He looked them over and then muttered half to himself, "Perhaps it was a mistake to put the paper into print so fast. Perhaps I should have waited until I had it right." Then, he went over to his personal collection of reprints, took from it a paper he had presented to the Danish Academy of Sciences six months after the publication of his original paper, and handed it to me with the words, "It's alright there, isn't it?" About the earlier model not a word was ever forthcoming.[5]

These stories typify the autobiographical reports of the participants in a discovery and often of their contemporaries as well. Not always but quite usually, scientists will strenuously resist recognizing that their discoveries were the product of beliefs and theories incompatible with those to which the discoveries themselves gave rise. Similar resistance is encountered among later generations, but memory and its distortion are no longer involved.

Later accounts of a discovery typically redescribe it, but again in the conceptual vocabulary of the period in which they are prepared. The result is the linearized or cumulative histories familiar from science textbooks and from the introductory chapters of specialized monographs. Those accounts, however, almost never withstand detailed comparison with documents from the period of the discovery. And when discrepancies are pointed out, one or another variant of a second standard technique for resisting the past is often deployed. Faced with apparent anomalies in the work of the discoverer, scientists and at least an occasional historian protect their version of the discovery by invoking the discoverer's "confusion" during the early stages of its emergence. It is only because he was confused, they explain, that his words

---

5. The words in quotation marks are paraphrases from memory. A confidential memorandum I prepared for the file later in the year cannot now be found. Search for it has, however, turned up a second memorandum, one I had entirely forgotten. Written in haste the day *before* the third interview, it recounts a report from Bohr's assistant that Bohr now said "he should not have published so soon. Parts II and III," he went on, "should have been entirely suppressed and Part I should have been withheld until recast in the form of the lecture delivered in Copenhagen in December."

fail to fit their story.

These appeals to confusion are damaging, but not because discoverers are never confused. Typically, they are, and Bohr's discovery of the Bohr atom is a clear example. When he wrote the paper announcing his discovery, he had two incompatible models in mind, and he occasionally confused them, mixed the two up. No reading of his first reports on his invention will eliminate the resulting contradictions, and those contradictions, which testify to his confusion, provide essential clues to the reconstruction of his route to the discovery. The standard appeal to confusion dismisses those clues, rejects them as challenges to historical reconstruction, and permits the attribution of confusion to stand as the end of the story. That is the first part of the damage.

For the second, more serious part, compare the case of Planck. Again there are anomalies in the early papers; again they provide clues to an unsuspected state of mind; and, again, dismissing them discards evidence essential for historical reconstruction. Thus far the damage is the same. But in Planck's case, unlike Bohr's, the anomalies do not take the form of internal contradictions, and they therefore provide no reason to suppose that Planck himself felt or had reason to feel confused. If it is nevertheless appropriate to apply the term to him, that is by virtue of a second standard use of the word "confused," one independent of the state of mind of the person to whom it is applied.

Consider, for example, the case of a student who, having read a textbook derivation of the black-body distribution law, then wrote it up in a way like the one found in Planck's early papers. That student would be confused, not in the sense of being pulled about by conflicting elements in his thought, but in the sense of having seen only dimly or confusedly the structure of the derivation that had been set before him by the text. That, I believe, is the sense of "confused" in the minds of the people, mostly scientists, who complain, for example, that I try too hard to make the thought of a Planck or a Boltzmann logical and coherent. Why, I am repeatedly asked, can I not simply acknowledge that they were confused?

That way of talking about a discoverer makes no sense. Taken literally, it suggests that the discovery, of which its author is said to have had only a confused view, had already been made, was somehow already there, in the discoverer's mind. Occasionally that implication is explicit. The discoverer, I am then told, was relying on intuition; his view of his discovery was still so clouded that he could only grope his way to it; that is why he described what he had in mind in such odd and inconsistent ways, appeared so much a sleepwalker as he proceeded towards his discovery.

Doubtless, few of those who explain anomaly by resort to confusion would go quite so far, but all must encounter the identical difficulty.

What licensed our calling the student confused was our knowledge of the concepts he brought to the text and of the proper way to fit the two together. If only he had clearly seen that much himself, he would not, any more than we, have described the derivation as he did. When, in the absence of internal contradiction, we apply the label "confused" to Planck, we are again using ourselves as measure. We assume that Planck brought to his problem the same concepts as we do, and we explain his anomalous behavior as we would explain similar behavior of our own. But the concepts we bring to the black-body problem are themselves products of the discovery Planck had not yet made. To claim for them a role in the emergence of his discovery is again to make him a sleepwalker or else clairvoyant. That is an incoherent notion of discovery—one that makes discovery dependent on prior grasp of what is to be discovered. No other result of the resort to confusion is so damaging.

So much for manifestations of resistance. What can possibly be their cause? Why should so many scientists and also an occasional historian fight so hard, often without quite realizing they are doing so, to recast past developments in the language of modern concepts? Why should attempts to reconstruct a conceptual past in its own terms so often be regarded as subversive? Answers to those questions are to be sought in two directions. The first is independent of any special characteristics of science, and I shall here merely identify it. The second involves the sciences uniquely, and about it I shall say a bit more.

I have already suggested that the past of science should be approached as an alien culture, one that the historian strives first to enter and then to make accessible to others. Entry into another culture, scientific or not, is regularly resisted, however, and the standard form of resistance is to carry one's own culture with one and assume that the world conforms. Remember the not-so-mythical English tourist who thinks that English-spoken-loudly will suffice for life in France. Motives for such resistance are familiar to anyone who has experienced the anxieties of foreign travel. Cultures shape life forms and with them the world in which the participants in the culture live. Entry into another culture does not simply expand one's previous form of life, open new possibilities within it. Rather, it opens new possibilities at the expense of old ones, exposing the foundations of a previous life form as contingent and threatening the integrity of the life one had lived before. Ultimately the experience can be liberating, but it is always threatening.

That is the general case, and it needs much exploration. The transition between cultures always threatens something, but what is at stake is not always clear. Where the cultures are scientific, however, one object of the threat stands out. In the socialization or professionalization of scientists the concept of the unit discovery plays a profound and generally functional role. The knowledge they acquire in school and university is transmitted to them in such units, both experimental and theoretical; unit discoveries are the bricks from which, in a familiar image, the edifice of science is piecemeal built. When the student later enters the profession, it is with the understanding that success is measured by the size and number of bricks the individual is able to put in place. That is why, since at least the seventeenth century, attempts to establish priority in discovery have played so large a part in scientific development. The concept of the unit discovery is constitutive of the scientific life as we know it.

Among historians, however, it is by now virtually a truism that that concept will not do. Discoveries are extended processes, seldom attributable to a particular moment in time and sometimes not even to a single individual. Usually there is little doubt who was responsible for them—who brought things to a point from which there could be no retreat. Planck himself is a clear case in point: given his distribution law and his technique for deriving it, the recognition of discontinuity was bound soon to occur. But, as we have seen, it then typically turns out that the person responsible for a discovery did not believe in it himself, at least not if belief in the discovery is belief in its subsequent canonical formulation. Even when the discoverer's own contribution was substantially complete, he still thought of what he had done in a way significantly different from the way in which that contribution came later to be described.

The priority controversies for which science is known are again especially revealing. The description of the work an individual did or the date at which he did it is rarely a issue. Matters of that sort are usually easily settled. Instead, what participants debate is whether one or another piece of work can properly be said to constitute *the* discovery. Most historians would insist that the question has no answer, that the debate itself rests on a misconception. But scientists can continue the argument in perpetuity, insisting that one party or the other must be *the* discoverer. Implicitly or explicitly, what is at stake for them is the concept of the unit discovery, a concept that will not withstand application to actual practice, but on which much of the reward system of science as well as important elements of the scientist's conception of self are nevertheless based.[6] Doubtless, other things

---

6. The following fragment, atypical only in having reached the published record, is illustrative. Early in 1979 I presented to an Einstein centennial symposium a brief account of Einstein's role in the genesis of the concept of quantization. Eugene Wigner, who was not convinced, commented on my paper as follows: "In spite of all my admiration of Professor Kuhn, I would like to contradict him a little bit. I think Planck did wonderful work, and even the discovery of his equation is wonderful. I do not believe he believed

besides the unit discovery are threatened by the resurrection of older modes of scientific thought. But I shall stop with this single example which, at the very least, illustrates the form that explanations of resistance must take.

I conclude with a few brief remarks on a question I do not quite understand. "Suppose you were right," I have been asked, "suppose that the first suggestion of a restricted energy spectrum did come, not from Planck, but from Einstein and Ehrenfest; what difference could it possibly make, and to whom?" For me there are two sorts of answer. I wrote the book as an historian of ideas, and my primary object was just to get the facts straight. Different specialties differ in the facts they must get straight, and no one is obliged to engage in this one. But a person who assigns no importance to such questions ought not even make the attempt.

Most historians would, I think, be content to stand on that answer alone, and the ground it provides is bedrock. For me, however, something else is also involved. I believe that history can provide data to the philosopher of science and the epistemologist, that it can influence views about the nature of knowledge and about the procedures to be deployed in its pursuit. Part of the appeal of the standard account of Planck's discovery is, I think, the closeness with which it matches a still cherished view of the nature of science and its development. Though I appreciate both the charms and the functions of that view, understanding requires that it be recognized as myth.

---

in the details of any derivation of it, and this was natural since the physics of that time was full of contradictions.... So that I feel the quantum was discovered, at least from what I read, by Planck and even though we admire Einstein, this is not for what we admire him most." (Cf. Harry Woolf, ed., *Some strangeness in the proportion* (Reading, MA, 1980), p. 194.) A few weeks later the following paragraph appeared in *Science news* (30 Mar 1979, p. 112): "Indeed, at the centennial symposium the historian of science Thomas S. Kuhn proposed to credit Einstein with the very concept of the quantum of radiation, an idea usually attributed to Max Planck. This provoked a rebuke from Eugene Wigner (from Wigner a not-so-gentle rebuke) to the effect that people ought to be allowed to keep the credit they have justly earned. Not everything in physics need be annexed to Einstein."

INDEX

(Reference to pages with numbers 255 or greater are to the notes. These are indexed only when the resulting entry contains information more substantial than the citation of a paper.)

Adams, E. P., 253
Agassi, Joseph, 278
α-rays, 245, 284
Arrhenius, Svante, 17
Atomic models, 206f., 221, 227-230, 247f., 253, 311, 315f.
Atomic theory, 22-44, 29-31; *see also* Bohr atom; Energeticists
Avogadro's law, 42, 111
Avogadro's number, 176, 179, 181, 211

Balmer formula, 247, 316
Bavink, Bernard, 285
Behn, U., 310
Bernoulli, Daniel, 261
β-rays, 226
Bjerrum, Niels, 219f.
Black-body problem, 3-10
Black-body theory (*see* Wien displacement law; Distribution laws; Ehrenfest, route to; Einstein, route to; Electron theory and; Experimental black cavities; Experimental verification of radiation distribution laws; Lorentz, route to; Planck's theory)
Blackmore, J. T., 279
Bohr, Niels, 112, 185, 227, 247f., 253; on Planck's second theory, 247-249
Bohr atom, 221, 247f., 251, 253, 256
Bolometer, 8, 10
Boltzmann, Ludwig, 5, 17, 32, 148, 152
Boltzmann and: Burbury, 63f., 65-67; Ehrenfest, 152; Kirchhoff, 63, 66, 273f.; Loschmidt, 47, 52, 57f., 273f.; Maxwell, 20, 41, 44; O. E. Meyer, 48, 272. *See also* Einstein and Boltzmann; Planck and Boltzmann.
Boltzmann on: Clausius's definition of entropy, 269; the combinatorial definition of entropy, 38f., 46-60, 109f., 128, 134, 278f., 282; the H-theorem, 38-46, 61, 123, 269, 271; irreversibility, 38-66, 73, 77; permutation numbers, 48-54, 101, 111, 282; the second law, 21, 24f., 51-54, 269 (*see also* Molar disorder; Molecular disorder)
Boltzmann equation, the, 41, 43f., 46, 69f., 275, 282
Boltzmann's *Lectures on Gas Theory*, 21, 39, 41-44, 53, 56, 64f., 67-71, 98, 123, 269f., 298
Born, Max, 205
Bosscha, Johannes, 110
Boyle's law, 19, 42, 88
Brillouin, Marcel, 252
British Association Meetings: Belfast (1902), 137; Birmingham (1913), 139, 231
Brush, S. G., 261, 268f., 272f.
Bryan, G. H., 60, 71, 140, 263, 273, 275
Burbury, S. H., 60f., 64f., 68f., 71, 136, 138, 159, 273; and Boltzmann, 63f., 65f.; on Condition A, 61, 63f., 66, 68f., 273; and Culverwell, 60; and Ehrenfest, 295; and Jeans, 69
Byk, A., 320

Canal rays, 222f.
Carnot, Sadi, 12-15
Cathode rays, 221, 224
Charles's law, 19, 42, 88
Classen, Alexander, 309
Clausius, Rudolph, 12-15, 19-21, 24-26, 40, 85, 259f.; *Mechanical Theory of Heat*, 13, 26; and Planck, 13-16, 21, 259f.; on the second law of thermodynamics, 13-15, 24-26, 260f., 269
Combinatorial definition of entropy, 46-54, 67-71, 105-109, 115f.
Combinatorials: Boltzmann's use of, 38, 60, 109f., 128, 134, 153; Ehrenfest's use of, 166f.; Einstein's use of, 134, 139, 181; Planck's use of, 98-110, 115-121, 124f., 128-130, 134, 156, 188, 243, 279, 282f.
Complexions: Boltzmann's use of, 128-130; Ehrenfest's use of, 154, 156, 158f., 167-169; Einstein on, 186; Lorentz's use of, 109f.; Planck's use of, 105, 120f., 128f., 184
Compton effect, 222
Congresses (*see* International Congress of Mathematicians; International Physics Congress; Solvay Congress)
Constants, universal: Larmor on, 136f.; Planck on, 90, 104-106, 111, 113, 117, 243, 278; Thiesen on, 111. *See also* Einstein's theory of: the universal constant, χ; Planck's theory: constant, h, *and* constant, k.
Correspondence principle, 240, 248f., 320
Crova, A. P. P., 7
Culverwell, E. P., 60, 273

Damping by radiation, 33f., 120, 237, 267
Davy, Humphrey, 261
Day, A. L., 290
Debye, Peter, 209f., 308
∇^2 V Club at Cambridge University, 317
Discontinuity: acceptance of, 188-207; development of, 143f.; origins of, 125-130. *See also* Energy discontinuity; Planck on; Planck's second theory; Quantization.
Dispersion theory, 203
Displacement law (*see* Wien displacement law)
Distribution laws (*see* Maxwell's; Planck's theory; Rayleigh; Rayleigh-Jeans; Thiesen's distribution formula; Wien; Experimental verification of radiation)
Doppler effect, 6, 223
Doran, B. G., 266
Drude, Paul, 136, 210
Dulong-Petit law, 211-215, 310

Ehrenfest, Paul, 132, 138, 143f., 152-170, 188f., 195, 294; route to black-body theory, 152-158, 294f.
Ehrenfest and: Boltzmann, 153, 167-169; Burbury, 295; Lorentz, 139, 152f., 169; Planck, 159f., 163f., 166, 168, 294
Ehrenfest on: "Abstract entropy theory," 166; Boltzmann's H-theorem, 153, 157f.; electron theory, 153, 160f., 168; entropy, 167f.; Planck's distribution law, 143, 154, 158, 168, 188, 210; Planck's electro-magnetic H-theorem, 153f.; Planck's model for radiation distribution, 158-160, 163, 166; Planck's physical hypothesis, 154f., 161f.; Planck's uniqueness proof, 155f.; the quantum of energy, 138f., 168f., 188, 201f., 209, 220; "quasi-entropies," 156, 158, 160f.; "quasi-H-theorems," 157f., 165
Ehrenfest's "Begriffliche Grundlagen der statistischen Auffassung in der Mechanik" (1912), 287
Eidgenössische Technische Hochschule, 215
Einstein, Albert, 21, 134, 136, 139, 143f., 150, 152; route to

black-body theory, 170-178
Einstein and: Boltzmann, 171, 262; Hopf, 303; Laub, 214; Laue, 188f.; Lorentz, 195; Planck, 170, 179f., 182-188, 196-198
Einstein on: Boltzmann's definition of entropy, 181; Boltzmann's H-theorem, 298; electron theory, 182, 210; energy fluctuation, 178; entropy, 172-178, 181-183, 299; Planck's distribution law, 143, 170, 180, 183, 185f., 188f., 202, 321; Planck's model for radiation distribution, 170, 185; Planck's second theory, 246, 252, 321; the quantum of energy, 170f., 183-185, 187f., 200-202, 210f., 219-222, 230, 278; thermodynamics, 171; Wien distribution law, 180f., 221f.
Einstein's light particle hypothesis, 143, 182, 221, 300; Compton effect and, 222; general reception of, 222; Laue on, 189; Lorentz on, 195, 205; Planck on, 197, 199f., 246; Sommerfeld on, 225; Stark on, 223f.; J. J. Thomson on, 229
Einstein's theory of: Brownian motion, 171f., 265, 299; relativity, 112, 186 (Planck on, 196; Sommerfeld on, 226); specific heats, 210-216, 245f.; statistical ensembles, 173f., 177, 263, 299; statistical thermodynamics, 171-176; 262f.; the universal constant, χ, 175-178
Electron theory and black-body theory, 132-135, 195, 211, 246f.; Ehrenfest on; 153, 160f., 168f.; Einstein on, 182, 210; Haas on, 227; Lorentz on, 132f., 135, 189f., 193f., 195.; Nernst on, 215; Planck on, 131-134, 198f.; J. J. Thomson on, 227f.; Wien on, 203
Electronic charge, 111f., 284f.; see also Quantum of electricity
Energeticists, 17, 24-26
Energy discontinuity, 125, 128, 137-140, 143f., 184f., 186-189, 193, 197f.; Planck's acceptance of, 235-238; proofs of, 210. See also Quantization.
Energy fluctuation, 176-179, 184, 187, 299
Entropy (see Combinatorial definition of; Ehrenfest on; Einstein on; Planck on)
Equipartition theorem, 146-152, 168, 191f., 203, 216, 286, 297
Ergodic hypothesis, 56, 172, 286f., 297
Eucken, Arnold, 247
Experimental black cavities, 11, 93f.
Experimental verification of equations for: free energy, 213f.; mean free path, 246f.; optical dispersion, 211; photoelectric effect, 223; specific heat of solids, 211-216, 309
Experimental verification of radiation distribution laws, 111, 114f., 122f., 127, 135, 147, 150, 152, 179, 191f., 221, 254, 281f., 292
Experimentalists, 7-11, 147f., 150-152, 193
Experiments (see Fizeau's, Michelson-Morley, Stern-Gerlach)

Fitzgerald, G. F., 273
Fizeau's experiment, 112
Fokker, A. D., 321
Forman, P. L., xi
Fox, Robert, 261
Franck, James, 222, 313

γ-radiation, 225, 315
Gans, Richard, 312
Garber, Elizabeth, 290
Gas constant, R, 111, 176, 180f.
Gas laws (see Avogadro's law; Boyle's law; Charles's law; Dulong-Petit law; Gas constant, R; Gas theory)
Gas theory, 18-28, 75-79, 82, 88-90, 123, 261f.; Ehrenfest on, 156-158; Einstein on, 171-173, 176, 298; Planck on, 25-28. See also Boltzmann's Lectures on Gas Theory; Gas laws; Jeans's Dynamical Theory of Gases;

Mean free path; Specific heats: of gases; Thermometer.
Geiger, H., 255, 284
German Physical Society, 92-98, 102-104, 110, 119-121, 147, 236, 281
Gibbs, J. W., 21, 171-173, 176f., 191, 263, 270, 298; *Statistical Mechanics*, 171, 176, 262, 275
Goldberg, Stanley, 260
Graetz, Leo, 27, 52-54, 270

Haas, A. E., 227f., 315f.
Haber, Fritz, 231
Hamiltonian differential equations, 200, 205, 236
Hamilton-Jacobi theory, 251
Hasenöhrl, Fritz, 216f.
Heilbron, J. L., xi
Helmholtz, Hermann von, 14, 30
Hermann, Armin, 278, 285, 301, 316
Herschel, William, 7
Hertz, Gustav, 222, 313
Hertz, Heinrich, 5, 32-34
Herzfeld, K. F., 316
Hevesy, George, 214
Hiebert, E. N., 259-261
Hirosige, T., 256, 320
Hopf, Ludwig, 303
H-theorem, 39-42; Ehrenfest on, 153-158; Einstein on, 298. *See also* Boltzmann on the; Ehrenfest on quasi-H-theorems; Planck's theory: electromagnetic H-theorem.

International Congress of Mathematicians, Fourth meeting, Rome (1908), 190, 193; *see also* Lorentz's: Rome lecture
International Physics Congress, Paris (1900), 99
Irreversibility (*see* Boltzmann on; Planck on)
Ishiwara, Jun, 251, 321

Jahnke, E., 147
Jammer, Max, 320
Jeans, J. H., 68-71, 133, 137-139, 168, 180, 185, 188, 194; and Burbury, 68f.; and Lummer and Pringsheim, 204; on molecular disorder, 68f.; on Planck's distribution law, 143, 188, 204; on quantum theory, 202, 205f., 209, 231f.; route to black-body problem, 148-150. *See also* Rayleigh-Jeans distribution law.
Jeans's: *Dynamical Theory of Gases*, 21, 148, 150, 275, 287; *Report on Radiation and the Quantum Theory*, 232, 254
Joffé, A. F., 221

Kamerlingh-Onnes, Heike, 319
Kangro, Hans, xi, 256-259, 279, 287, 291
Kayser, Heinrich, 135
Keesom, W. H., 319
Kim, Yung Sik, 260
Kinetic theory of gases (*see* Gas theory)
Kirchhoff, Gustav, 4f., 14, 17, 63, 66, 71; and Boltzmann, 62f., 66f., 273f.; and Planck, 14, 17
Kirchhoff's *Lectures on the Theory of Heat*, 21, 61, 100, 270
Kirchhoff's radiation law, 4-6, 8, 10, 29, 35f., 63, 72, 116f., 133-135, 150, 153, 191, 257f.
Klein, M. J., xi, 171, 256, 264, 278f., 289, 291, 299
Kopp, Hermann, 212
Kries, J. von, 121, 286
Kurlbaum, Ferdinand, 11, 17, 147

Ladenburg, Rudolph, 221
Lagrangian equations, 172
Landé, Alfred, 311
Lange, Victor, 293
Langevin, Paul, 210
Langley, S. P., 7f., 258
Larmor, Joseph, 136f., 230, 273, 275, 284
Laub, Jakob, 214
Laue, Max von, 143, 188f., 201
Lenard, Philipp, 221
Lindemann, F. A., 316
Liouville's theorem, 55, 122, 153, 171, 286
Lorentz, H. A., 32, 102, 112f., 132f., 135, 138-140, 144, 150, 152, 155,

INDEX

169, 179, 283; proof of Planck's distribution law, 102-104; route to black-body theory, 189-191
Lorentz and: Ehrenfest, 138f., 152, 160; Planck (*see* Planck and: Lorentz); Wien, 190, 192-194, 203
Lorentz on: Einstein's light particle hypothesis, 195, 205; electron theory and black-body theory, 32f., 132f., 135, 189f., 193, 196, 245, 319; the quantum, 189-200, 202, 204-206, 209-211, 220, 230
Lorentz's: *Les Théories statistiques en thermodynamique*, 263; "Old and New Questions of Physics" (1910), 205, 216; Rome (1908) lecture, 190f., 195-197, 216
Loschmidt, Josef, 46f., 52, 57f., 273; and Boltzmann, 47, 52, 57, 273f.
Loschmidt's number, 111
Loschmidt's reversibility paradox, 44, 54, 58, 60, 70, 273
Lummer, Otto, 11, 17, 93-96, 99, 111, 147, 192f., 204, 281

Mach, Ernst, 257, 259
Magneton (*see* Weiss's magneton)
Maxwell, J. C., 19f., 22-26, 31-33, 38; and Boltzmann, 20, 41, 44; and Clausius, 19f.
Maxwell's demon, 24, 29, 31f., 44, 264
Maxwell's distribution law, 7, 10f., 20f., 38, 41-43, 50f., 54f., 63, 68, 71, 158; Einstein on, 184; Kirchhoff-Planck derivation of, 62, 100; Meyer's derivation of, 271
Maxwell's equations, 4, 7, 12, 29, 31f., 34, 77, 132, 204; Boltzmann and, 139; Ehrenfest and, 153, 155; Einstein and, 139, 180-184, 186; Lorentz and, 195f.; Planck and, 112, 116, 118, 125, 139, 197; Wien and, 203
Maxwell's: *Theory of Heat*, 23f., 44; *Treatise on Electricity and Magnetism*, 31
McCormmach, Russell, 285, 309

Mean free path, 19f., 40, 246, 273
Meyer, O. E., 48, 70f., 271f.
Michelson, W. A., 8, 11
Michelson-Morley experiments, 112
Millikan, R. A., 222
Molar vs. molecular disorder, 57-60, 64-67
Molecular disorder, 39, 43-45, 48, 57, 60-71, 121; Burbury on, 68f., Jeans on, 68f., and natural radiation, 43, 318; Planck on, 67f., 124; and randomness, 67, 242
Müller, J. H. J., 7

Natural radiation, 77f., 82-84, 120-125, 132, 154f., 235, 276f.; and molecular disorder, 43, 318; and randomness, 242
Naturforscherversammlung, 22, 96, 186f., 216, 219, 221, 226, 229, 231, 281
Needell, Allan, 283
Nernst, Walther, 134; and Bjerrum, 219; on electron theory, 215; and Planck, 230f.; on Planck's calculation of e, 284; on quantum theory, 214f., 230f.; on specific heats, 213-215, 219-221, 231, 243, 309f.
Nernst's *Theoretische Chemie*, 213
Netto, E., 282f.
Newton's laws, 31, 57
Nicholson, J. W., 220, 229, 316
Nisio, S., 256, 315, 320

Olesko, Kathryn, 298
"Oscillators," Planck's use of term, 200, 305; *see also* Resonators

Paschen, Friedrich, 10f.
Peierls, R. E., 275
Pelseneer, Jean, 317
Phase space formulations, 129, 237f., 249-251, 297, 318
Photochemistry, 223f., 231
Photoelectric effect, 207, 221f., 226, 245f.
Photon (*see* Einstein's light particle hypothesis)
Physical chemistry, 213-216, 220, 253

INDEX

Physical Society of London, 232, 317
Physikalisch-Technische Reichsanstalt, 93
Planck, Erwin, 113, 278, 285
Planck, Max, 13-18, 21-24, 30f., 97f., 113, 133f., 200, 254, 266f., 278f.
Planck and Boltzmann, 20-22, 36-39, 61-63, 67, 270, 277; Comparison of views on: combinatorials, 91, 98-105, 110, 125, 128, 134, 139, 279, 284; complexions, 129f.; disorder, 119; electrodynamics, 118, 130f.; entropy, 101f., 111, 238; general techniques, 37, 73, 77f., 91, 101, 106, 125, 127; molecular disorder, 43, 77f., 80, 82, 91, 121-125; natural radiation, 121f.; probability, 48-60; radiation and gas theory, 20-26, 75, 78f., 82, 88f., 123
Planck and: Clausius, 13-16, 21f., 259f.; Ehrenfest, 132, 155, 162f., 166, 196-198, 253, 305f.; Gibbs, 270; Graetz, 27, 52, 54, 270; Helmholtz, 14, 30; Kirchhoff, 14, 17 (*Lectures on the Theory of Heat*, 21, 61, 100, 270); Lorentz, 102, 105f., 194-202, 235f., 289, 304f., 319; Mach, 259; Maxwell, 22; Nernst, 230f.; Sommerfeld, 315; Wien, 246, 270, 289, 305f.; Zermelo, 26f., 72, 271
Planck on: atomic theory, 22f., 29-31; discontinuity, 196-201, 235-237, 288, 304 (*see also* Planck's second theory); Einstein's light-particle hypothesis, 197, 199f., 246; Einstein's relativity theory, 196-198; electron theory, 132-134, 198f.; entropy, 85-90, 95, 100, 119, 136f., 281f. (combinatorial forms for, 105-109, 115f.; oscillator entropy, 93-96, 238f., 280f.); irreversibility, 26-36, 72f., 78, 82, 85-88, 90, 115, 130f., 154-156, 161f., 174, 267; kinetic theory of gases, 18-21; molecular disorder, 66f., 73, 124;

Rayleigh-Jeans law, 151f., 194-198, 235f., 240f., 243; thermodynamics, 15-17, 24-26, 53, 71, 98, 125, 259, 264, 275; the Thiesen distribution, 94-96; universal constants (*see* Constants); the Wien distribution law, 87-90, 92, 96, 108
Planck's *Lectures on the Theory of Thermal Radiation* (1906), 102, 110, 114-133, 136, 139, 143, 151, 162f., 167, 195f., 198, 254, 286; second edition (1913), 235f., 240-243, 246; third through fifth editions, 253f.; Ehrenfest on, 166, 169; Einstein on, 296; reviews of, 139f.
Planck's second theory, 144, 200f., 207, 236-254, 318; multi-quantum emission, 247, 318; reception of, 321, 340-352
Planck's theory: calculation of e, 111, 284; calculation of Loschmidt's number, 111; constant, h, 104, 111, 115, 131f., 135, 181f., 198-202, 206-232, 278, 305; constant, k, 105, 111, 113f., 132f., 151 (and Boltzmann's "permutability measure," 111; and Einstein's χ, 175); distribution laws, 97, 100, 107-109, 115, 135f., 151, 202, 205, 238-240, 246f., 281f. (Ehrenfest on, 143, 154, 158f., 168, 188, 210; Einstein on, 143, 170, 180, 183-185, 188, 320; Jeans on, 143, 188, 204; Lorentz on, 102-104, 190-196; others on, 143, 209f., 218f.); electromagnetic H-theorem, 77-84, 117f., 120-125, 241-243 (Ehrenfest on, 154); energy element, $h\nu$, 107-109, 125-132, 139, 143, 198-200 (Ehrenfest on, 168f., 188; Einstein on, 170, 183-185, 187f., 211); "fundamental equation," 82-84, 95; resonator energy, U, 79f., 82-84, 87, 95, 98, 107f., 290; resonator equations, 167, 200f., 237-239; resonators, 35f., 79f.,

119, 129, 160, 164, 169, 198f., 218-220 (analyzing resonators, 80, 159, 241f., 277); uniqueness theorem, 92-100, 115f. (Ehrenfest on, 153-156)
Pohl, R. W., 284f.
Poincaré, Henri, 26, 210, 231
Pringsheim, Ernst, 11, 17, 93f., 96, 111, 281; *see also* Lummer

Quantization, 125-127, 139, 143f., 169, 183f., 218-220; *see also* Energy discontinuity
Quantum conditions, 218-220
Quantum of electricity, e, 132f., 201
"Quantum of energy," use of term, 201f.
Quantum publications, growth of, 206, 209, 216, 229f., 308
Quantum theory: acceptance of (by Lorentz, 194-196, 216; by Planck, 197-201, 215f.; in general, 206f., 228-232); birth of, 170-172, 182, 188f., 278; national character of, 209f., 231f.; of spectra, 216, 221-225, 229, 247f., 252f. *See also* α-rays; Atomic models; β-rays; Canal rays; Cathode rays; Photochemistry; Photoelectric effect; Radioactivity; Specific heats; X-rays.

Radiation distribution mechanism: Ehrenfest on, 153f., 157-166, 196, 302f.; Einstein on, 170, 180f., 185, 303; Jeans on, 148-150, 165, 185; Planck on, 151, 162f., 196; Rayleigh on, 145-147, 165f., 185
Radiation laws (*see* Black-body theory: Distribution laws; Wien displacement law)
Radioactivity, 245; *see also* α-rays; β-rays
Radiometer, 5, 258
Rayleigh, Lord, 137, 143-152, 180, 188, 263; distribution law, 145-147, 152, 291f.; and Jeans, 137, 144, 148f.; model for cavity radiation, 145-147, 165f., 185
Rayleigh-Jeans distribution law, 143, 145-152, 167-169; Ehrenfest on, 158f., 167f., 188; Einstein on, 170, 180f., 183-185, 188; experimentalists on, 152, 192; Lorentz on, 189-194; Planck on, 151f., 195-198, 235f., 240, 243
Recurrence paradox (*see* Zermelo's recurrence paradox)
Relativity (*see* Einstein's theory of)
Resonators, 158-167, 194, 296; *see also* Oscillators; Planck's theory
Reversibility paradox (*see* Loschmidt's paradox)
Roscoe, H. E., 309
Rosenfeld, Léon, 100
Rubens, Heinrich, 11, 17, 147, 214, 281
Rumford, Count, 261
Rutherford, Ernest, 112, 247f., 284
Rydberg constant, 247

Sackur, Otto, 320
Schaefer, Clemens, 140
Scheel, Karl, 255
Schidlof, Arthur, 227
Schuster, Arthur, 273
Second law: Boltzmann on, 21, 24f., 51-55, 269f.; Carnot on, 12f., 15; Clausius on, 13-15, 24-26, 269; Einstein on, 175-177; Maxwell on, 24; Planck on, 15-17, 25-27, 53, 71f.; Tait on, 24; W. Thomson on, 13, 24
Siegel, Daniel, 257
Solvay, Ernest, 215
Solvay Congress (1911), 204, 216, 219, 226-228, 230, 245f., 252f.
Sommerfeld, Arnold, 144, 192, 203, 216, 285; on Einstein's light particle hypothesis, 225; and Planck, 314f.; on Planck's quantum of action, 225f.; on Planck's second theory, 251, 321; on quantum theory, 225f., 311; on relativity theory, 226; and Stark, 224f.; and Wien, 192, 203
Specific heats: Behn on, 310; Einstein on, 210-213, 246; of gases, 147-151, 204, 215; Hasenöhrl on, 216; Nernst on, 213-216; Planck on, 246f.; and

INDEX

the quantum, 206, 209-221, 245;
Sommerfeld on, 216
Spectra (see Quantum theory of)
Stark, Johannes, 222-226
Statistical ensembles, 173-177, 191, 263, 299
Statistical thermodynamics, 98, 171-177, 262f., 275
Stefan, Josef, 6
Stefan-Boltzmann law, 5f., 10f., 116, 150, 153, 155, 179, 191
Stern-Gerlach experiment, 254
Stewart, Balfour, 258
Stirling's approximation, 49f., 101, 106
Stokes's rule, 221f.
Strutt, John William (see Rayleigh, Lord)

Tait, P. G., 25f.
Thermocouple, 8
Thermodynamics, development of, 12-15, 171-176, 207; see also Planck on; Second law; Statistical thermodynamics
Thermometer, 172, 174-176, 299
Thiesen, M. F., 93f., 110, 132
Thiesen's distribution formula, 93-96
Thomson, J. J., 227, 315f.; on Einstein's light particle hypothesis, 228; on electron theory, 227f.
Thomson, William, 13, 24-26, 71
Tyndall, John, 7

Ultraviolet catastrophe, 152, 195

Valentiner, Siegfried, 253
van der Waals, J. D., 270
van der Waals's equation, 264
Van't Hoff, J. H., 17
Vienna Academy of Science, 152, 154, 227
Voigt, Woldemar, 135f.

Waterston, J. J., 262
Watson, H. W., 55, 272; *Treatise on the Kinetic Theory of Gases*, 20, 55, 263
Weber, H. F., 8
Weinstein, Bernhard, 309
Weiss, J., 308
Weiss, Pierre, 220, 229
Weiss's magneton, 229, 231; Einstein on, 312; Gans on, 312; Langevin on, 229
Wertheimer, Eduard, 227
Wheaton, Bruce, 298, 312
Wien, Wilhelm, 5-11, 17, 99, 139, 144f., 189, 270
Wien and: Lorentz, 190, 192f., 203; Lummer and Pringsheim, 192f.; Planck, 246, 270, 289, 305f.; Sommerfeld, 192, 203; Stark, 246
Wien on: Einstein's light-particle hypothesis, 246; electron theory, 203; Planck's resonators, 99, 118; the quantum, 202-204, 206, 210, 222f., 225, 230
Wien displacement law, 6-8, 10, 116, 133, 135, 150, 153, 191, 202
Wien distribution law, 11, 87-89, 93, 95, 111, 115f., 135f., 145; alternatives to, 93-96, 144-152; Ehrenfest on, 154f.; Einstein on, 180f., 220f.; Planck on, 92f., 115, 145
Wilson, William, 251, 321
Wilson-Sommerfeld quantum conditions, 219, 251
Wise, Norton, 271

X-rays, 206, 222-225, 245

Zermelo, E. F. F., 26f., 72, 270f.
Zermelo's recurrence paradox, 26, 54, 270; Boltzmann on, 270, 274
Zero-point energy, 246-248, 290, 319f.